# Hydrology and Sustainable Water Resources

# Hydrology and Sustainable Water Resources

Edited by **William Sobol**

R CALLISTO
REFERENCE

New York

Published by Callisto Reference,
106 Park Avenue, Suite 200,
New York, NY 10016, USA
www.callistoreference.com

**Hydrology and Sustainable Water Resources**
Edited by William Sobol

International Standard Book Number: 978-1-63239-426-2 (Hardback)

Printed in the United States of America.

# Contents

# Preface

Every book is a source of knowledge and this one is no exception. The idea that led to the conceptualization of this book was the fact that the world is advancing rapidly; which makes it crucial to document the progress in every field. I am aware that a lot of data is already available, yet, there is a lot more to learn. Hence, I accepted the responsibility of editing this book and contributing my knowledge to the community.

This book brings forth both significant challenges in the sphere of water resource management and various approaches to resolving them. From drinking to agriculture, and sanitation to industrial needs, people depend on freshwater reserves for a myriad of activities essential for the smooth and efficient functioning of their daily lives. Hence, understanding the threats and problems pertaining to freshwater resources is critical. There is also a growing need to work towards enhancing the availability of water resources to all. This book is an important tool to address the need for improvements in resources and reduction of threats by giving an assessment of available reviews and case studies that shed light on methods of contamination in the environment. Additionally, it provides details on the sustenance of groundwater and surface-water resources across the globe. This book is meant for students and professionals working in fields which are closely associated with water management and hydrology.

While editing this book, I had multiple visions for it. Then I finally narrowed down to make every chapter a sole standing text explaining a particular topic, so that they can be used independently. However, the umbrella subject sinews them into a common theme. This makes the book a unique platform of knowledge.

I would like to give the major credit of this book to the experts from every corner of the world, who took the time to share their expertise with us. Also, I owe the completion of this book to the never-ending support of my family, who supported me throughout the project.

**Editor**

# Contaminant Hydrology: Surface Water

# Environmental Factors that Influence Cyanobacteria and Geosmin Occurrence in Reservoirs

Celeste A. Journey, Karen M. Beaulieu and
Paul M. Bradley

Additional information is available at the end of the chapter

## 1. Introduction

Phytoplankton are small to microscopic, free-floating algae that inhabit the open water of freshwater, estuarine, and saltwater systems. In freshwater lake and reservoirs systems, which are the focus of this chapter, phytoplankton communities commonly consist of assemblages of the major taxonomic groups, including green algae, diatoms, dinoflagellates, and cyanobacteria. Cyanobacteria are a diverse group of single-celled organisms that can exist in a wide range of environments, not just open water, because of their adaptability [1-3]. It is the adaptability of cyanobacteria that enables this group to dominate the phytoplankton community and even form nuisance or harmful blooms under certain environmental conditions [3-6]. In fact, cyanobacteria are predicted to adapt favorably to future climate change in freshwater systems compared to other phytoplankton groups because of their tolerance to rising temperatures, enhanced vertical thermal stratification of aquatic ecosystems, and alterations in seasonal and interannual weather patterns [7, 8]. Understanding those environmental conditions that favor cyanobacterial dominance and bloom formation has been the focus of research throughout the world because of the concomitant production and release of nuisance and toxic cyanobacterial-derived compounds [4-6, 7-10]. However, the complex interaction among the physical, chemical, and biological processes within lakes, reservoirs, and large rivers often makes it difficult to identify primary environmental factors that cause the production and release of these cyanobacterial by-products [9].

### 1.1. Hydrologic controls

Hydrologic processes control the delivery and retention of nutrients and suspended sediments to lakes and reservoirs, which influence the composition of phytoplankton and zooplankton communities within those lakes and reservoirs [11-16]. One of the major hydrologic processes that constrain nutrient retention and availability to the phytoplankton community is the flushing rate, related to the lake or reservoir volume and inflow from streams and rivers [17]. Residence times range from several months to many years in natural lakes and from days to weeks in reservoirs [18]. The shorter residence times in reservoirs tend to lower phytoplankton abundances when compared to natural lakes [12,16]. Conversely, climatically or anthropogenically induced lengthening of residence times in reservoirs can promote episodic cyanobacterial dominance [14,16]. Further clouding the relation between hydrologic processes and cyanobacterial abundances, stream or river inflows control the transport of suspended sediment into lakes and reservoirs and affect the phytoplankton community structure and physiology whereby excessive turbidity attributed to the suspended sediment loadings can shift the community towards cyanobacteria [19].

## 1.2. Environmental risk associated with cyanobacteria

While most lakes and reservoirs have multiple uses, about two-thirds of the United States population drinks water treated from surface-water sources and, of those, the majority of the largest public utilities obtain their drinking water from lakes and reservoirs [20]. Taste-and-odor episodes are common in lakes and reservoirs used for drinking water throughout the United States [1,5-6, 21-23]. Taste-and-odor episodes are often sporadic, and intensities vary spatially [6,23]. Cyanobacterial production of trans-1, 10-dimethyl-trans-decalol (geosmin), and 2-methylisoborneol (MIB), which produce musty, earthy tastes and odors in drinking water, represents one of the primary causes of taste-and-odor complaints to water suppliers [24]. Compounds that produce taste and odor in drinking water are not harmful; therefore, taste-and-odor problems are a palatability, rather than health, issue for drinking water systems.

Geosmin and MIB can be produced by cyanobacteria and certain other bacteria. Three genera of actinomycetes, a type of bacteria found ubiquitously in soils and also present in the aquatic environment, are important sources of geosmin and MIB: *Microbispora, Nocardia*, and *Streptomycetes* [6,22]. Genera of cyanobacteria that contain known geosmin- and MIB-producing species include *Anabaena, Planktothrix, Oscillatoria, Aphanizomenon, Lyngba, Symploca* [6,25-26] and *Synechococcus* [21]. Geosmin and MIB are problematic in drinking water because the human taste-and-odor detection threshold for these compounds is extremely low (10 nanograms per liter (ng/L)) [27-29], and conventional water-treatment procedures (particle separation, oxidation, and adsorption) typically do not reduce concentrations below the threshold level [24].

If cyanobacteria are identified as the source of geosmin and MIB, human health concerns arise because these cyanobacterial-dervied compounds frequently co-occur with cyanobacterial-derived toxins (cyanotoxins). Although many species of cyanobacteria capable of producing geosmin or MIB are also capable of producing toxins, most species are not capable of producing taste-and-odor compounds and cyanotoxins simultaneously [4,30-32]. Cyanotoxins generally are associated with a bloom formation of a toxin-producing cyanobacterial species. Less frequently, cyanobacterial releases of geosmin and MIB that are not associated with cyanotoxins, have been linked to seasonal periods of high transparency (clear-water phase) attributed to heavy zooplankton grazing [6, 33-36].

The biological function of these cyanobacterial-derived compounds is not well known, but production of geosmin and MIB are reported to occur during active growth and extracellular release during stationary periods, cellular senescence, or cell lysis [26]. The release of cyanotoxins and taste-and-odor compounds by cyanobacteria simply may be a mechanism for the removal of excess metabolites during periods of environmental stress [1,3]. The possibility that cyanotoxins, geosmin, and MIB may contribute to the distribution, abundance, and survival of cyanobacteria in the environment has been investigated [1,35,37]. Secondary metabolites may deter herbivore grazing and shift grazing pressure toward chemically undefended cyanobacterial and algal species, but that allelopathic role is more often attributed to cyanotoxins [37]. If, however, the availability of chemically undefended algae and cyanobacteria is limited (for example, during seasonally heavy zooplankton grazing events that can produce a clear-water phase), it is possible that a shift could occur in the herbivore community toward species that consume chemically defended cyanobacteria [34-39].

### 1.3. Purpose

Cyanobacterial blooms can be stimulated by human activities that introduce excessive nutrients or modifies the flushing rate in a lake or reservoir [5,9,12-16,40]. Therefore, focus of most research has been on nutrient-enriched, eutrophic to hypereutrophic lake systems that experience cyanobacteria blooms at least seasonally, including agriculturally dominated watersheds of the Midwestern United States [14, 31-32,41-44] and Florida [40,45-47]. Production and release of geosmin often have been reported to occur during periods when cyanobacteria dominated the phytoplankton community and often produced species-specific blooms [5,9,25,40-49]. Environmental factors that have been reported to enhance the ability of cyanobacteria to dominate the phytoplankton community include decreased availability of nitrogen, increased phosphorus concentrations, low total nitrogen to phosphorus ratios, reduced light availability (turbidity), warmer water temperatures, greater water column stability, and longer residence times [1,4-6,9,31-32,44-50].

Two cascading reservoirs that serve as drinking water supplies experienced periodic taste-and-odor problems although the reservoirs were not excessively enriched in nutrients, did not experience observable blooms [51-54], and, therefore, did not appear to fit the existing chemical models for cyanobacterial-dominated systems [9,14,31-32,41-44]. The two reservoirs are located in a rural watershed in the Piedmont region of Spartanburg County, South Carolina. Three synoptic surveys and a 2-year seasonally intensive study of limnological conditions in Lake William C. Bowen (Lake Bowen) and Municipal Reservoir #1 (Reservoir #1) were conducted from 2005 to 2009 to assess the chemical, physical, and biological processes that influenced the occurrence of cyanobacteria and cyanobacterial-derived compounds geosmin, MIB, and microcystin, a common cyanotoxin [52,54].

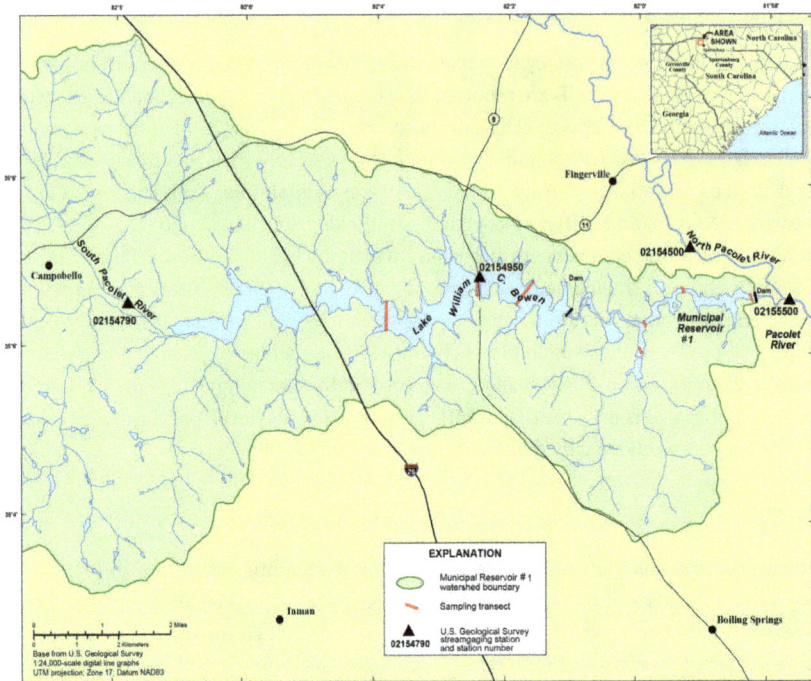

**Figure 1.** Location of sampling transects and U.S. Geological Survey gaging stations in the Lake William C. Bowen and Municipal Reservoir #1 watershed in Spartanburg County, South Carolina.

## 2. Methods

Water quality and phytoplankton community structure in Lake Bowen and Reservoir #1 were monitored synoptically in 2005 and 2006 and intensively from May 2007 to June 2009 to assess the conditions associated with cyanobacterial, geosmin, MIB, and microcystin occurrence. Water samples were collected near the surface (1-meter (m) depth) and near the bottom (6-m depth, where sufficiently deep) at selected transects (fig. 1). Euphotic-zone composite samples were collected during winter, spring, and summer 2009 only to compare to the corresponding surface samples. Samples were collected and processed using U. S. Geological Survey (USGS) protocols and guidelines described in [29]. Discrete depth samples were collected at three locations across each transect (25, 50, and 75 percent of transect width) using Van Dorn samplers and were composited to create a depth-specific sample. Transparency (Secchi disk depth) and light attenuation (to determine euphotic zone depth) were measured at the time of sampling. Lake profile measurements of fluorescence (an estimate of chlorophyll), specific conductance, pH, dissolved oxygen concentrations, and water temperature were collected at 1-m depth intervals along each transect. Reservoir sampling frequency varied seasonally with 67 percent of the samples collected during the peak algal growth period (spring to late summer) [54]. Samples were analyzed for nutrients, major ions, organic carbon, phytoplankton biomass, chlorophyll $a$, pheophytin $a$, dissolved geosmin and MIB, total microcystin, and actinomycetes concentrations. Analytical methods for chemistry and algal taxonomy are described in detail in [54]. The degree of stratification was quantified for each sampling event by computing the relative thermal resistance to mixing (RTRM) at 1-m depth intervals from the lake profile of water temperature at the time of sampling at each site. The RTRM for each 1-m depth interval was computed as the ratio of the density difference between water at the top and the bottom of the 1-m depth interval divided by 0.000008 (the density difference between water at 5 and 4 °C; [18,55-56]. The maximum RTRM for that lake profile was used as a measure of the degree of stratification at that site for that sampling event.

Bivariate and multivariate nonparametric techniques were utilized for data exploration. Data preparation included assigning censored values the same rank and ranked below estimated and quantitative (detections above the laboratory reporting limit or LRL) values [57-58]. Estimated values that are semiquantitative detections below the LRL were assigned the same rank, that is, above censored values but below detected values [57-58]. For biotic data, analyses were done using cell biovolumes (in cubic micrometers per milliliter); preliminary analyses of cell densities (cells per milliliter) yielded similar results but are not provided in this chapter. For chemical, physical, and a subset of phytoplankton data, the Kruskal-Wallis (KW) test was applied to the data to determine if a statistical difference existed (alpha level = 0.05) among groups of data, and the Tukey's Studentized Range test was used to identify which group or groups were different [57-58]. Data from selected sites were evaluated by the Kendall tau correlation procedure to measure the strength of the monotonic bivariate relation between the environmental factors and geosmin concentrations, microcystin concentrations, and cyanobacteria biovolumes [57-58].

Phytoplankton assemblage data were evaluated using multivariate techniques described in [59]. Prior to evaluation, assemblage data were transformed using a fourth root transformation. Hypotheses of temporal (seasonal, annual) and spatial (depth, reservoir location) similarities in the taxonomic composition and biovolumes of phytoplankton communities were examined with a series of non-metric multi-dimensional scaling (NMDS) and one-way analysis of similarity (ANOSIM) tests [59-60]. Prior to plotting the sample patterns of the complex relations among phytoplankton groups and species in 2- and 3-dimensional (2-D, 3-D, respectively) space using NMDS, between-sample similarity (or dissimilarity) coefficients were computed and a triangular matrix constructed [60]. The goodness-of-fit of the NMDS was measured as a stress value, whereby stress < 0.1 corresponds to an effective ordination in 2-D space and < 0.2 is useful but should be superimposed with grouping from a hierarchical cluster analysis to verify [60]. Therefore, the hierarchical Cluster analysis with the SIMPROF option of the cyanobacterial assemblages was superimposed on the NMDS for this study. The statistical test that was used to determine differences among phytoplankton groups and species was ANOSIM, based on a Global R statistic [60]. Global R statistic falls between -1 and 1 whereby R equal to zero indicates completely random grouping of phytoplankton assemblages while R equal to 1 indicates that all assemblages of a group are more similar to each other than to assemblages from another group [60]. Significant R values (p-value < 0.05) indicate the R value is significantly different from zero. [60]. The last step of the multivariate analysis was to link the cyanobacterial assemblage data to the environmental data using RELATE and the global BEST test procedures [59,60]. RELATE provided a Spearman rho correlation analysis between the similarity matrices of the cynaobacterial assemblage and environmental data. Prior to analysis, a logarithmic (base 10) transformation was applied to environmental data. Subsets of explanatory environmental variables and a fixed resemblance matrix computed from the cyanobacterial data were used as input into the BEST model to determine the best combination of variables that explains the observed cyanobacterial assesmblages [59,60]. The stepwise procedure Bio-ENV was employed [60].

On the basis of the results of the exploratory statistical analysis, one site in Lake Bowen and one site in Reservoir #1 were selected to develop a regression model to estimate geosmin concentrations by using environmental factors as explanatory variables. The Lake Bowen site was located nearest the dam and represented the quality of water released to Reservoir #1 (fig. 1). The Reservoir #1 site was located near its dam and represents the quality of water near the raw water intake for the R.B. Simms Water Treatment Plant (fig. 1). Because of the high percentage of censored geosmin concentrations at both sites, ordinary least squares regression was not used to develop a multiple linear regression model. Instead, the multiple logistic regression approach was used to identify environmental factors that best explained the likelihood of geosmin concentrations exceeding the human detection threshold of 10 ng/L [57]. Variables selected for input into the multiple logistic regression analysis included those identified in the Kendall tau correlation analysis and those that could be easily measured by Spartanburg Water as part of their watershed monitoring program. The best equation for each reservoir was selected on the basis of the Pearson Chi-square Statistic (goodness of fit greater at lower statistics and higher p-values), the Hosmer-Lemeshow Statistic (goodness of fit greater at lower statistics and higher p-values), and the minimum Likelihood Ratio Test Statistic, which tests how well an equation fits the data by summing the squares of the Pearson residuals (goodness of fit greater at lower p-values). Model output provided a Logit P result, whereby Logit P results greater than 0.5 resulted in a positive response (geosmin concentrations exceeded the human detection threshold of 10 ng/L) and less than 0.5 resulted in a reference response (geosmin concentrations were below the human detection threshold of 10 ng/L).

## 3. Reservoir hydrology

Lake Bowen and Reservoir #1 are relatively small, shallow (4.8 and 2.3 m depths, respectively) cascading impounds of the South Pacolet River in Spartanburg County, South Carolina (fig. 1) [52,61]. At the full-pool elevation, Lake Bowen has a surface area of 621 hectares (ha) and has 53.2 kilometers (km) of shoreline. Lake Bowen releases spillage (overflow) at the dam directly into Reservoir #1 and by controlled releases at depth from gated conduits (flowthrough). Reservoir #1 is substantially smaller and older than Lake Bowen. At the full-pool elevation, Reservoir #1 has a surface area of 110 ha and 21.1 km of shoreline. Water from these reservoirs is treated by Spartanburg Water at the R. B. Simms Water Treatment Plant. The water-treatment facility and raw water intake are located on Reservoir #1. Recreational activities are allowed on Lake Bowen but are prohibited on Reservoir #1.

## 3.1. Hydrologic conditions in the reservoirs

Inflow to Lake Bowen from the South Pacolet River at the USGS streamgaging station 02154790 generally represented 70 to 85 percent of the total inflow into Lake Bowen on the basis of the synoptic streamflow measurements of minor tributaries (fig. 1) [53]. Unit-area streamflows (computed as the measured synoptic streamflow divided by the drainage areas at the site) were similar among the South Pacolet River and its minor tributaries. Therefore, daily mean unit-area streamflow at the South Pacolet River streamgaging station was multiplied by the total drainage area above Lake Bowen (207 km$^2$) to extrapolate the total inflow to Lake Bowen from 2005 to 2009. Daily residence times in Lake Bowen were computed by dividing the daily mean inflow to Lake Bowen by the daily mean water volume in Lake Bowen (determined using the stage-volume curve developed by the USGS [54] and the daily mean water level at the USGS streamgaging station 02154950). Outflow from Lake Bowen was considered the major inflow to Reservoir #1 and was computed by the summation of daily volume of water associated with spillage (overflow) and controlled releases. However, the estimated Lake Bowen inflow did not represent all the inflow to Reservoir #1 because it did not include contributions from a tributary to the south of Reservoir #1. Therefore, Reservoir #1 outflow also was considered. Determinng outflow from Reservoir #1 was further complicated by the confluence of the South and North Pacolet Rivers to form the Pacolet River just below the Reservoir #1 dam. So, outflow from Reservoir #1 was estimated by computing a 30-day moving window average of differences in streamflow between the Pacolet River USGS streamgaging station (02155500) and North Pacolet River USGS streamgaging station (02154500).

## 3.2. Climate controls on hydrology

The climate of the Pacolet River Basin is classified as temperate [62-63], with 30-year mean annual precipitation of 127.6 centimeters (cm) and mean annual temperature of 15.6 degrees Celsius (°C) [64]. The study area was under severe to extreme drought conditions from September 2007 through December 2008, with annual precipitation amounts that were less than 38 percent of the long-term mean (48.7 and 31.0 cm for 2007 and 2008, respectively) [64]. Annual mean precipitation in 2009 was near the 30-year mean annual precipitation.

As is common for reservoirs, computed residence times were relatively short and averaged 0.60 year in Lake Bowen (estimated from inflow) and 0.26 year in Reservoir #1 (estimated from outflow) for the 5-year study period (2005 – 2009) (fig. 2A). Maximum residence times of 1.04 years in Lake Bowen and 0.80 year in Reservoir #1 occurred in 2008 when drought conditions were prevalent in Spartanburg County (fig. 2A). The maximum residence times were concomitant with major changes in chemical and biological conditions in both reservoirs.

Longer residence times in 2008 were attributed to lower water contributions from the South Pacolet River to Lake Bowen and less water being released from Reservoir #1. This lower water contribution affected the nutrient transport from the watershed to the reservoirs, with much lower annual loads of nitrogen and phosphorus being delivered in 2008 (fig. 2B,C) [53]. Annual total nitrogen loads decreased an order of magnitude from 117,600 kilograms per year (kg/yr) in 2005 (wet period) to 14,000 kg/yr in 2008 (drought period). In 2008, the annual total phopshorus load of 795 kg/yr was only 5 percent of the 2005 total phosphours load. While annual nitrate plus nitrite loads decreased four-fold from 35,380 kg/yr in 2005 to 8,740 kg/yr in 2008, the 2008 nitrate plus nitrite load represented a greater percentage of the total nitrogen load (62 percent) than in 2005 (30 percent).

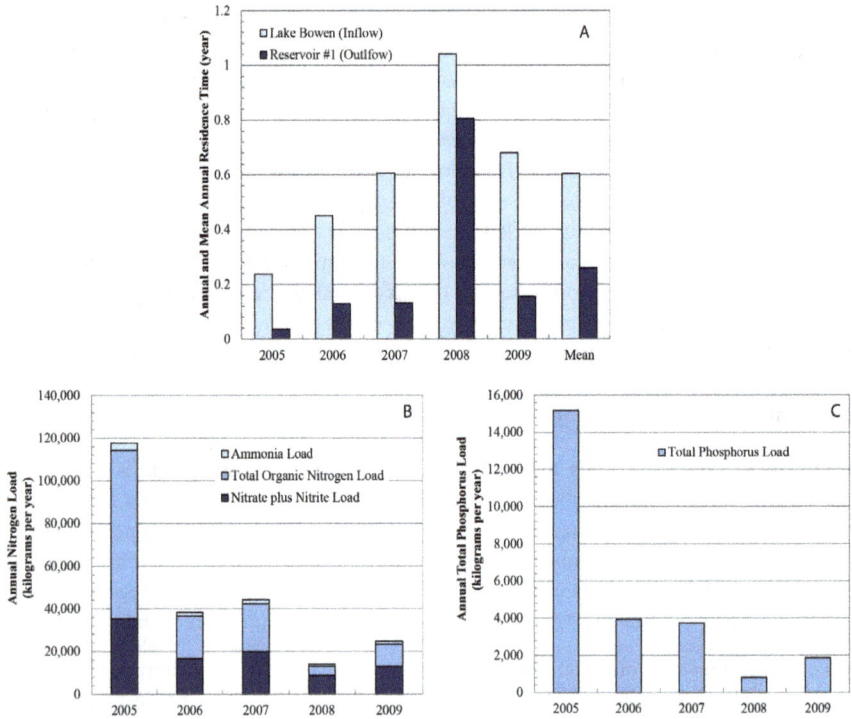

**Figure 2.** (A) Annual and mean annual residence times in Lake William C. Bowen (Lake Bowen) and Municipal Reservoir #1 (Reservoir #1) and annual (B) nitrogen and (C) phosphorus loadings from the South Pacolet River into Lake Bowen, from 2005 to 2009, Spartanburg County, South Carolina.

# 4. General limnological conditions

Overall, Lake Bowen and Reservoir #1 can be classified as warm monomictic reservoirs that stratify from early June to early-to-late October in locations where depths exceed 5 m (for detailed bathymetry, see [61]). From 2005 to 2009, thermal stratification occurred seasonally in the deeper, downgradient regions of the reservoirs [52,54]. Maximum RTRMs that were representative of strongly stratified conditions (greater than 80) were prevalent during the summer season, but were near zero from November through the winter season for these sites [18]. During periods of stratification, the hypolimnion became anoxic (dissolved-oxygen concentrations below 1 milligram per liter (mg/L)) in both reservoirs.

Water column transparencies were greatest at deep, downgradient sites in Lake Bowen and Reservoir #1, but lower at the shallower, upgradient sites in Reservoir #1 (KW p < 0.001). Statistically significant seasonal patterns in transparencies were identified in Lake Bowen and Reservoir #1, with the greatest transparencies in the spring and the least in the fall (KW p < 0.001). Median transparencies were about 1.6 m in the downgradient regions and 1.0 m in the upgradient regions of the reservoirs; however, maximum transparencies of 3.7 m in Lake Bowen and 2.0 m in Reservoir #1 were measured in May 2008.

Overall, Lake Bowen and Reservoir #1 had statistically similar chlorophyll $a$ and nutrient concentrations that were indicative of mesotrophic conditions (KW p > 0.10) [65-66]. Maximum nutrient concentrations remained below 0.034 mg/L for total phosphorus and 0.70 mg/L for total nitrogen in the reservoirs for the study period [54]. Maximum chlorophyll $a$ concentration was 26 micrograms per liter ($\mu$g/L) and occurred in Lake Bowen, but median concentrations were below 9 $\mu$g/L at all sites in both reservoirs. Maximum chlorophyll $a$ concentrations in both reservoirs were indicative of eutrophic conditions [18,65]. Chlorophyll-to-phosphorus ratios (Chl:TP) that are substantially less than 1 indicate phytoplankton are not phosphorus limited [41,67]. Median chlorophyll-to-total-phosphorus ratios ranged from 0.5 to 0.7 for all sites in Lake Bowen and Reservoir #1 and indicated phosphorus concentrations were approaching limiting conditions. Median TN:TP ratios ranged from 20 to 24 in Lake Bowen and from 13 to 18 in Reservoir #1. Based on these median TN:TP ratios, potential phosphorus limitation (TN:TP >17) was common in Lake Bowen and co-limitation (17 < TN:TP > 10) by phosphorus and nitrogen was common in Reservoir #1 [65]. Low total-nitrogen-to-total-phosphorus (TN:TP) ratios (generally below 29:1) were consistent with environmental conditions reported to favor nitrogen-fixing cyanobacteria (fig. 3C, 3D) [40,45,47].

Seasonal variation in dissolved inorganic nitrogen (DIN; sum of dissolved nitrate plus nitrite and ammonia) concentrations indicated periods of limitation and enrichment in readily bioavailable forms of nitrogen for most phytoplankton groups (fig. 3A, 3B). DIN concentrations were consistently low in the reservoirs near the surface (1 m) during the spring and summer of 2008, when compared to other spring periods. Elevated DIN concentrations were observed near the bottom relative to the surface (not shown in fig. 3) and were attributed to increased dissolved ammonia in the anoxic hypolminion [54]. Statistically, dissolved ammonia concentrations were highest in the fall and lowest in the spring and summer (KW p < 0.001), and dissolved nitrate plus nitrite concentrations were highest in the winter and spring and lowest in the summer and fall (KW p < 0.001). However, the winter 2009 sample period also had dissolved ammonia concentrations comparable to the fall sample periods (fig. 3A). Increases in total phosphorus concentrations in the hypolimnion relative to the epilimnion during summer were evident in both reservoirs (KW p < 0.001), and especially was pronounced during the summer of 2008. Hypoliminetic increase in total phosphorus concentrations was concurrent with the development of strong stratification (KW p < 0.001; maximum RTRMs greater than 100) and anoxic conditions in the summer and early fall, indicating a source of phosphorus from the bed sediment during that seasonal period [66]. Hypolimnetic increases in total phosphorus concentrations were as high as 0.017 mg/L in Lake Bowen and 0.012 mg/L in Reservoir #1. Dissolved nitrate plus nitrite concentrations were highest in the winter (KW p < 0.001), while total nitrogen and TN:TP ratios were statistically lower during the summer (KW p < 0.001) (fig. 3). For all depths and sites, total organic nitrogen concentrations were statistically greater during the spring and winter than during the summer and fall (KW p = 0.003) (fig. 3).

Overall, chlorophyll $a$, total nitrogen, and phosphorus concentrations observed in Lake Bowen and Reservoir #1 were not indicative of reservoir systems that would experience cyanobacterial dominance of the phytoplankton community, based on empirical models developed in [9] (fig. 3). Instead, the empirical models predicted that cyanobacteria would represent less than 20 percent of the phytoplankton community based on the observed chlorophyll and nutrient levels. One exception based on the empirical model was for observed TN:TP ratios that indicated greater likelihood of cyanobacterial dominance [9] (fig. 3C, 3D), whereby TN:TP ratios were generally below 29:1 and were consistent with environmental conditions reported to favor nitrogen-fixing cyanobacteria [40,45,47].

## 5. Phytoplankton community structure

Overall, no statistical differences were identified in total phytoplankton and cyanobacterial biovolumes among sites and depths in Lake Bowen and Reservoir #1 (KW p > 0.20). Median total phytoplankton biovolumes ranged from 1,853,000 to 2,480,000 cubic micrometers per milliliter ($\mu m^3$/mL) in Reservoir #1 and from 1,869,000 to 2,292,000 $\mu m^3$/mL in Lake Bowen. Cyanobacterial biovolumes were an order of magnitude lower, with median biovolumes ranging from 127,700 to 180,000 $\mu m^3$/mL in Reservoir #1 and from 111,800 to 123,700 $\mu m^3$/mL in Lake Bowen. Maximum cyanobacterial biovolumes occurred during the summer (July to August), but did not always result in cyanobacteria dominating the phytoplankton community (KW p < 0.001). In fact, median percentages of the cyanobacterial fraction of the total phytoplankton community were less than 10 percent in all samples from all sites, indicating cyanobacteria rarely dominated the phytoplankton community and verifying the empirical model results described in the last section [9] (fig. 4). However, during drought conditions in July 2008 in Lake Bowen, cyanobacteria represented as much as 59 percent of the total phytoplankton community in Lake Bowen (fig. 4). During the same time period, cyanobacteria dominance was less pronounced in Reservoir #1 (31 to 44 percent of the total phytoplankton

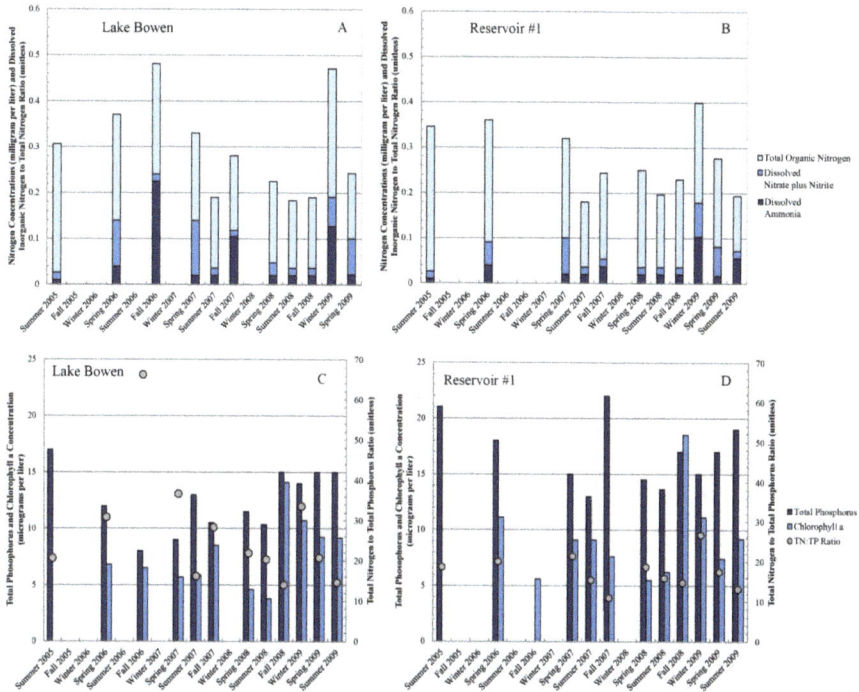

**Figure 3.** Seasonal variation in nitrogen species concentrations near the surface in (A) Lake Bowen and (B) Reservoir #1 and in total phosphorus and chlorophyll a concentrations and TN:TP ratios in (C) Lake Bowen and (D) Reservoir #1, Spartanburg County, South Carolina, 2005 to 2009.

community) than Lake Bowen. Additionally, surface algal blooms were not observed during the study period.

Within the cyanobacteria group, genera that contained known geosmin-producing species and known toxin-producing species were present at all sites. Biovolumes of these genera, however, varied seasonally and annually. Known geosmin-producing genera identified in the two reservoirs included *Planktolyngbya, Aphanizomenon, Synechococcus, Psuedoanabaena, Oscillatoria,* and *Anabaena,* many of which also can produce toxins. Other known toxin-producing genera identified in the two reservoirs included *Cylindrospermopsis, Synechocystis, Microcystis,* and *Aphanacapsa.* Median percentages of known geosmin-producing genera in the cyanobacterial group ranged from 48 to 59 percent among all sites. Fractions of all known gesomin-producing genera in the cyanobacterial group were similar among sites and depths (KW p > 0.90). However, seasonal differences were identified that indicated greater fractions of known geosmin-producing genera occurred during the spring and winter and the least fractions occurred during the fall (KW p < 0.001).

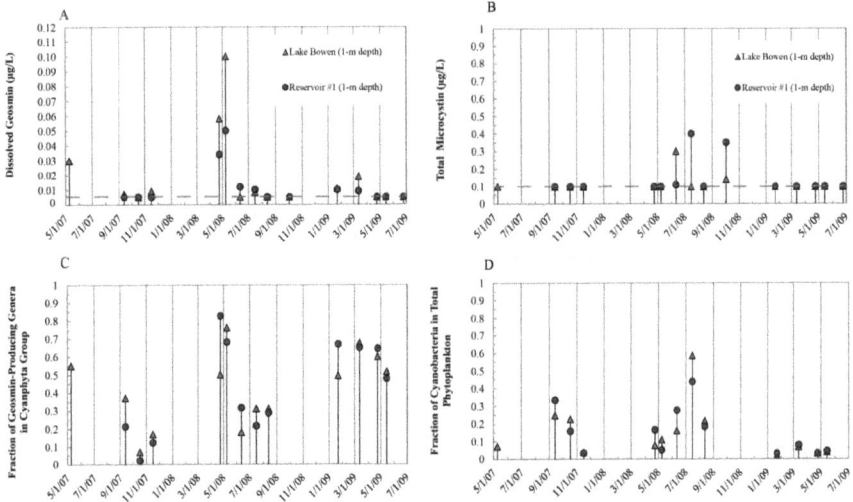

**Figure 4.** Seasonal variation in (A) dissolved geosmin concentrations, (B) total microcystin concentrations, (C) fraction of geosmin-producing genera within the Cyanophyta (cyanobacteria) group, and (D) fraction of cyanobacteria in the total phytoplankton for sites near the surface in Lake William C. Bowen and Municipal Reservoir #1, Spartanburg County, South Carolina, 2005 to 2009. Red dashed line represents the laboratory reporting level for dissolved geosmin (< 0.005 μg/L) total microcystin (< 0.010 μg/L).

Results of the ANOSIM tests for the differences among samples grouped by reservoir, season, year, and depth position indicated significant and relatively strong temporal variation (particularly seasonal variation) in phytoplankton assemblages in terms of biovolumes of algal divisions, all Cyanophyta genera, and the 6 known geosmin-producing Cynaophyta genera. Overall, phytoplankton assembages of all divisions and genera within the Cyanophyta division had low Global R values (< 0.22) when grouped by reservoir and depth position, indicating relatively random grouping (table 1). Conversely, the same assemblages had the highest Global R values ( > 0.40) when grouped by season (table 1). The greatest seasonal differences were for genera within the Cyanophyta genera (including genera with known geosmin-producing species). Once the taxonomic assemblage of known-geosmin producing Cyanophyta genera became restricted by season (spring only) and depth position (spring only, surface depth only), no differences among samples were identified (table 1). In summary, multivariate statistical analysis indicated that taxonomic assemblages of the phytoplankton community (represented by the major algal divisions) and the cyanobacterial community varied season to season and, to a lesser degree, year to year. The above-mentioned taxonomic assemblages between the two reservoirs and between the two depth positions were similar. This seasonal pattern of variability is consistent with the pattern identified in the Kruskal-Wallis analysis of the chemical and physical data.

| Analysis of Similarity (Nonparametric One-Way ANOSIM) | | | | | | | |
|---|---|---|---|---|---|---|---|
| Reservoir Group | | Season | | Year | | Depth Position | |
| Global R | p-value | Global R | p-value | Global R | p-value | Global R | p-value |
| All Phytoplankton Divisions | | | | | | | |
| 0.009 | 0.740 | *0.401* | *0.001* | 0.231 | 0.001 | 0.078 | 0.006 |
| All Genera in Cyanophyta Division (all seasons, all depths) | | | | | | | |
| 0.005 | 0.027 | *0.723* | *0.001* | *0.331* | *0.001* | 0.045 | 0.051 |
| Cyanophyts Genera with known geosmin-producing species only (all seasons, all depths) | | | | | | | |
| 0.017 | 0.110 | *0.595* | *0.001* | *0.280* | *0.001* | 0.086 | 0.010 |
| Cyanophyta Genera with known geosmin-producing species only (spring only, all depths) | | | | | | | |
| 0.051 | 0.056 | *NA* | *NA* | 0.146 | 0.017 | 0.220 | 0.003 |
| Cyanophyta Genera with known geosmin-producing species only (spring only, surface only) | | | | | | | |
| 0.030 | 0.236 | *NA* | *NA* | 0.113 | 0.083 | NA | NA |

**Table 1.** Nonparametric Analysis of Similarity (one-way) results for 5 phytoplankton taxonomic assemblages in relation to 4 factors of reservoir group, season, year, and depth position for Lake William C. Bowen and Municpal Reservoir #1, Spartanburg County, South Carolina, May 2007 to June 2009. Bold italized text highlights most significant relation between groups and phytoplankton assemblages. Global R statistic falls between -1 and 1 whereby R equal to zero indicates completely random grouping of phytoplankton assemblages whereas R equal to 1 indicates that all assemblages of a factor are more similar to each other than to assemblages from another factor [60]. Significant R values (p-value < 0.05) indicate the R value is significantly different from zero.

Four NMDS 2-D plots of the cyanobacterial assemblages as Cyanophyta genera were constructed and samples were grouped by reservoir (fig. 5A), season (fig. 5B), depth (fig. 5C), and season by year (fig. 5D) to illustrate the ANOSIM results. The hierarchical Cluster analysis with the SIMPROF option of the cyanobacterial assemblages was superimposed on the NMDS, producing 5 distinct assemblages at 60 percent similarity [59, 60]. Comparison of the cyanobacterial assemblages in Lake Bowen and Reservoir #1 and at the 3 depths identifed samples that were fairly evenly distributed throughout the 5 cluster groups (fig. 5A,5C). Euphotic-zone composite samples were collected during winter, spring, and summer 2009 only. Therefore, as determined by ANOSIM and illustrated by NMDS, reservoir and depth were not considered to be major factors that explain the differences in cyanobacterial assemblages (fig. 5A, 5C). Seasonal changes explained most of the differences in cyanobacterial assemblages among the 5 cluster groups, however, spring and fall samples were split into two distinct groups (fig. 5B). The best explanatory factor was the combined season and year that identified the spring 2008 cyanobacterial assembage as a separate group from the spring 2009 and 2007 samples. Additionally, the two distinct fall 2007 groups were determined to be cyanobacterial assemblages associated with pre- and post-overturn periods, with the post-overturn samples plotting to the upper left of the pre-overturn samples shown in Figure 5D.

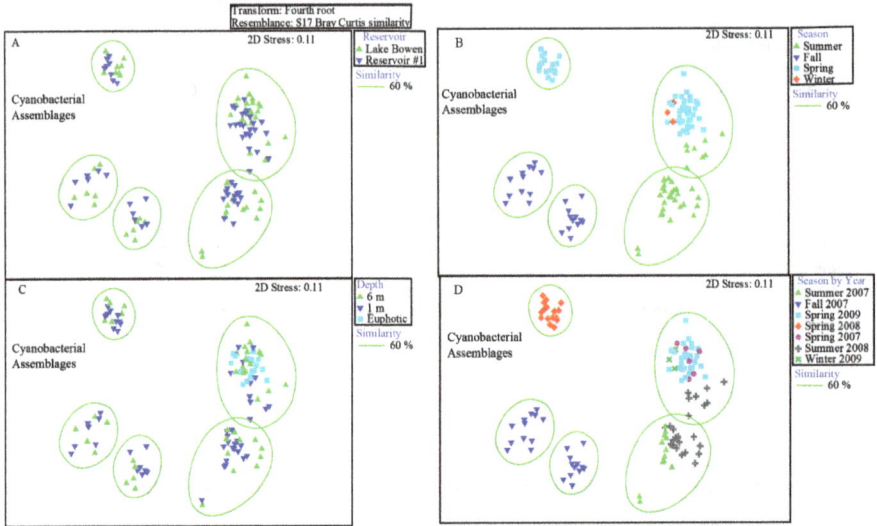

**Figure 5.** Non-metric multi-dimensional scaling (NMDS) ordinations of cyanobacterial assemblages in surface, euphotic zone, and bottom depths at sites in Lake William C. Bowen and 3 sites in Municipal Reservoir #1, Spartanburg County, South Carolina, May 2007 to June 2009, based on (A) reservoir, (B) season, (C) depth, and (D) season and year. Fourth-root transformation was applied to the cyanobacterial assemblage data prior to construction of the Bray Curtis resemblance matrix for the NMDS ordination. Goodness-of-fit of the NMDS is measured by the 2-dimensional stress of 0.11, which is indicative of relatively good ordination but to verify, cluster groups, determined by Hierarchical Cluster analysis program with the SIMPROF option in PRIMER, were superimposed [60].

# 6. Geosmin, MIB, and microcystin occurrence

Geosmin was the most commonly detected taste-and-odor compound in Lake Bowen and Reservoir #1 during the study period from May 2006 to June 2009 [51]. However, about 35 percent of the samples in Lake Bowen and more than 44 percent of the samples in Reservoir #1 had non-detectable geosmin concentrations (< 0.005 µg/L). Median geosmin concentrations ranged from < 0.005 to 0.006 µg/L at the study sites. MIB rarely was detected (median concentrations were less than the laboratory reporting level of 0.005 µg/L at all sites). When present, MIB was at very low concentrations with maximum MIB concentrations ranging from 0.005 to 0.014 µg/L.

As is often observed in surface-water systems, geosmin concentrations in Lake Bowen and Reservoir #1 exhibited strong annual and seasonal variability during the study, with peak geosmin concentrations occurring in the spring (KW p < 0.001; fig. 4A). Maximum geosmin concentrations occurred in spring 2008 (April and May 2008) at all sites in both reservoirs. During the spring of 2008, maximum geosmin concentrations ranged from 0.060 to 0.100 µg/L near the surface in Lake Bowen, which were 6 to 10 times greater than the human detection threshold of 10 ng/L (0.010 µg/L). Maximum geosmin concentrations near the surface in Reservoir #1 were about half (0.035 to 0.050 µg/L) the maximum concentration observed in Lake Bowen during the same spring 2008 period. Annual maximum geosmin concentrations tended to re-occur in the April–May period, but in 2009, the annual maximum geosmin concentrations occurred in March (fig. 4). Nonetheless, the spring (March-to-May) period of annual maximum geosmin concentrations was not concurrent with the period of annual maximum cyanobacterial biovolume (fig. 4A, D). Cyanobacteria were present in both reservoirs during the peak geosmin period in the spring of 2008, but represented less than 20 percent of total phytoplankton biovolume (fig. 4D). The peak geosmin period was concurrent with a peak in the fraction of known geosmin-producing genera in the cyanobacteria group (fig. 4A, C). Microcystin rarely was detected. Maximum microcystin concentrations of 0.30 and 0.40 µg/L were observed in Lake Bowen and Reservoir #1, respectively, during the summer of 2008 immediately following the peak geosmin period (fig. 4B). Microcystin concentrations were below concentrations of concern of 1 µg/L for drinking water or 20 µg/L for recreational activities [4]. No statistical differences existed in geosmin, MIB, and microcystin concentrations and in cyanobacterial biovolume between Lake Bowen and Reservoir #1, among sites within each reservoir, and between depths at a site (KW p > 0.70).

## 7. Environmental factors influencing geosmin concentrations and cyanobacterial biovolumes

In Lake Bowen, elevated geosmin concentrations near the surface were correlated significantly (p < 0.05) with deep (greater than 6 m) mixing zone ($Z_m$) depths (Kendall tau ($\tau$) = 0.60), lower euphotic zone-to-mixing zone depth ($Z_{eu}$: $Z_m$) ratios ($\tau = -0.47$), and elevated dissolved-oxygen concentrations ($\tau = 0.42$), which are environmental conditions indicative of unstratified conditions in Lake Bowen. In relation to phytoplankton community structure, elevated geosmin concentrations in Lake Bowen were correlated to greater *Oscillatoria* biovolumes ($\tau = 0.60$), reduced algal biomass (dry weight) ($\tau = -0.61$), and lower total phytoplankton biovolumes ($\tau = -0.4$). This disparate pattern of increased cyanobacterial biovolumes and overall reduced phytoplankton biovolumes during periods of elevated geosmin was not in line with past geosmin occurrence research [25,41,43,44]. Conversely, geosmin concentrations were not correlated to residence times, nutrient concentrations, chlorophyll *a* concentrations, actinomycetes concentrations, other known geosmin-producing genera of cyanobacteria, fraction of known geosmin-producing genera in the cyanobacteria group, or cyanobacterial biovolumes in Lake Bowen.

In Reservoir #1, elevated geosmin concentrations near the surface were correlated to elevated dissolved-oxygen and chloride concentrations ($\tau = 0.55$ and 0.58, respectively). Elevated geosmin concentrations were correlated to lower pheophytin *a* concentrations ($\tau = -0.61$) and to reduced algal biomass (dry weight) ($\tau = -0.53$). Elevated geosmin concentrations were correlated to greater fraction of geosmin-producing genera in the cyanobacteria group ($\tau = 0.56$); however, total cyanobacterial biovolumes were not correlated to geosmin concentrations. As was determined for Lake Bowen, geosmin concentrations were not correlated to residence times, nutrient, chlorophyll $a$, or actinomycetes concentrations in surface samples.

Sites in Lake Bowen had geosmin concentrations correlated to the known geosmin-producing genus *Oscillatoria* within the cyanobacteria group but not to actinomycetes concentrations, suggesting that cyanobacteria were the probable source of geosmin. Sites in Reservoir #1 also had geosmin concentrations that correlated to known geosmin-producing genera as a fraction of the overall cyanobacteria group. Again, no significant correlation existed between geosmin and actinomycetes concentrations in either reservoir.

Cyanobacterial biovolumes tended to be greatest during the summer when the reservoirs were stratified (KW p < 0.001). In contrast to elevated geosmin concentrations, environmental factors indicative of stratified or stable water column conditions, including warmer water temperatures ($\tau = 0.86$), lower $Z_m$ ($\tau = -0.84$), greater $Z_{eu}:Z_m$ ratios ($\tau = 0.88$), lower dissolved-oxygen concentrations ($\tau = -0.52$), and maximum RTRM ($\tau = 0.91$) correlated strongly with elevated cyanobacterial biovolumes in Lake Bowen. Additionally, in contrast to elevated geosmin concentrations, cyanobacterial biovolumes in Lake Bowen correlated with changes in nutrient levels in the reservoirs. Elevated cyanobacterial volumes were correlated to lower nitrogen concentrations (as ammonia, nitrate plus nitrite, total organic nitrogen, and TN) and lower TN:TP ratios. Elevated cyanobacteria biovolume in both reservoirs correlated strongly to elevated hypolimnetic to epilimnetic TP ratios ($\tau = 0.55$) that was considered indicative of release of phosphorus from the sediment by biotic or abiotic processes during anoxic conditions. Elevated cyanobacterial biovolumes in surface samples from Reservoir #1 had similar correlative relations to environmental factors as cyanobacterial biovolumes in Lake Bowen, with the exception of no correlation to TN:TP ratios and a negative correlation to TP ($\tau = -0.51$).

On the basis of 30-day moving window averages, elevated cyanobacterial biovolumes in Lake Bowen and Reservoir #1 were correlated to longer residence times (computed from inflow; $\tau = 0.69$ and 0.50, respectively) and lower overflow volumes ($\tau = -0.61$ and $-0.44$, respectively), indicative of low-flow or drought conditions. Biovolumes of known geosmin- and toxin-producing cyanobacteria genera, including *Cylindrospermopsis, Planktolyngbya, Synechococcus, Synechocystis,* and *Aphanizomenon* (Lake Bowen only), correlated with greater cyanobacteria biovolumes and were the dominant taxa in the cyanobacteria group.

The RELATE and BEST multivariate analyses also were used to link cyanobacterial assemblages to environmental factors by testing the hypothesis of no relation between multivariate pattern of two datasets (table 2) [59,60]. The RELATE procedure was used to evaluate potential correlations between three assemblage measures (all phytoplankton divisions, all genera and genera with known-geosmin producing species only in the Cyanophyta division,) and 14 potential explanatory factors. Specifically, resemblance matrices for phytoplankton division, Cyanophyta genera, and known geosmin-producing Cyanophyta genera assemblages for all seasons and for spring only were correlated to the resemblance matrices for 14 selected environmental variables using Spearman rho correlation analysis (table 2) [59]. Strongest correlations were identified for the potential explanantory variables and Cyanophyta genera ($\varrho = 0.526$) and known geosmin-producing Cyanophyta genera ($\varrho = 0.436$) during the 3 spring seasons (2007, 2008, 2009), which indicated that differences in cyanobacterial assesmblages can be explained by differences in environmental factors during the spring period (table 2). Statistically signficant (p < 0.001) but weak correlations were identified between environmental factor and phytoplankton divisions and Cyanophyta genera when all seasons were considered (table 2). The BEST analysis, which incorporated the stepwise search algorithm (BVSTEP), was used to select the environmental variables that best explain the cyanobacterial assemblages during the spring season. BEST analysis was performed by comparing fixed resemblance matrices of springtime Cyanophyta assemblages (all genera and genera with known geosmin-producing species only) with resemblance matrices of a combination of up to 14 variables, with the final result being the best set of explanatory variables that explain structure in the assemblage data. A six-variable model was selected that best explained the assemblages of all Cyanophyta genera in the two reservoirs during the spring ($\varrho = 0.601$) (table 2). The explanatory variables included specific conductance, the two forms of DIN (ammonia and nitrate plus nitrite), chlorophyll $a$, pheophytin $a$ (degradate of chlorophyll a), and iron (table 2). A four-variable model was selected to best explain the assemblages of only known geosmin-producing genera of Cyanophyta ($\varrho = 0.663$) and included magnesium, nitrate plus nitrite, chlorophyll $a$:pheophytin $a$ ratio (undegraded to degraded pigment), and geosmin (table 2).

| Phytoplantkon Taxonomic Assemblages | RELATE Analysis on Matched Resemblance Matrices | | BEST Analysis on Phytoplankton Fixed Resemblance Matrix and Selected Explanatory Variables | | |
|---|---|---|---|---|---|
| | Spearman Rho | P-value | Spearman Rho | P-value | Explanatory Variables |
| Phytoplankton Divisions, all season | 0.275 | 0.001 | -- | -- | -- |
| Cyanophyta genera, all seasons | 0.197 | 0.001 | -- | -- | -- |
| Cyanophyta genera, spring only | 0.526 | 0.001 | 0.601 | 0.001 | Specific conductance, ammonia, nitrate plus nitrite, chlorophyll a, pheophytin a, iron |
| Cyanophyta genera with known geosmin-producing genera, spring only | 0.436 | 0.001 | 0.663 | 0.001 | Magnesium, nitrate plus nitrite, chlorophyll a:pheophytin a ratio; geosmin |

**Table 2.** RELATE and BEST tests for linkages between springtime cyanobacterial assemblages and environmental factors for Lake William C. Bowen and Municpal Reservoir #1, Spartanburg County, South Carolina, May 2007 to June 2009.

Factor-specific relations were depicted using NMDS ordination of cyanobacterial assemblages superimposed with bubble plots of selected environmental factors (fig. 6). Maximum geosmin concentrations occurred during the spring of 2008 when the distinct cyanobacterial assemblage was present in Lake Bowen and Reservoir #1 (fig. 6A, 6C). During that same period, nitrate plus nitrite concentrations were reduced when compared to other spring periods but not as reduced as fall and summer periods (fig. 6E). Minimum iron concentrations were associated with the spring 2008 cyanobacterial assemblage when compared to spring 2007 and 2009 assemblages (fig. 6B). Less evident differences in chlorophyll a and pheophytin a concentrations were identified in the bubble plots for all cyanobacterial assemblages (fig. 6 D, F).

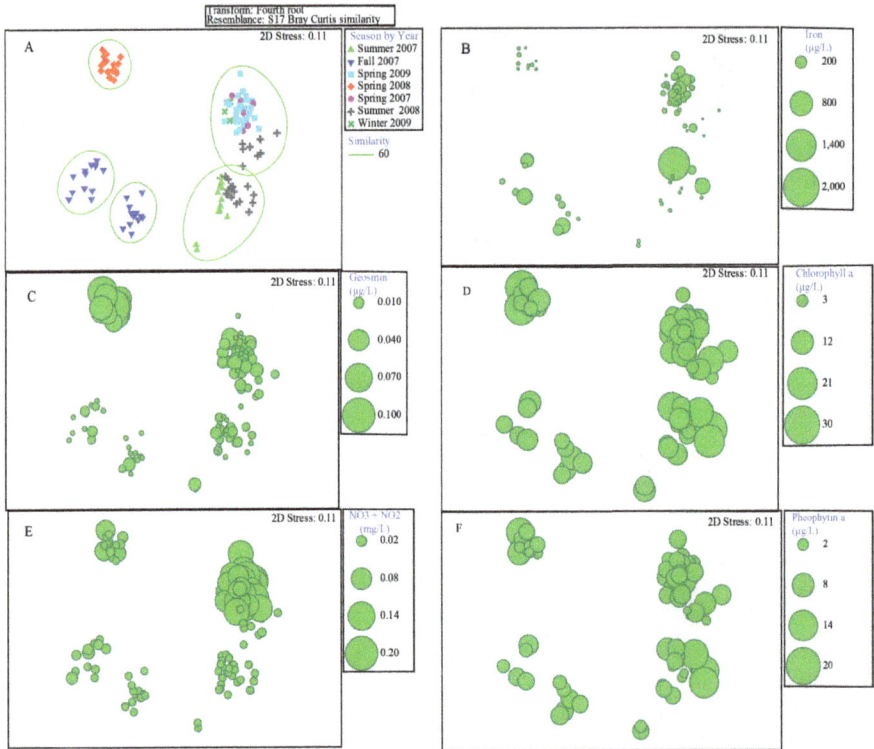

**Figure 6.** Non-metric multi-dimensional scaling (NMDS) ordinations of (A) cyanobacterial assemblages in surface, euphotic zone, and bottom depths at 2 sites in Lake William C. Bowen and 3 sites in Municipal Reservoir #1, Spartanburg County, South Carolina, May 2007 to June 2009 and superimposed NMDS Bubble plots of (B) iron, (C) geosmin, (D) chlorophyll a, (E) nitrate plus nitrite (NO3+NO2), and (F) pheophytin a concentrations. Fourth-root transformation was applied to all cyanobacterial assemblage data prior to construction of the Bray Curtis resemblance matrix for the NMDS ordination. Goodness-of-fit of the NMDS is measured by the 2-dimensional stress of 0.11, which is indicative of relatively good ordination but to verify, cluster groups, determined by Hierarchical Cluster analysis program with the SIMPROF option in PRIMER, were superimposed [60].

Production and release of geosmin often have been reported to be related to periods when cyanobacteria dominated the phytoplankton community and often produced species-specific blooms [5,9,25,40-49]. In turn, cyanobacterial dominance has been attributed to several environmental factors that allow the cyanobacteria to thrive more successfully than other phytoplankton groups, including decreased availability of nitrogen, increased phosphorus concentrations, reduced light availability, warmer water temperatures, greater water column stability, and longer residence times (1,4-6,9,31-32,44-50]. In Lake Bowen and Reservoir #1, cyanobacterial biovolumes were related to many of these reported environmental factors as well as increased total phosphorus in the hypolimnion attributed to sediment release during anoxic conditions (internal phosphorus cycling). Elevated dissolved geosmin concentrations, however, were not related to increased cyanobacterial biovolumes, and cyanobacteria rarely dominated the total phytoplankton community. Therefore, another mechanism was needed to explain the increased production and release of geosmin in April and May of 2008 for Lake Bowen and Reservoir #1.

One plausible explanation of elevated geosmin concentrations in 2008 is related to a pattern of greater transparencies observed in both reservoirs (more pronounced in Lake Bowen) concurrent with annual maximum dissolved geosmin concentrations in the spring. In fact, during the spring of 2008, maximum transparencies (3.7 m in Lake Bowen and 2.0 m in Reservoir #1) coincided with the period of peak dissolved geosmin production when maximum geosmin concentrations were 100 ng/L in Lake Bowen and 50 ng/L in Reservoir #1. That springtime pattern of greater transparency is consistent with a clear-water phase usually attributed to heavy zooplankton grazing of the phytoplankton community and has been reported commonly in mesotrophic reservoirs [33, 35]. Because zooplankton data were not available, the occurrence of this process could not be confirmed directly. Nonetheless, the relation between elevated dissolved geosmin concentrations and environmental factors other than deeper transparencies also suggested zooplankton grazing could be a mechanism for the direct or indirect release of geosmin from cyanobacteria into the dissolved phase. In both reservoirs, elevated dissolved geosmin concentrations were correlated to environmental factors not only indicative of greater light penetration (greater euphotic zone depths) but also reduced algal biomass and total phytoplankton biovolumes. Geosmin release from cyanobacterial cells has been associated with cellular senescence or cell lysis [26] and with periods of environmental stress [1,3], all of which are consistent with heavy zooplankton grazing. While cyanobacteria generally are not considered favorable food for zooplankton, a shift in the zooplankton community toward less selective predation is probable because of the increased competition for prey that would accompany a heavy zooplankton grazing event [35-39]. Another indication of heavy zooplankton grazing was the coincidence of the lowest total phytoplankton and cyanobacterial biovolumes near the surface (1-m depth) during the period of maximum dissolved geosmin concentrations in the spring of 2008.

Predator-driven natural selection of cyanobacterial genera with chemical-defense capabilities could further contribute to elevated dissolved geosmin concentrations. During the period of maximum geosmin concentrations, three genera with known geosmin-producing strains in the cyanobacteria group (*Oscillatoria, Aphanizomenon,* and *Synechococcus*) were the dominant cyanobacteria taxa in both reservoirs. Increasing grazing pressure with decreasing prey alternatives would be expected to trigger chemical defenses in this surviving population. While grazing of *Oscillatoria* and *Aphanizomenon* by zooplankton is not well documented, the genus *Synechococcus* consists of picoplankton that are important contributors to pelagic freshwater ecosystems and have been shown to be grazed by zooplankton [34,68].

In combination with or separate from heavy springtime zooplankton grazing, nutrient (N, P) limitation also could explain seasonal changes to the phytoplankton communities [12,69,70]. Limited supply of the biologically available nitrogen has been proposed as a major factor for the presence of nitrogen-fixing cyanobacteria [9,14,47]. During the 2008 drought period, Lake Bowen received reduced DIN loading (predominately in the form of nitrate plus nitrite) from the watershed (fig. 2) and experienced lower-than-normal springtime DIN concentrations in both reservoirs. Concurrently, total phosphorus concentrations remained relatively consistent in both reservoirs (fig. 3).

## 8. Multiple logistic regression model of geosmin concentrations

Two multiple logistic regression models (MLogModel) were developed that estimated the occurrence of dissolved geosmin concentrations above the human detection threshold of 10 ng/L (0.010 µg/L) on the basis of the multivariate and Kendall tau correlation analyses. Even though significant correlation existed between geosmin and several phytoplankton taxonomic variables (for example, total phytoplankton and *Oscillatoria* biovolumes), explanatory variables used as input for the MLogModels were limited to more easily or quickly measured chemical constituents and hydrodynamic properties. These selected, easily measured variables would be included in the watershed management and monitoring strategy for the reservoirs. Two models were labeled MLogRModel 1—developed from data at the surface (1 m) depth at Lake Bowen and MLogRModel 2—developed from data at the surface depth (1 m) in Reservoir #1. Although many of the same environmental factors were used as explanatory variables in initial model runs, the final explanatory variables that provided the best fit model varied among the two sites.

The best fit model for MLogModel 1 for Lake Bowen was the following:

$$\text{Logit P} = -4.691 + \left(2.184 * \left[Z_m\right]\right) - \left(24.419 * \ddot{O}\left[TN\right]\right) + \left(0.0351 * \left[\text{Overflow - Flowthrough}\right]\right), \qquad (1)$$

where $Z_m$ is mixing zone depth in meters, $\sqrt{TN}$ is the square root of the total nitrogen concentration in milligrams per liter, and Overflow – Flowthrough is the 30-day prior moving window average of overflow minus the 30-day prior moving-window average of flowthrough at Lake Bowen dam, in million gallons per day. Overall, MLogModel 1 determined that the likelihood of dissolved geosmin concentrations exceeding the human detection threshold in Lake Bowen occurred during greater mixing zone depths (unstratified conditions), greater spillage or overflow at Lake Bowen dam rather than controlled releases or flowthrough (indicative of higher water levels), and by lower total nitrogen concentrations. Of the three explanatory variables, only mixing zone depth was correlated significantly to geosmin concentrations in the exploratory data analysis. The MLogModel 1 correctly estimated the likelihood of geosmin concentrations exceeding the human detection threshold 83 percent of the time and not exceeding the human detection threshold 100 percent of the time, resulting in an overall sensitivity of 94 percent (table 3).

The best fit model for MLogModel2 for Reservoir #1 was the following:

$$\text{Logit P} = -20.098 - \left(5.970 * \text{Log10}\big[\text{MR1 Outflow}\big]\right) + \left(4.444 * \big[Z_{eu}\big]\right), \tag{2}$$

where Log10[MR1 Outflow] is the logarithm of the 30-day prior moving window average of outflow from Reservoir #1, in million gallons per day, and $Z_{eu}$ is the euphotic zone depth, in meters. Euphotic zone depth extended from the surface of the water downward to a depth where light intensity fell to 1 percent of that at the surface. In Reservoir #1, greater euphotic zone depth and reduced outflow from Reservoir #1 provided the best estimation of dissolved geosmin concentrations above the human detection threshold in MLogModel2. Of the two explanatory variables, only euphotic zone depth was correlated significantly to geosmin concentrations. The MLogModel 2 was less sensitive than MLogModel 1. When applied to the dataset, MLogModel 2 correctly estimated the likelihood of geosmin concentrations exceeding the human detection threshold 67 percent of the time and not exceeding the human detection threshold 90 percent of the time, resulting in an overall sensitivity of 85 percent (table 3). The reduced sensitivity of the MLogModel2 for Reservoir #1 relative to the other model was attributed, in part, to the fact that the site had a much lower number of observed positive responses (geosmin concentrations above the human detection threshold) than the other site, which decreased the ability of the regression model to accurately estimate that response.

| Classification Table | Predicted Reference | Predicted Positive | Totals | Diagnostic | Percent | Hosmer-Lemshow (p-value) | Pearson Chi-Squared (p-value) | Likelihood Ratio Test (p-value) |
|---|---|---|---|---|---|---|---|---|
| MLogModel 1 for Lake Bowen | | | | | | | | |
| Actual Reference Responses | 11 | 0 | 11 | Specificity | 100 | 9.07 (0.336) | 15.49 (0.216) | 12.12 (0.007) |
| Actual Positive Responses | 1 | 5 | 6 | Sensitivity | 83 | | | |
| Totals | 12 | 5 | 17 | Overall | 94 | | | |
| MLogModel 2 for Reservoir #1 | | | | | | | | |
| Actual Reference Responses | 9 | 1 | 10 | Specificity | 90 | 2.014 (0.981) | 5.09 (0.827) | 8.593 (0.014 |
| Actual Positive Responses | 1 | 2 | 3 | Sensitivity | 67 | | | |
| Totals | 10 | 3 | 13 | Overall | 85 | | | |

**Table 3.** Classification tables for the multiple logistic regression models developed to estimate the likelihood of geosmin concentrations exceeding the human detection threshold of 10 ng/L in Lake William C. Bowen (MLogModel 1) and in Municipal Reservoir #1 (MLogModel 2), Spartanburg County, South Carolina, 2005 to 2009.

Overall, MLogModel 1 indicated greater likelihood for both dissolved geosmin to exceed the human detection threshold during periods of lower nitrogen concentrations. MLogModel 1 indicated a geosmin exceedence during periods of higher water levels in Lake Bowen (greater spillage or overflow compared to controlled releases or flowthrough at Lake Bowen dam). Conversely, MLogModel 2 indicated a greater likelihood of threshold exceedences by dissolved geosmin concentrations during periods of reduced outflow from Reservoir #1 and greater light penetration. It also should be noted that the calibration dataset for each logistic model had a small sample size (less than 20 samples) and was collected during a hydrologic period of extremely low- to average-flow conditions. The small sample size and extreme hydrologic conditions potentially may limit the applicability of these models for above average and, especially, high-flow conditions.

## 9. Conclusions

The occurrence of dissolved geosmin was studied in two reservoirs in Spartanburg County, South Carolina, from August 2005 to June 2009. Lake Bowen and Reservoir #1 are relatively shallow, meso-eutrophic, warm monomictic, cascading impoundments on the South Pacolet River. Overall, water-quality conditions and phytoplankton community assemblages were similar between the two reservoirs but differed seasonally and annually. Median dissolved geosmin concentrations in the reservoirs ranged from 0.004 to 0.006 µg/L, below the human detection threshold of 0.010 µg/L. Annual maximum dissolved geosmin concentrations tended to occur between March and May. In this study, peak dissolved geosmin production occurred in April and May 2008, ranging from 0.050 to 0.100 µg/L at the deeper reservoir sites near the dams. The peak geosmin period coincided with drought conditions that extended the water residence time in both reservoirs and reduced the nutrient inputs. In situ production of geosmin by cyanobacteria was the most probable source of elevated geosmin concentrations in Reservoir #1 and Lake Bowen. Elevated cyanobacterial biovolumes in the reservoirs that were present during the summer of 2008 were related to many environmental factors that have been previously reported to enhance cyanobacterial dominance of the phytoplankton community, including decreased availability of nitrogen, increased phosphorus concentrations, reduced light availability, warmer water temperatures, greater water column stability, and longer residence times. However, unlike previous research, elevated dissolved geosmin concentrations did not coincide with increased cyanobacterial biovolumes and cyanobacteria were not dominating the total phytoplankton community. Therefore, another mechanism was needed that could explain the increased production and release of geosmin in April and May of 2008. Heavy springtime zooplankton grazing and nutrient (N, P, iron) limitation were two plausible mechanisms that could explain seasonal changes to the phytoplankton communities and associated geosmin production in the spring period [12,69,70].

In Lake Bowen, elevated geosmin concentrations near the surface were correlated to environmental conditions indicative of unstratified conditions (higher dissolved-oxygen concentrations and greater $Z_m$). In relation to phytoplankton community structure, elevated geosmin concentrations were correlated to greater Oscillatoria biovolumes, a genus of cyanobacteria with known geosmin-producing species. Elevated geosmin concentrations were correlated to reduced algal biomass and lower total phytoplankton biovolumes. In Reservoir #1, elevated geosmin concentrations near the surface were correlated to greater dissolved-oxygen concentrations and to reduced algal biomass. Rather than a specific genus of cyanobacteria, elevated geosmin concentrations in Reservoir #1 were correlated to a greater fraction of geosmin-producing genera in the cyanobacteria group. However, total cyanobacterial biovolumes were not correlated to geosmin concentrations.

In contrast to elevated geosmin concentrations in surface samples from Lake Bowen and Reservoir #1, environmental factors indicative of stratified or stable water column conditions, including warmer water temperatures, lower $Z_m$, greater $Z_{eu}:Z_m$ ratios, lower dissolved-oxygen concentrations, and maximum RTRM, correlated strongly with elevated cyanobacterial biovolumes. Additionally, in contrast to elevated geosmin concentrations in surface sample, elevated cyanobacterial biovolumes correlated with changes in nutrient levels in the reservoirs, including lower nitrogen concentrations (as ammonia, nitrate plus nitrite, total organic nitrogen, and TN) and elevated hypolimnetic TP concentrations relative to the epilimnetic TP concentrations (considered indicative of release of phosphorus from the sediment during anoxic conditions). In both reservoirs, elevated cyanobacterial biovolumes were correlated to longer residence times, indicative of low-flow or drought conditions, and lower overflow volumes, indicative of lower water levels. A greater fraction of cyanobacteria in the total phytoplankton community and biovolumes of known geosmin- and toxin-producing cyanobacteria genera, including *Cylindrospermopsis, Planktolyngbya, Synechococcus, Synechocystis*, and *Aphanizomenon* correlated with the greater cyanobacteria biovolumes and were the dominant taxa in the cyanobacteria group. During the summer 2008 when these genera dominated the phytoplankton community, low-level (< 0.5 µg/L ) microcystin concentrations also were observed in the two reservoirs.

The BEST analysis selected a six-variable model that best explained the assemblages of all Cyanophyta genera in the two reservoirs during the spring ($\varrho$ = 0.601). The explanatory variables included specific conductance, the two forms of DIN (ammonia and nitrate plus nitrite), chlorophyll *a* and its degradate, pheophytin *a*, and iron (table 2). NMDS bubble plots indicated less nitrate plus nitrite and iron concentrations during the spring 2008 when compared to spring 2007 and 2009 that may be attributed to the change in the assemblage of Cyanophyta genera and greater geosmin concentrations. A four-variable model was selected to best explain the assemblages of only known geosmin-producing genera of Cyanophyta ($\varrho$ = 0.663) and included magnesium, nitrate plus nitrite, chlorophyll *a*:pheophytin *a* ratio (undegraded to degraded pigment), and geosmin.

Logistic regression models indicated geosmin concentrations had the greatest probability (83 percent model sensitivity) of exceeding 10 ng/L during periods of greater overflow (higher water levels in Lake Bowen) relative to flowthrough releases at the dam, lower total nitrogen, and unstratified conditions (greater mixing zone depths) in Lake Bowen. Conversely, in the source water in Reservoir #1, geosmin concentrations above 10 ng/L were more probable (only a 67 percent model sensitivity) during periods of lower outflow but greater light penetration (euphotic zone depth, that correlated to transparency). Fewer periods of geosmin concentrations exceeding 10 ng/L (only 3 compared to 6 at the other site) could have produced the reduced sensitivity, poorer fit, and apparent inverse relations of elevated geosmin concentrations to hydrodynamic conditions relative to Lake Bowen and the raw water. It also should be noted that the calibration dataset for the logistic model had a small sample size (less than 20 samples) and was collected during a hydrologic period of extremely low-flow to average conditions. The small sample size and extreme hydrologic conditions potentially may limit the applicability of these models for above average and, especially, high-flow conditions.

## Author details

Celeste A. Journey[1*], Karen M. Beaulieu[2] and Paul M. Bradley[1]

*Address all correspondence to: cjourney@usgs.gov

1 U.S. Geological Survey, Columbia, SC, USA

2 U.S. Geological Survey, Hartford, CT, USA

## References

[1]  Watson, S. B. Cyanobacterial and eukaryotic algal odour compounds: Signals or by-products? A review of their biological activity. Phycologia (2003). , 42(4), 332-350.

[2]  Krogman, D. W. Cyanobacteria (blue-green algae): Their evolution and relation to other photosynthetic organisms. Bioscience (1981). , 31(2), 121-124.

[3]  Paerl, H. W, & Millie, D. F. Physiological ecology of toxic aquatic cyanobacteria. Phycologia (1996). S):, 160-167.

[4]  Chorus, I, & Bartram, J. Toxic cyanobacteria in water-A guide to their public health consequences, monitoring and management. London, E & FN Spon /Chapman & Hall: (1999).

[5]  Paerl, H. W, Fulton, R. S, Moisander, P. H, & Dyble, J. Harmful freshwater algal blooms, with an emphasis on cyanobacterial. Science World (2001). , 1, 76-113.

[6]  Jüttner, F, & Watson, S. B. Biochemical and ecological control of geosmin and 2-methylisoborneol in source waters. Applied and Environmental Microbiology (2007). , 73(14), 4395-4406.

[7]  Paerl, H. W, & Huisman, J. Climate change: a catalyst for global expansion of harmful cyanobacterial blooms. Environmental Microbiology Reports (2009). , 1(1), 27-37.

[8]  Paerl, H. W. (1988). Nuisance phytoplankton blooms in coastal, estuarine, and inland waters: Limnology and Oceanography, , 33, 823-847.

[9] Downing, J. A, Watson, S. B, & Mccauley, E. Predicting cyanobacteria dominance in lakes. Canadian Journal of Fishery and Aquatic Sciences (2001). , 58, 1905-1908.

[10] Dodds, W. K, Bouska, W. W, Eitzmann, J. L, Pilger, T. J, Pitts, K. L, Riley, A. J, Schloesser, J. T, & Thornbrugh, D. J. Eutrophication of U.S. freshwaters: analysis of potential economic damages. Environmental Science & Technology (2009). , 43, 12-19.

[11] Kimmel, B. L, & Groeger, A. W. (1984). Factors controlling primary production in lakes and reservoirs: a perspective. Lake and Reservoir Management, 1, , 1(1), 277-288.

[12] Bukaveckas, P. A, & Crain, A. S. Inter-annual, seasonal and spatial variability in nutrient limitation of phytoplankton production in a river impoundment. Hydrobiologica (2002). , 481, 19-31.

[13] De Silva, C. A, Train, S, & Rodrigues, L. C. (2005). Phytoplankton assemblages in Brazilian subtropical cascading reservoir system. Hydrobiologica, , 537, 99-109.

[14] Wang, S. H, Dzialowski, A. R, Meyer, J. O, Denoyelles, F, Lim, N. C, Spotts, W. W, & Huggins, D. G. (2005). Relationships between cyanobacterial production and the physical and chemical properties of a Midwestern Reservoir, USA. Hydrobiologica, , 541, 29-43.

[15] Burford, M. A, Johnson, S. A, Cook, A. J, Packer, T. V, Taylor, B. M, & Townsley, E. B. Correlations between watershed reservoir characteristics and algal blooms in subtropical reservoirs. Water Resources (2007). , 41, 4105-4114.

[16] Lehman, J. T, Platte, R. A, & Ferris, J. A. (2007). Role of hydrology in development of vernal clear water pahse in an urban impoundment. Freshwagter Biology, , 52, 1773-1781.

[17] Dillon, P. J. A critical review of Vollenweider's nutrient budget model and other related models. Water Resources Bulletin (1974). , 10(5), 969-989.

[18] Wetzel, R. G. Limnology-Lake and reservoir ecosystems (3d ed.). New York: Academic Press; (2001).

[19] Burkholder, J. M. Phytoplankton and episodic suspended sediment loading-Phosphate partitioning and mechanisms for survival. Limnology and Oceanography (1992). , 37(5), 974-988.

[20] Cooke, G. D, Welch, E. B, Peterson, S. A, & Nichols, S. A. Restoration and management of lakes and reservoirs: Boca Raton, Florida: Taylor and Francis Group; (2005).

[21] Taylor, W. D, Losee, R. F, Torobin, M, Izaguirre, G, Sass, D, Khiari, D, & Atasi, K. Early warning and management of surface water taste-and-odor events. American Water Works Association Research Foundation Reports; (2006).

[22] Zaitlin, B, & Watson, S. B. Actinomycetes in relation to taste and odour in drinking water-Myths, tenets, and truths. Water Research (2006). , 40, 1741-1753.

[23]  Peters, A, Köster, O, Schildknecht, A, & Von Gunten, U. Occurrence of dissolved and particle-bound taste and odor compounds in Swiss lake waters. Water Research (2009)., 43, 2191-2200.

[24]  Suffet, I. H, Corado, A, Chou, D, & Butterworth, S. MacGuire M.J. AWWA taste and odor survey. Journal of American Water Works Association (1996)., 88(4), 168-190.

[25]  Izaguirre, G, Hwang, C. J, Krasner, S. W, & Mcguire, M. J. Geosmin and 2-methyios-borneol from cyanobacteria in three water supply systems. Applied and Environmental Microbiology (1982)., 43(3), 708-714.

[26]  Rashash, D, Hoehn, R, Dietrich, A, Grizzard, T, & Parker, B. Identification and control of odorous algal metabolites. American Water Works Association (AWWA) Research Foundation and AWWA;(1996).

[27]  Young, W. F, Horth, H, Crane, R, Ogden, T, & Arnott, M. Taste and odor threshold concentrations of potential potable water contaminants. Water Research (1996)., 30(2), 331-340.

[28]  Wnorowski, A. U. Tastes and odors in the aquatic environment-A review. Water SA (1992)., 18(3), 203-214.

[29]  Graham, J. L, Loftin, K. A, Ziegler, A. C, & Meyer, M. T. Guidelines for design and sampling for cyanobacterial toxin and taste-and-odor studies in lakes and reservoirs. U.S. Geological Survey Scientific Investigations Report (.2008)., 2008-5038.

[30]  Carmichael, W. W. The toxins of cyanobacteria. Scientific American (1994)., 270, 78-86.

[31]  Graham, J. L, & Jones, J. R. Microcystin in Missouri reservoirs. Lake and Reservoir Management (2009)., 25, 253-263.

[32]  Graham, J. L, Loftin, K. A, Meyer, M. T, & Ziegler, A. C. Cyanotoxin mixtures and taste-and-odor compounds in cyanobacterial blooms from the Midwestern United States. Environmental Science and Technology (2010)., 44(19), 7361-7373.

[33]  Durrer, M, Zimmermann, U, & Jüttner, F. Dissolved and particle-bound geosmin in a mesotrophic lake (Lake Zurich)- Spatial and seasonal distribution and the effect of grazers. Water Research (1999)., 33(17), 3628-3636.

[34]  Callieri, C, Balseiro, E, Bertoni, R, & Modenutti, B. (2004). Picocyanobacterial photo-synthetic efficiency under Daphnia grazing pressure. Journal of Plankton Research, 26, , 12(12), 1471-1477.

[35]  Scheffer, M. Ecology of shallow lakes. The Netherlands, Kluwer Academic Publishers; (2004).

[36]  Boyer, J, Bollens-rollwagen, G, & Bollens, S. M. Microzooplankton grazing before, during, and after a cyanobacterial bloom in Vancouver Lake, Washington, USA. Aquatic Microbial Ecology (2011)., 641, 63-174.

[37] Sterner, R. W. Resource competition during seasonal succession toward dominance by cyanobacteria. Ecology (1989). , 70, 229-245.

[38] Sarnelle, O, & Wilson, A. E. Local adaptation of Daphnia pulicaria to toxic cyanobacteria. Limnology and Oceanography (2005). , 50, 1565-1570.

[39] Hansson, L, Gustafsson, A, Rengefors, S, & Bomark, K. L. Cyanobacterial chemical warfare affects zooplankton community composition: Freshwater Biology (2007). , 52, 1290-1301.

[40] Havens, K. E, James, R. T, East, T. L, & Smith, V. H. N. P ratios, light limitation, and cyanobacteria dominance in a subtropical lake impacted by non-point source nutrient pollution. Environmental Pollution (2003). , 122, 379-390.

[41] Graham, J. L, Jones, J. R, Jones, S. B, Downing, J. A, & Clevenger, T. E. Environmental factors influencing microcystin distribution and concentration in the Midwestern United States. Water Research (2004). , 38, 4395-4404.

[42] Graham, J. L, Loftin, K. A, & Kamman, N. Monitoring recreational freshwaters. LakeLine (2009). , 29(2), 18-24.

[43] Christensen, V. G, Graham, J. L, Milligan, C. R, Pope, L. M, & Ziegler, A. C. Water quality and relation to taste-and-odor compounds in North Fork Ninnescah River and Cheney Reservoir, south-central Kansas, U.S. Geological Survey Scientific Investigations Report 2006-5095; (2006). , 1997-2003.

[44] Dzialowski, A.R, & Smith, .J.H. Development of predictive models for geosmin-related taste and odor in Kansas, USA, drinking water reservoirs. Water Research 2009; 43: 2829-2840.

[45] Smith, V. H. Low nitrogen to phosphorus ratios favor dominance by blue-green algae in lake phytoplankton. Science (1983). , 221, 669-671.

[46] Smith, V. H, Bierman, V. J, Jones, B. L, & Havens, K. E. Historical trends in the Lake Okeechobee ecosystem IV-Nitrogen:phosphorus ratios, cyanobacterial dominance, and nitrogen fixation potential. Archiv fur Hydrobiologie, Monographische Beitrage (1995). , 107, 71-78.

[47] Smith, V. H, Sieber-denlinger, J, & Denoyelles, F. Jr., Campbell S., Pan S., Randke S.J., Blain G.T., Strasser, V.A. Managing taste and odor problems in a eutrophic drinking water reservoir. Lake and Reservoir Management (2002). , 18(4), 319-323.

[48] Jacoby, J. M, Collier, D. C, Welch, E. B, Hardy, F. J, & Crayton, M. Environmental factors associated with a toxic bloom of Microcystis aeruginosa: Canadian Journal of Fishery and Aquatic Sciences (2000). , 57, 231-240.

[49] Downing, J. A, & Mccauley, E. The nitrogen:phosphorus relationship in lakes. Limnology and Oceanography (1992). , 37(5), 936-945.

[50] Dokulil, M. T, & Teubner, K. Cyanobacterial dominance in lakes. Hydrobiologica (2000). , 438, 1-12.

[51] South Carolina Department of Health and Environmental ControlWater classifications and standards. South Carolina Department of Health and Environmental Control Code of Regulations, State Register Regulation (2008). http://www.scdhec.net/environment/water/regs/r61-68.docaccessed 28 August 2012)., 61-68.

[52] Journey, C. A, & Abrahamsen, T. A. Limnological conditions in Lake William C. Bowen and Municipal Reservoir #1, Spartanburg County, South Carolina, August to September 2005, May 2006, and October 2006. U.S. Geological Survey Open-File Report (2008). http://pubs.usgs.gov/of/2008/1268/accessed 28 August 2012)., 2008-1268.

[53] Journey, C. A, Caldwell, A. W, Feaster, T. D, Petkewich, M. D, & Bradley, P. M. Concentrations, loads, and yields of nutrient and suspended sediment in the South Pacolet, North Pacolet, and Pacolet Rivers, northern South Carolina and southwestern North Carolina, October 2005 to September 2009. U.S. Geological Survey Scientific Investigations Report (2010). http://pubs.usgs.gov/sir/2010/5252/accessed 28 August 2012)., 2010-5252.

[54] Journey, C. A, Arrington, J. M, Beaulieu, K. M, Graham, J. L, & Bradley, P. M. Limnological conditions and occurrence of taste-and-odor compounds in Lake William C. Bowen and Municipal Reservoir #1, Spartanburg County, South Carolina, U.S. Geological Survey Scientific Investigations Report 2011-5060; (2011). http://pubs.usgs.gov/sir/2011/5060/accessed 28 August 2012)., 2006-2009.

[55] Vallentyne, J. R. Principles of modern limnology: American Science (1957)., 45, 218-244.

[56] Welch, E. B. Ecological effects of wastewater: London, Chapman & Hall; (1992).

[57] Helsel, D. R. Nondetects and data analysis-Statistics for censored environmental data. New York Wiley (2005).

[58] Helsel, D. R, & Hirsch, R. M. Statistical methods in water resources. Amsterdam, The Netherlands, Elsevier Science Publishers (1992).

[59] Clarke, K. R, & Ainsworth, M. A method of linking multivariate community structure to environmental variables. Marine Ecology Progress Series (1993)., 92, 205-219.

[60] Clarke, K. R, & Warwick, R. M. Change in marine communities-An approach to statistical analysis and interpretation (2d ed.). Plymouth, U.K.: PRIMER-E; (2001).

[61] Nagle, D. D, Campbell, B. G, & Lowery, M. A. Bathymetry of Lake William C. Bowen and Municipal Reservoir #1, Spartanburg County, South Carolina, 2008. U.S. Geological Survey Scientific Investigations Map 3076; (2009). http://pubs.usgs.gov/sim/3076/accessed on 28 August 2012).

[62] Purvis, J. C, Tyler, W, & Sidlow, S. F. Climate Report G-General Characteristics of South Carolina's Climate. Columbia, South Carolina Department of Natural Resources Water Resources Division; (1990)., 5.

[63] Kronberg, N, & Purvis, J. C. Climates of the States-South Carolina, in Climatography of the States. Washington, D.C.: U.S. Department of Commerce; (1959).

[64] National Climatic Data CenterClimatography of the United States for Greenville-Spartanburg Airport Station near Greer, S.C.; (2004). http://www.ncdc.noaa.gov/oa/climate/normals/usnormals.htmlaccessed 26 July 2010).(20-1971)

[65] Forsberg, C, & Ryding, S. Eutrophication parameters and trophic state indices in 30 Swedish waste-receiving lakes. Archives of Hydrobiology (1980). , 89, 189-207.

[66] Nürnberg, G. K. Trophic state of clear and colored, soft- and hardwater lakes with special consideration of nutrients, anoxia, phytoplankton and fish. Lake and Reservoir Management (1996). , 12, 432-447.

[67] White, E. Utility of relationships between lake phosphorus and chlorophyll-a as predictive tool in eutrophication control studies. New Zealand Journal of Marine and Freshwater Research (1989). , 23, 25-41.

[68] Fahnenstiel, G. L, Carrick, H. J, & Iturriaga, R. Physiological characteristics and food-web dynamics of Synechococcus in Lakes Huron and Michigan. Limnology and Oceanography (1991). , 36(2), 219-234.

[69] Vanni, M. J, & Temte, J. Seasonal patterns of grazing and nutrient limitation of phytoplankton in a eutrophic lake. Limnology and Oceanography (1990). , 35, 697-709.

[70] Smith, V. H, & Bennett, S. J. Nitrogen:phosphorus supply ratios and phytoplankton community structure in lakes. Archives of Hydrobiology (1999). , 146, 37-53.

# Chapter 2

# Managing the Effects of Endocrine Disrupting Chemicals in Wastewater-Impacted Streams

Paul M. Bradley and Dana W. Kolpin

Additional information is available at the end of the chapter

## 1. Introduction

A revolution in analytical instrumentation circa 1920 greatly improved the ability to characterize chemical substances [1]. This analytical foundation resulted in an unprecedented explosion in the design and production of synthetic chemicals during and post-World War II. What is now often referred to as the 2nd Chemical Revolution has provided substantial societal benefits; with modern chemical design and manufacturing supporting dramatic advances in medicine, increased food production, and expanding gross domestic products at the national and global scales as well as improved health, longevity, and lifestyle convenience at the individual scale [1, 2]. Presently, the chemical industry is the largest manufacturing sector in the United States (U.S.) and the second largest in Europe and Japan, representing approximately 5% of the Gross Domestic Product (GDP) in each of these countries [2]. At the turn of the 21st century, the chemical industry was estimated to be worth more than $1.6 trillion and to employ over 10 million people, globally [2].

During the first half of the 20th century, the chemical sector expanded rapidly, the chemical industry enjoyed a generally positive status in society, and chemicals were widely appreciated as fundamental to individual and societal quality of life. Starting in the 1960s, however, the environmental costs associated with the chemical industry increasingly became the focus, due in part to the impact of books like "Silent Spring" [3] and "Our Stolen Future" [4] and to a number of highly publicized environmental disasters. Galvanizing chemical industry disasters included the 1976 dioxin leak north of Milan, Italy, the Love Canal evacuations in Niagara, New York beginning in 1978, and the Union Carbide leak in Bhopal, India in 1984 [2].

Understanding the environmental impact of synthetic compounds is essential to any informed assessment of net societal benefit, for the simple reason that any chemical substance that is in commercial production or use will eventually find its way to the environment [5]. Not surprisingly given the direct link to profits, manufacturers intensely investigate and routinely document the potential benefits of new chemicals and chemical products. In contrast, the environmental risks associated with chemical production and uses are often investigated less intensely and are poorly communicated.

An imbalance in the risk-benefit analysis of any synthetic chemical substance or naturally occurring chemical, which presence and concentration in the environment largely reflects human activities and management, is a particular concern owing to the fundamental link between chemistry and biology. Biological organisms are intrinsically a homeostatic balance of innumerable internal and external chemical interactions and, thus, inherently sensitive to changes in the external chemical environment.

## 1.1. Environmental contamination: historical emphases

Much of the focus on environmental contamination in the decades since the institution of the 1970 Clean Air and 1972 Clean Water Acts in the U.S. and comparable regulations in Europe and throughout the world has been on what are now frequently referred to as conventional "priority pollutants" (so-called legacy contaminants). These include two primary groups: 1) wastewater nutrients and pathogens, and 2) a small subset of anthropogenic chemicals with relatively well-recognized toxicological risks, most notably "persistent bioaccumulative toxicants" (PBT) or "persistent organic pollutants" (POP). For example, the wastewater treatment infrastructure primarily reflects the early-recognized need to manage the environmental release of nutrients and human pathogens associated with human and animal waste. Likewise, the second driver of environmental regulation primarily concerns the relatively small number of known toxins or toxin-containing contaminant groups that, at least historically, were widely used in industry, frequently released accidentally or intentionally to the environment, are typically observed at part per billion (ppb) to part per million (ppm) concentrations, and are often well above recognized toxicological impact thresholds including carcinogenic thresholds. Managing the environmental impacts of these chemicals was the original motivation for and continues to be the primary focus of wastewater and hazardous waste regulations in the U.S.

## 1.2. Environmental contamination: expanding emphasis

The contaminants of historical environmental focus (conventional priority pollutants) are but a small fraction of the known and unknown chemicals that are potential environmental contaminants. As of September 2012, the Chemical Abstracts Service (CAS) has registered more than 68 million organic and inorganic chemical substances (not including proteins, etc.) [6]. While this chemosphere of known anthropogenic chemicals is impressive, the actual number of potential anthropogenic contaminants is incalculably larger, due to the continuing research, development, and marketing of novel chemical products and to the countless, unmanaged chemical transformations that occur following release to the environment [5].

The numbers and quantities of anthropogenic chemicals continue to increase rapidly [6]. In March 2004, the number of CAS registered organic and inorganic chemical substances was approximately 23 million [5, 6]. Thus, the current estimate of approximately 68 million indicates a three-fold increase in the number of known chemicals between 2004 and 2012 [6]. To put this issue in perspective, Bohacek et al. [5, 7] provided a glimpse of the magnitude of the potential anthropogenic contaminant pool. Conservatively limiting the candidate atoms to C, N O, and S and the total number of structural atoms to 30 or less, Bohacek et al. estimated over $10^{60}$ distinct possible structures [7]. Obviously, inclusion of additional common constituent atoms (e.g. phosphorous and halogens) or increasing the numbers of atoms per molecule would greatly increase this estimate [5].

The environmental impact of any anthropogenic chemical can be amplified due to the formation of numerous unidentified daughter products resulting from subsequent chemical and biological transformation processes in the environment [5]. A common example among the contaminants of historical focus is the reductive dechlorination of trichloroethene (TCE) and its intermediate daughter products (dichloroethenes, DCE) to form vinyl chloride (VC) [8]. Historically, TCE has been widely employed in dry cleaning and as a degreasing agent in industry. TCE has an MCL of 5 $\mu$g/L and a $10^{-4}$ Cancer Risk level of 300 $\mu$g/L [9, 10]. In contrast, VC is a demonstrated human carcinogen with an MCL of 2 $\mu$g/L and a $10^{-4}$ Cancer Risk level of 2 $\mu$g/L [9, 10]. An example among the contaminants of more recent concern is the transformation of 4-nonylphenol polyethoxylate compounds (primarily used as nonionic surfactants) to 4-nonylphenol (4-NP) and nonylphenol mono- and di-ethoxylates. The aquatic toxicity of 4-nonylphenol 16-ethoxylate (NP16EO) is 110 mg/L for fish, while that of 4-nonylphenol is 1.4 mg/L [11]. The 4-nonylphenol polyethoxylates are not estrogenically active. In contrast, 4-nonylphenol is a demonstrated xenoestrogen with a relative binding affinity of 2.1 $\times$ $10^{-4}$ relative to the natural estrogen, 17$\beta$-estradiol (E2) [11].

Thus, considering just the inventoried substances, only about 0.4% (>295000) of the more than 68 million (as of Sept 08 2012) commercially available organic and inorganic chemical substances registered in CAS are government inventoried or regulated worldwide [6]. Thus, even considering only these registered commercial chemicals, each of which are or may become environmental contaminants, the vast majority are unregulated and largely unmonitored in the environment. Environmental contaminants, which are currently unregulated, are often referred to as a group as "emerging contaminants," in an effort to distinguish them from the conventional priority pollutants (legacy contaminants).

### 1.3. Emerging concern versus emerging contaminants

The term, "emerging contaminant," is misleading in the unintended implication that these chemicals are collectively new to the environment. In fact, large fractions of these emerging contaminants have been in use and, by extension, have been present in the environment for many years. However, many of these compounds occur in the environment at concentrations well below historical ppb to ppm analytical detection limits. The environmental threat associated with these contaminants has gone largely unrecognized or undefined, due to a lack of analytical methods of sufficient sensitivity and resolution to allow detection at environmentally relevant concentrations. Thus, while newly synthesized and produced commercial chemicals would in fact fit the perception; the "emerging" characteristic for the majority of these unregulated compounds is not recent environmental release, but a nascent and growing appreciation of their real and potential impacts in the environment.

To illustrate the magnitude of the problem, consider just the pharmaceutical compounds, chemicals synthesized specifically to affect a biological impact. Pharmaceuticals were estimated to be approximately 23% of the global chemical production in 2000 [2]. More than 12000 approved prescription and "over the counter" (non-prescription) drug products and formulations are currently listed by the U.S. Food and Drug Administration, along with more than 5000 discontinued products [12, 13]. More than 80 new drug products or formulations were approved in 2011 [12, 13]. In contrast, analytical methods for detection and quantification in environmentally relevant matrices (e.g. sediment and water) exist for only a small fraction of the pharmaceuticals approved for use in the U.S. For example, the U.S. Geological Survey (USGS) has developed one of the more comprehensive analytical methods for the monitoring of pharmaceuticals in the environment [14]. However, the currently available USGS direct aqueous injection liquid chromatography/mass spectrometry/mass spectrometry (LC/MS/MS) method for filtered water includes only approximately 112 pharmaceutical compounds [14]. Similarly, the U.S. Environmental Protection Agency (USEPA) method for pharmaceutical and personal care products in water, soil, sediment, and biosolids by LC/MS/MS covers only about 60 pharmaceutical compounds [15].

Using these methods as a measure of the analytical coverage of pharmaceutical compounds in the environment and not including environmental transformation products, the vast majority of pharmaceutical chemicals, which have been in use and, consequently, may reasonably be expected to occur in the environment, are not currently monitored in the environment. From this perspective, these contaminants are more appropriately viewed as emerging concerns.

## 1.4. Contaminants of emerging concern

The potential impacts of contaminants of emerging concern (CEC) on the environment, in general, and on natural surface-water and riparian ecosystems, in particular, are a critical environmental management issue in the U.S. and Europe [11, 16]. CEC is a "catch-all" phrase that refers to a wide range of chemicals, which occurrence in and potential impacts on the environment have long been suspected but only recently validated with the advent of sensitive modern analytical capabilities. The CEC umbrella covers several broad classes of contaminants that are loosely categorized according to source, original intended use, and/or primary mode of ecological impact and which include: pharmaceuticals and personal care products, organic wastewater compounds, antimicrobials, antibiotics, animal and human hormones, as well as domestic and industrial detergents.

## 1.5. Endocrine disrupting chemicals (EDC)

Many CEC interact with animal endocrine systems and, consequently, are classified as endocrine disrupting chemicals (EDC). The endocrine system, sometimes referred to as the hormone system, is present in all vertebrate animals and consists of glands, hormones, and receptors that regulate all biological functions including metabolism, growth, behavior, and reproduction [see for example, 11, 17, 18, 19]. Endocrine hormones include the estrogens, androgens, and thyroid hormones. The USEPA defines an EDC as:

---

"An exogenous agent that interferes with the synthesis, secretion, transport, binding, action, or elimination of natural hormones in the body that are responsible for the maintenance of homeostasis, reproduction, development, and/or behavior."[17]

---

Because the common conceptualization of "endocrine systems" is typically associated with vertebrates, much of the attention on environmental EDC has been focused on endocrine disruption impacts in vertebrate animals, particularly aquatic vertebrates [11, 18-24] and associated terrestrial food webs [25]. It is important to realize, however, that invertebrates (molluscs, insects, etc.) also have hormone systems that regulate biological function and maintain homeostasis [26-29]. Thus, many invertebrates are also susceptible to the impacts of EDC [26-29]. Because invertebrates account for approximately 95% of all animals on earth and are critical elements of freshwater environments, the potential impacts of EDC on these organisms cannot be overlooked [26].

EDC threaten the reproductive success and long-term survival of sensitive aquatic populations. The impacts of EDC in the environment are detectable at multiple ecological endpoints, including induction of male vitellogenin (egg yolk protein) expression [30], skewed sex ratios and intersex characteristics [31], degraded predator avoidance behavior [23, 24], as well as reproductive failure and population collapse in sensitive fish species [22]. All of these impacts have been observed at concentrations that have been widely documented in wastewater effluent and effluent-impacted surface-water systems [16, 23, 24, 30, 31]. The widespread co-occurrence of EDC [see for example, 16] and intersex characteristics in black basses (Micropterus species) [20, 21] in U.S. streams suggests endocrine disruption may be pervasive in aquatic populations and emphasizes the potential EDC threat to high value, sensitive surface-water and riparian ecosystems.

### 1.5.1. Natural and xenobiotic EDC

EDC can be divided into two general classes: endocrine hormones and endocrine mimics (xenobiotics including xenoestrogens, xenoandrogens, phytoestrogens, etc.).

Endocrine hormones are natural or synthetic chemicals produced specifically to interact with the hormone binding sites of animal endocrine systems. The release of endocrine hormones, including estrogens and androgens, is a particular concern owing to their high endocrine activity/potency and additive effects. These hormones have been identified as primary estrogenic agents in wastewater effluent [22, 32-39]. Examples of reproductive hormones that are commonly detected in effluent-affected ecosystems are 17β-estradiol (E2), estrone (E1), testosterone (T), and the synthetic birth control compound, 17α-ethinylestradiol (EE2).

Other endocrine disrupting chemicals share sufficient structural similarity with the endocrine hormones to interact with animal endocrine receptors sites and trigger organ- and organism-level endocrine responses. These endocrine mimics generally exhibit less endocrine reactivity, but are essentially ubiquitous in wastewater, are often reported at concentrations 3-5 orders of magnitude higher than the endocrine hormones, and have been detected in the majority of investigated surface-water systems. Examples of these structural analog EDCs include organic wastewater compounds like the ubiquitous detergent metabolite, nonylphenol and naturally-occurring phytoestrogens.

### 1.5.2. Environmental EDC sources

Numerous potential sources of EDC to the environment have been documented, including: pharmaceutical industry, other industry and manufacturing, land application of municipal biosolids, landfills and associated leachates, livestock and aquaculture operations, domestic septic systems, latrine and vault toilets, and municipal and industrial wastewater treatment plants (Fig. 1) [16].

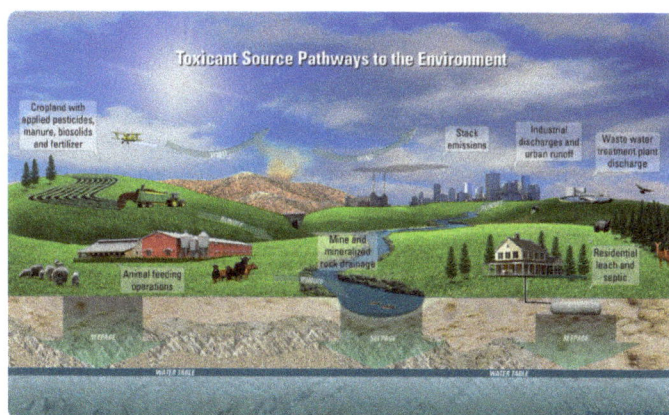

**Figure 1.** Potential sources of EDC in the environment (figure by E.A. Morrissey, USGS).

Among these, wastewater treatment plants (WWTP) discharge directly to surface waters and are often a particular concern for downstream surface-water and riparian ecosystems [11, 16, 23, 30, 40].

### 1.6. Chapter focus

Recent research indicates that a substantial and potentially protective capacity for in situ EDC biodegradation exists in the sediments and water columns of effluent-affected, surface-water systems in the U.S. However, the efficiency and circumstances of biodegradation can vary substantially between stream systems and between compound classes. Likewise, the potentials for in situ biodegradation of a large number of EDC remain untested. Improved understanding of the extent of contaminant occurrence and of the tendency of surface-water receptors to degrade or to accumulate these wastewater contaminants is needed to support development of regulatory contaminant criteria and maximum load polices for the release of EDC to the environment. This chapter focuses on the impacts of wastewater EDC on downstream surface-water and riparian ecosystems and on the potential importance of the natural assimilative capacity of surface-water receptors as a mechanism for managing these EDC impacts.

## 2. EDC risk in wastewater-impacted surface-water and riparian ecosystems

The environmental or ecological risk associated with EDC can be defined in a number of ways. In one approach (Fig. 2), environmental EDC risk can be viewed as the net result of the interaction of three conceptual drivers:

- Environmental EDC occurrence and distribution

- EDC impact thresholds of species in downstream ecosystems

- EDC attenuation capacity of the surface-water receptor

The first two drivers are widely recognized and, currently, are the focus of a majority of investigations of environmental EDC risk. By comparison, relatively little is known about the environmental fate, transport and persistence of EDC.

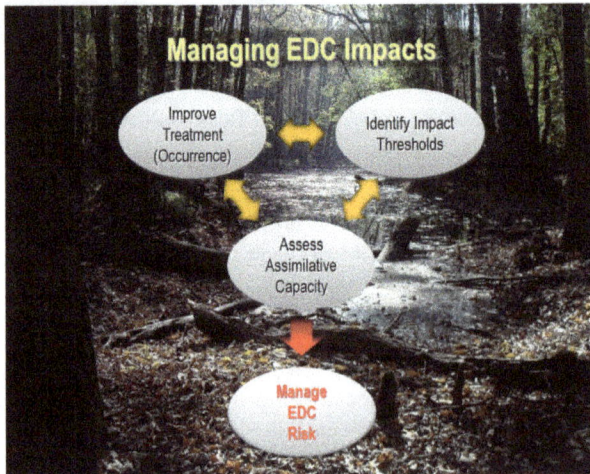

**Figure 2.** Interaction of occurrence and distribution, adverse impact thresholds and site-specific assimilative capacity as drivers of EDC environmental risk.

## 2.1. EDC occurrence and distribution

The risks of EDC are clearly predicated on their presence, concentration, matrix of occurrence, and bioavailability in the environment. Thus, developing analytical methods to detect and quantify EDC in water, sediment, and other environmental matrices has been a primary focus of field investigations over the past two decades. Current approaches to assessing EDC occurrence and distribution in the environment fall into two primary categories, selective and non-selective methods.

Selective methods have traditionally been the cornerstone of contaminant monitoring and this general approach has been critical to the documentation of EDC in the environment, identification of potential EDC sources, and the establishment of EDC as a fundamental environmental threat. Full scan, high-resolution Liquid Chromatography/Mass Spectrometry (LC/MS) is the mainstay of environmental EDC analysis, due primarily to the fact that many of these compounds are not volatile in the inlet of gas chromatography (GC) systems [41]. The complexities of environmental matrices and environmental EDC mixtures have led to wide use of LC/TOF/MS (time of flight, TOF) combined with isotopically labeled internal standards in order to achieve full spectral mass sensitivity, required analytical resolving power, and high mass-measurement accuracies sufficient to estimate elemental composition [41-43]. The fundamental limitation to these methods is the requirement for clean-up and separation methods tailored to selected target analytes and chemically-related unknowns. In essence, in analytical chemistry "what you see is largely dictated by what you look for."

In light of the largely unknown nature of environmental EDC mixtures, using selective analytical methods to assess the total endocrine disrupting impact in a given environmental setting is not straightforward [32]. To address this general screening need, a number of biologically based assays (BBA) have been developed to assess the total amount of a specific endocrine activity (e.g. estrogenicity) that is present in the environment [32]. For example, a number of assays have been developed and successfully employed to assess total estrogenic activity, including the Yeast Estrogen Screen (YES) [44] and the bioluminescent version (BLYES) [45]. BBA are sensitive, cost-effective tools for assessing total estrogenicity of water samples. A priori knowledge of individual estrogenic compounds is unnecessary, because the assay measures target (estrogen) receptor binding. Thus, BBA can add considerable ecological relevance to selective analytical chemical results.

Current areas of active research include application of these analytical improvements to quantify the distribution of EDC between matrices. While a number of studies have demonstrated EDC impacts at concentrations observed in wastewater-impacted surface waters, the tendency of aromatic and polyaromatic contaminants to partition to the sediment phase is well recognized and sediment concentrations can exceed water concentrations by several orders of magnitude [46-48].

## 2.2. EDC environmental impact thresholds of aquatic populations

As noted earlier, the impacts of EDC in the environment involve multiple ecological endpoints. The adverse impact threshold for each of these ecological endpoints may differ substantially. Moreover, the threshold for each of these ecological endpoints can vary substantially among organisms within a specific setting and among environmental settings.

EDC present fundamental challenges to the traditional toxicological assessment approach. Historically, toxicological assessments have been based on a "dose alone determines the poison" maxim [49-51] and the use of a generalized monotonic dose response curve (threshold or linear nonthreshold models) for estimating adverse impact thresholds for individual toxins [51, 52]. However, a number of EDC, including several hormones, show nonmonotonic U-shaped and inverted U-shaped dose response curves for different biological endpoints [52-55]. In fact, the compelling argument has been made that threshold assumptions do not apply to EDC because these compounds are endogenous molecules or mimic endogenous molecules (like estrogen) that are critical to development. Thus, homeostatic balance is disrupted and the "threshold" is automatically exceeded with exposure to the EDC.

While the viewpoint that EDC do not have an acceptable "No Observable Effect Level" (NOEL) is compelling, practical management of EDC risk will depend on establishment of regulatory adverse impact thresholds ("acceptable risk" thresholds). The several challenges to a comprehensive understanding of environmental EDC risk and development of "acceptable risk" thresholds for EDC include the facts that: (1) these compounds generally occur in the environment as complex chemical mixtures, not single compounds, (2) many EDC exhibit trans-generational (epigenetic) impacts, (3) EDC impacts can vary substantially over the life-cycle of an organism and are often particularly severe during gestation and early development, and (4) EDC impacts can occur long after exposure. Development and implementation of appropriate methods for assessing EDC adverse impacts at multiple endpoints are environmental priorities.

In the U.S., regulatory adverse impact thresholds for EDC are under development and not currently available for implementation. Although thresholds for acceptable risk remain undefined, a number of studies have demonstrated that EDC concentrations currently observed in the environment often exceed levels known to cause adverse effects in aquatic populations. To illustrate, consider again E2, E1, and 4-NP.

Both E2 and E1 induce vitellogenesis and feminization in fish species [35, 39, 56-61] at dissolved concentrations as low as 1-10 ng/L [35, 39]. Municipal wastewater treatment plant (WWTP) effluent concentrations of 0.1-88 ng/L and 0.35-220 ng/L have been reported for E2 and E1, respectively. More common detections are in the range of 1-10 ng/L [see for review, 46]. E2 and E1 concentrations above 100 ng/L have been reported in surface waters [16], but are typically in the range of <0.1-25 ng/L [see for review, 46]. Because sensitive fish species are affected by concentrations as low as 1 ng/L and because the effects of reproductive hormone and non-hormonal EDCs are often additive [62], such dissolved concentrations are an environmental concern. Furthermore, estrogen concentrations in surface-water sediment can be up to 1000 times higher per volume than in the associated water column, ranging from 0.05-29 ng/g dry weight [see for review, 46].

Alkylphenol contaminants, like 4-NP, exhibit less estrogenic reactivity [36, 38] than E2, but are ubiquitous in WWTP effluent [11, 16], have been reported at concentrations up to 644 μg/L [40, 48] and have been detected in the majority of investigated surface-water systems [see for review, 63]. Nonylphenol-based compounds are the primary alkylphenol contaminants detected in WWTP-impacted stream systems [16], because nonylphenol ethoxylates constitute approximately 82 % of the world production of alkylphenol ethoxylate [11]. The widespread occurrence of 4-NP in stream systems is attributable to WWTP effluents and microbial transformation of effluent-associated nonylphenol ethoxylates to 4-NP in anoxic, surface-water sediments [47]. Short-chain nonylphenol ethoxylates and 4-NP are produced within WWTP from biodegradation of ubiquitous, nonylphenol ethoxylate nonionic surfactants [47]. 4-NP that is released to the stream environment, rapidly and strongly adsorbs to the sediments suspended in the water column and to the bedded sediments [47, 48].

### 2.3. EDC attenuation and persistence

In contrast to the focus on assessment of EDC occurrence and distribution and EDC adverse impact thresholds, comparatively little is known about the environmental attenuation or persistence of EDC. Environmental persistence, however, is a fundamental component of contaminant environmental risk.

Persistence can be viewed as the resistance of the contaminant molecule to biological or chemical transformations. Pseudo-persistence may also result in settings where the contaminant molecule is continually replenished (e.g. wastewater-impacted systems). Because the longer a contaminant persists in the environment the greater the chance that the contaminant will reach and eventually exceed an adverse impact threshold, improved understanding of the fate of EDC in the environment is essential to a comprehensive assessment of EDC environmental risk.

Conservative mechanisms of contaminant attenuation like dilution and sorption have been the historical foundation of wastewater management in surface-water systems. However, the fact that EDC may trigger organ-, organism-, and community-level responses at ng/L concentrations raises concerns about the ultimate reliability of attenuation mechanisms that do not directly degrade endocrine function [64]. Endocrine disruption at hormone concentrations (1-10 ng/L) [35, 39, 60, 61], which have become detectable only with recent analytical innovations, illustrates this concern and emphasizes the importance of characterizing non-conservative, contaminant attenuation processes. In the following section, recent findings on the potential for EDC biodegradation are presented to illustrate the potential importance of this environmental attenuation mechanism and identify existing data gaps that need to be addressed in order to employ natural attenuation for the management of EDC environmental risk.

# 3. Biodegradation of wastewater EDC in surface-water receptors

This section focuses on EDC biodegradation as an example of the potential importance of the natural assimilative capacity of surface-water receptors as a mechanism for managing EDC impacts in aquatic habitats. Recent results demonstrating the potential for EDC biodegradation in wastewater-impacted streams are discussed along with several environmental factors known to affect the efficiency of EDC biodegradation.

## 3.1. Methods

The potential for EDC biodegradation was assessed in microcosms using $^{14}C$-radiolabeled model compounds [see for example, 46, 63] representing the two general classes of EDC: endocrine hormones and endocrine mimics. Each $^{14}C$-model contaminant compound contained a cyclic (aromatic) ring structure that is considered essential to compound toxicity and biological activity. Consequently, the $^{14}C$-radiolabel of each model contaminant was positioned within the aromatic ring such that recovery of $^{14}C$-radioactivity as mineralization products ($^{14}CO_2$ and/or $^{14}CH_4$) indicated ring cleavage and presumptive loss of endocrine activity [see for example, 46, 63].

Headspace concentrations of $CH_4$, $^{14}CH_4$, $CO_2$, and $^{14}CO_2$ were monitored by analyzing 0.5 mL of headspace using gas chromatography/radiometric detection (GC/RD) combined with thermal conductivity detection. Compound separation was achieved by isocratic (80 C), packed-column (3 m of 13× molecular sieve) gas chromatography. The headspace sample volumes were replaced with pure oxygen (oxic treatments) or nitrogen (anoxic treatments). Dissolved phase concentrations of $^{14}CH_4$ and $^{14}CO_2$ were estimated based on Henry's partition coefficients that were determined experimentally as described previously [65, 66]. The GC/RD output was calibrated by liquid scintillation counting using $H^{14}CO_3^-$. To confirm the presence of oxygen (headspace $[O_2]$ = 2-21% by volume) in oxic treatments or the absence of oxygen (headspace $[O_2]$ minimum detection limit = 0.2 part per million by volume) in anoxic treatments, headspace concentrations of $O_2$ were monitored throughout the study using GC with thermal conductivity detection.

## 3.2. EDC biodegradation in surface water: environmental factors

While most investigations into the potential for EDC biodegradation continue to focus on WWTP, a growing number of studies address the potential for biodegradation of CEC, in general, and EDC, specifically, in a variety of environmental settings. For simplicity, we focus here on recent findings from USGS scientists, which illustrate that a substantial and potentially exploitable capacity for in situ biodegradation of a number of CEC, including known EDC, exists in the sediments and water columns of surface-water systems in the U.S. The efficiency and circumstances of biodegradation, however, vary substantially among stream locations, stream systems, environmental matrices, and EDC compounds. These findings illustrate the data gaps that need to be addressed in order to develop best management practices for individual surface-water systems and specific compound classes.

### 3.2.1. Between and within stream variation

Biodegradation of E2, E1, and testosterone (T) was investigated recently in three WWTP-affected streams in the U.S. [46]. Relative differences in the mineralization of [4-$^{14}C$] hormones were assessed in oxic microcosms containing saturated sediment from locations upstream and downstream of the WWTP outfall in each system. The results for E2 are shown in figure 3.

Sediment collected upstream from the WWTP outfall in each of the three surface-water systems demonstrated substantial aerobic mineralization of [4-$^{14}$C] E2 (Fig. 3), with initial linear rates of $^{14}CO_2$ recovery ranging from approximately 1% d$^{-1}$ (percent of theoretical) for E2 mineralization in Fourmile Creek (Iowa) sediment (Fig. 3) up to approximately 3 % d$^{-1}$ for E2 mineralization in Boulder Creek (Colorado) sediment. The recovery of $^{14}CO_2$ observed in this study was attributed to microbial activity, because no significant recovery of $^{14}CO_2$ (recovery less than 2% of theoretical) was observed in sterilized control microcosms. Recovery of $^{14}CO_2$ was interpreted as explicit evidence of microbial cleavage of the steroid "A" ring and loss of endocrine activity, as demonstrated previously using the YES assay [67, 68]. The results are consistent with previous reports of microbial transformation and "A" ring cleavage of [4-$^{14}$C] E2 in rivers in the United Kingdom [67] and Japan [69] and suggest that the potential for aerobic biodegradation of reproductive hormones may be widespread in stream systems.

Upstream sediment demonstrated statistically significant mineralization of the "A" ring of E2. This result indicated that, in combination with sediment sorption processes which effectively scavenge hydrophobic contaminants from the water column and immobilize them in the vicinity of the WWTP outfall, aerobic biodegradation of reproductive hormones can be an environmentally important mechanism for non-conservative (destructive) attenuation of hormonal endocrine disruptors in effluent-affected streams.

The E2 "A" ring mineralization was substantially greater in sediment collected immediately downstream from the WWTP outfall in the effluent-dominated Boulder Creek and South Platte River (Colorado) study reaches (Fig. 3). The recovery of $^{14}CO_2$ in the immediate downstream sediment was approximately twice that observed upstream of the outfall in Boulder Creek and the South Platte River. Effluent may enhance *in situ* biodegradation of hormone contaminants by introducing WWTP-derived degradative populations or by stimulating the indigenous microorganisms through increased supply of nutrients and co-metabolites. The fact that no difference in E2 "A" ring mineralization was observed between upstream and downstream locations in the less effluent-affected Fourmile Creek suggested that the stimulation of E2 mineralization observed in the Boulder Creek and South Platte River study reaches was attributable to some characteristic of the WWTP effluent and may be concentration dependent. These observations illustrate the substantial variation in EDC biodegradation that may occur at different locations within a stream system and the need to account for location, particularly proximity to recognized sources, when assessing the potential for biodegradation of EDC in the environment.

These results also demonstrate that substantial variation in EDC biodegradation may occur between different stream systems. In Fourmile Creek, location relative to the WWTP had little effect on E2 biodegradation rates. However, location was a major influence on E2 biodegradation in Boulder Creek and in the South Platte River. Similarly, initial linear rates of $^{14}CO_2$ recovery in sediment collected immediately downstream of the WWTP outfalls ranged from approximately 4 % d$^{-1}$ (percent of theoretical) for E2 mineralization in Fourmile Creek up to approximately 11 % d$^{-1}$ for E2 mineralization in Boulder Creek.

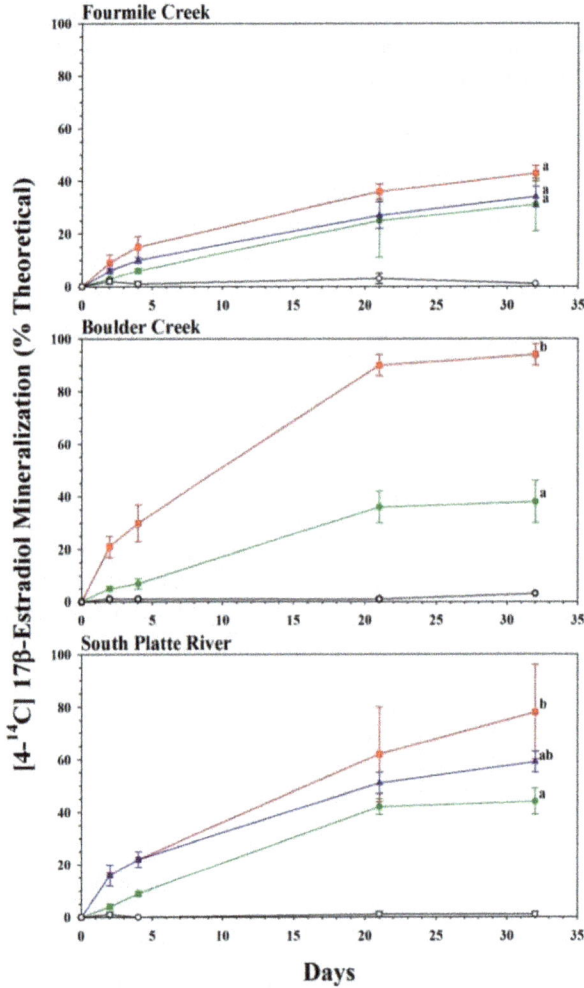

**Figure 3.** Mineralization of $^{14}$C-E2 to $^{14}$CO$_2$ in oxic microcosms containing sediment collected upstream (green), immediately downstream (red) and far downstream (blue) of the WWTP outfalls in Fourmile Creek, Boulder Creek and South Platte River. Black indicates sterile control.

### 3.2.2. Effects of environmental matrices

The effects of the environmental matrix on EDC biodegradation were evaluated for stream biofilm, sediment, and water collected from locations upstream and downstream from a WWTP outfall in Boulder Creek using E2, EE2, and 4-$n$-NP (linear chain isomer) as [14]C-model substrates [70] (Fig.4). Initial time intervals (0-7 d) evaluated biodegradation by the microbial community at the time of sampling. Later time intervals (70 and 185 d) provided insight into changes in EDC biodegradation potential as the microbial community adapted to the absence of light for photosynthesis (i.e. shifted from photosynthetic based community to a predominantly heterotrophic community).

No statistically significant mineralization ($p < 0.05$) of 4-$n$-NP or E2 was observed in the biofilm or water matrices during the initial time step (7 d), whereas statistically significant mineralization of 4-$n$-NP and E2 was observed in the sediment matrices. Mineralization was not observed in autoclaved matrices; therefore, mineralization observed in all matrices was attributed to biodegradation. After 70 d, mineralization of 4-$n$-NP and E2 was observed in the biofilm and sediment matrices, and after 185 d biodegradation of these compounds was observed in all matrices. Mineralization of EE2 was observed only in sediment treatments.

In this study [70], the sediment matrix was more effective than the biofilm and water matrices at biodegrading 4-NP, E2, or EE2. Biodegradation of all three EDC was generally least efficient in water only. These observations illustrate the substantial variation in EDC biodegradation that may occur in different environmental matrices from the same location within a stream system and the need to evaluate the potential for biodegradation of EDC in each.

### 3.2.3. EDC compound effects

The results of the study by Writer et al. [70] also demonstrated the substantial variation in biodegradation that may occur between different EDC compounds (Fig. 4). Biodegradation of EE2 typically is assumed to be slow in aquatic sediments, and limited direct assessments have been conducted [67].

Results from this study provided rare evidence that EE2 mineralization can occur in surface-water sediments, but EE2 mineralization was at least an order of magnitude lower than E2 or 4-$n$-NP mineralization. Because the $K_{om}$ values for E2 and EE2 were similar and about an order of magnitude lower than for 4-NP [70], the relative recalcitrance of EE2, compared to E2, was not due to sorption differences. These results illustrate the need to evaluate the location-specific potential for biodegradation of each environmentally important EDC.

### 3.2.4. Red-Ox effects

Microbial mechanisms for degradation of historical environmental contaminants and, by extension EDC, are fundamentally redox processes. Consequently, in situ redox conditions are expected to control the efficiency of EDC biodegradation. Environmental endocrine activity is dependent on the presence of an aromatic ring structure with an extended carbon backbone. All natural and synthetic hormones are aromatic compounds and the endocrine mimic EDC are generally expected to share this characteristic.

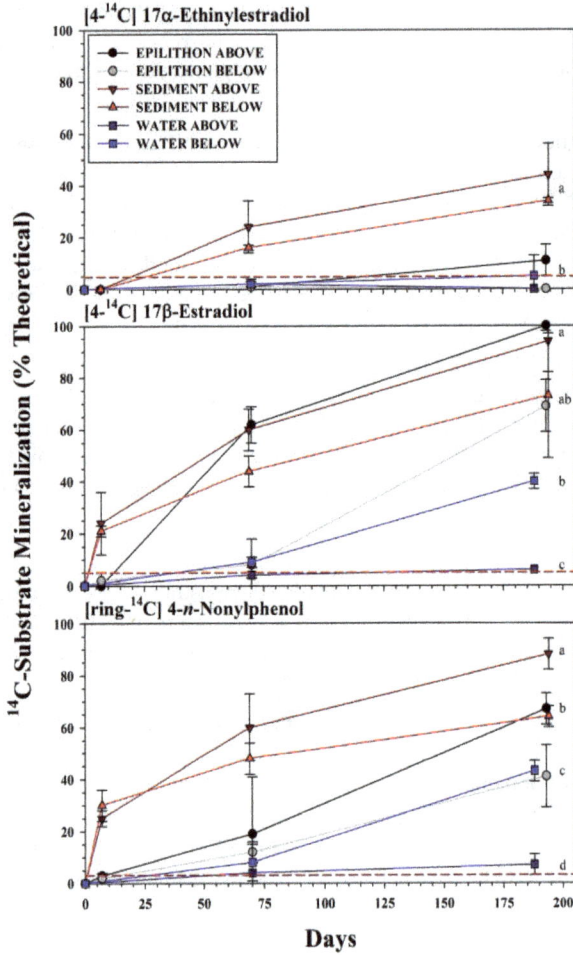

**Figure 4.** Mineralization of EE2, E2, and 4-*n*-NP in microcosms containing sediment, epilithon, or water only collected from upstream and downstream of the WWTP outfall in Boulder Creek.

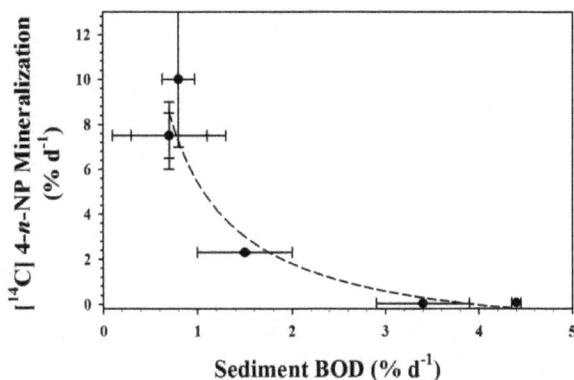

**Figure 5.** Inverse relation between 4-*n*-NP mineralization rate and sediment biological oxygen demand.

Because the energy available for microbial metabolism is a function of the potential differ-ence between the electron donor and the terminal electron acceptor, the theoretical energy yield from the biodegradation of a contaminant serving as an electron donor is greatest when coupled to oxygen, and decreases in the order of oxygen-reduction > nitrate-reduction > Fe(III)-reduction > sulfate reduction > methanogenesis. Experience from the remediation of legacy contaminants over several decades has demonstrated that microorganisms can de-grade aromatic contaminant compounds under a range of terminal electron accepting condi-tions. However, rates of aromatic contaminant biodegradation under oxic conditions are typically 1-2 orders of magnitude greater than under anoxic conditions.

Extending this experience to EDC, the efficiency of environmental EDC biodegradation would be expected to be greatest under oxic conditions and severely limited under anoxic conditions. The results of a recent assessment of the potential for 4-*n*-NP biodegradation in stream sedi-ments are consistent with this expectation [63]. While substantial mineralization of $^{14}$C-4-*n*-NP was observed in sediment microcosms incubated under oxic conditions, the rate of mineraliza-tion under oxic conditions was inversely related to the sediment biological oxygen demand (BOD)(Fig. 5) and no evidence of mineralization was observed under anoxic conditions [63].

The importance of redox conditions on EDC biodegradation is also demonstrated by results (Fig. 6) of a recent investigation of E2 and E1 biodegradation potential in manure-impacted stream sediments collected from a small stream in northcentral Iowa. An accidental spill from a manure lagoon raised concerns about the effect of oxygen-limited conditions on the fate of manure-derived EDC contaminants in the stream. Specific questions concerned the potential for continued EDC biodegradation under anoxic conditions and the potential that manure-derived nitrate (NO$_3$) amendment might stimulate EDC biodegradation by creating denitrifying conditions. Enhanced anaerobic biodegradation of environmental contaminants under denitrifying conditions has been reported previously [71].

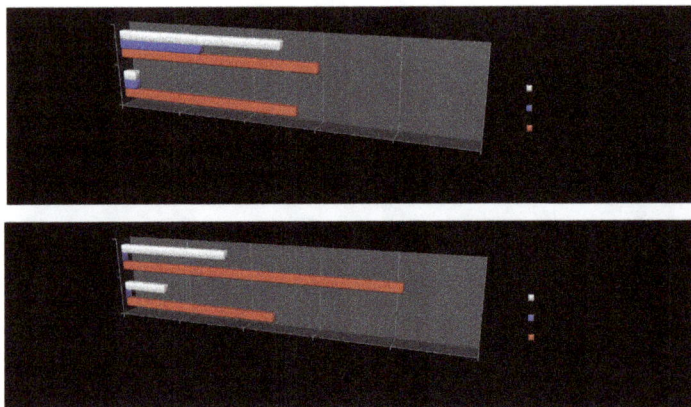

**Figure 6.** Effect of redox condition on E2 and E1 mineralization in microcosms containing sediment collected from upstream (A.) and downstream (B.) of the manure spill site at New York Branch.

The results (Fig. 6) demonstrated that significant biodegradation of E2 and E1 could occur in stream sediment under anoxic conditions. In general, biodegradation was substantially lower under anoxic conditions than under oxic conditions. However, comparable E2 biodegradation was observed in sediments collected upstream of the spill site under anoxic and oxic conditions (Fig. 6A). Somewhat surprisingly, rather than stimulating E2 and E1 biodegradation under anoxic conditions, the addition of $NO_3$ inhibited biodegradation in both E2 treatments and in the downstream sediment E1 treatment. These results and the results of the previous 4-$n$-NP biodegradation study (Fig. 5) [63] illustrate the substantial variation in EDC biodegradation that may occur in the same sediment under different redox conditions and the need to evaluate the potential for biodegradation of EDC under those redox conditions that predominate in situ.

# 4. Conclusion – Toward an integrated approach to EDC risk management in surface-water ecosystems

The risk associated with EDC in the environment may reasonably be viewed as the net result of the interaction of three conceptual drivers: (1) the occurrence and distribution of EDC in the environment, (2) site- and life-cycle-specific adverse-impact thresholds of EDC ecological endpoints, and (3) the potential for in situ EDC natural attenuation and persistence. The first two drivers are widely recognized and are the focus of numerous recent and ongoing investigations of environmental EDC risk. In contrast, environmental EDC persistence is a fundamental aspect of EDC environmental risk and relatively few studies have addressed the fate, transport and persistence of EDC in various environmental settings.

An understanding of the site-specific capacity for EDC attenuation is critical to EDC risk management. The longer contaminants or groups of contaminants persist in the environment the greater the chance that the contaminants will reach and eventually exceed an adverse impact threshold. The extent and rates of in situ contaminant biodegradation are key data needs for establishing total maximum daily load criteria for EDC, upgrading wastewater treatment infrastructure, and selecting protective treatment performance criteria. Ultimately, those EDC, which have little or no potential for biodegradation under environmentally relevant conditions, may need to be removed from commercial production, because any chemical substance that is in use will eventually find its way to the environment.

## Acknowledgements

The U.S. Geological Survey Toxic Substances Hydrology Program (http://toxics.usgs.gov/) supported this research.

## Author details

Paul M. Bradley and Dana W. Kolpin

U.S. Geological Survey, USA

## References

[1]  Cranor C. How should society approach the real and potential risks posed by new technologies? Plant Physiology. 2003;133 3-9.

[2]  Murmann J. Chemical industries after 1850. 2003. In: Oxford Encyclopedia of Economic History [Internet]. New York, NY: Oxford University Press; [22].

[3]  Carson R. Silent Spring. New York, NY: Houghton Mifflin; 1962.

[4]  Colborn T, Dumanoski D, Myers J. Our stolen future: are we threatening our fertility, intelligence, and survival? : a scientific detective story. New York, New York, USA: Penguin Books; 1996. 306 p.

[5]  Daughton C. Non-regulated water contaminants: emerging research. Environmental Impact Assessment Review. 2004;24 711-32.

[6]  American Chemical Society. CAS Content at a Glance http://www.cas.org/content/at-a-glance: American Chemical Society; 2012 [cited 2012 09 08 2012].

[7]   Bohacek R, McMartin C, Guida W. The art and practice of structure-based drug design: a molecular modeling perspective. Medicinal Research Reviews. 1996;16(1) 3-50.

[8]   Bradley P. History and ecology of chloroethene biodegradation: A review. Bioremediation Journal. 2003;7(2) 81-109.

[9]   U.S. Environmental Protection Agency. Basic Information about Regulated Drinking Water Contaminants and Indicators: U.S. Environmental Protection Agency 2012 [cited 2012 09 08 2012]. Available from: http://water.epa.gov/drink/contaminants/basicinformation/index.cfm.

[10]  U.S. Environmental Protection Agency. 2012 Edition of the Drinking Water Standards and Health Advisories. U.S. Environmental Protection Agency 2012 Contract No.: EPA 822-S-12-001.

[11]  Lintelmann J, Katayama A, Kurihara N, Shore L, Wenzel A. Endocrine disrupters in the environment. Pure and Applied Chemistry. 2003;75(5) 631-81.

[12]  U.S. Food and Drug Administration. Approved Drugs www.fda.gov: U.S. Food and Drug Administration; 2012 [cited 2012 September 12]. Available from: http://www.fda.gov/Drugs/InformationOnDrugs/ApprovedDrugs/default.htm.

[13]  U.S. Food and Drug Administration. Approved Drug Products with Therapeutic Equivalence Evaluations: U.S. Food and Drug Administration,; 2012. Available from: http://www.fda.gov/downloads/Drugs/DevelopmentApprovalProcess/UCM071436.pdf.

[14]  U.S. Geological Survey. National Water Quality Laboratory, 2012 [cited 2012 September 12]. Available from: http://wwwnwql.cr.usgs.gov/USGS/.

[15]  U.S. Environmental Protection Agency. Method 1694: Pharmaceuticals and Personal Care Products in Water, Soil, Sediment, and Biosolids by HPLC/MS/MS. Washington DC: U.S. Environmental Protection Agency,, 2007 EPA-821-R-08-002 Contract No.: EPA-821-R-08-002.

[16]  Kolpin D, Furlong E, Meyer M, Thurman E, Zaugg S, Barber L, et al. Pharmaceuticals, hormones and other organic wastewater contaminants in U.S. streams, 1999-2000: A national synthesis. Environmental Science and Technology. 2002;36(6) 202-11.

[17]  U.S. Environmental Protection Agency. Endocrine Primer: U.S. Environmental Protection Agency,; 2012 [cited 2012 September 12]. Available from: http://www.epa.gov/scipoly/oscpendo/pubs/edspoverview/primer.htm.

[18]  Fossi M, Marsilli L. Effects of endocrine disruptors in aquatic mammals. Pure and Applied Chemistry. 2003;75(11-12) 2235-47.

[19]  Jobling S, Tyler C. Endocrine disruption in wild freshwater fish. Pure and Applied Chemistry. 2003;75(11-12) 2219-34.

[20] Blazer V, Iwanowicz L, Iwanowicz D, Smith D, Young J, Hendrick J, et al. Intersex (testicular oocytes) in smallmouth bass from the Potomac River and selected nearby drainages. Journal Aquatic Animal Health. 2007;19(4) 242-53.

[21] Hinck J, Blazer V, Schmitt C, Papoulias D, Tillit D. Widespread occurrence of intersex in black basses (Micropterus spp.) from U. S. rivers, 1995-2004. Aquatic Toxicology. 2009;95(1) 60-70.

[22] Kidd K, Blanchfield P, Mills K, Palance V, Evans R, Lazorchak J, et al. Collapse of a fish population after exposure to a synthetic estrogen. Proceeding of the National Academy of Science. 2007;104(21) 8897-901.

[23] McGee M, Julius M, Vajda A, Norris D, Barber L, Schoenfuss H. Predator avoidance performance of larval fathead minnows (Pimephales promelas) following short-term exposure to estrogen mixtures. Aquatic Toxicology. 2009;91(4) 355-61.

[24] Painter M, Buerkley M, Julius M, Vajda A, Norris D, Barber L, et al. Antidepressants at environmentally relevant concentrations affect predator avoidance behavior of larval fatheaad minnows (Pimephales promelas). Environmental Toxicology and Chemistry. 2009;28(12) 2677-84.

[25] Giesy J, Feyk L, Jones P, Kurunthachalam K, Sanderson T. Review of the effects of endocrine-disrupting chemicals in birds. Pure and Applied Chemistry. 2003;75(11-12) 2287-303.

[26] Soin T, Smagghe G. Endocrine disruption in aquatic insects: a review. Ecotoxicology. 2007;DOI 10.1007/s10646-006-0118-9 11.

[27] Fingerman M, Jackson N, Nagbhushanam R. Hormonally regulated functions in crustaceans as biomarkers of environmental pollution. Comparative Biochemistry and Physiology Part C Pharmacology, Toxicology, and Endocrinology. 1998;120(3) 343-50.

[28] Gagne F, Blaise C. Effects of municipal effluents on serotonin and dopamine levels in the freshwater mussel *Elliptio complanata*. Comparative Biochemistry and Physiology Part C. 2003;136 117-25.

[29] Lagadic L, Coutellec M-A, Caquet T. Endocrine disruption in aquatic pomonate molluscs: few evidences, many challenges. Ecotoxicology. 2007;1 61-81.

[30] Barber L, Lee K, Swackhammer D, Schoenfuss H. Reproductive responses of male fathead minnows exposed to plant efflunet, effluent treated with XAD8 resin, and an environmentally relevant mixture of alkylphenol compounds. Aquatic Toxicology. 2007;82(1) 36-46.

[31] Vajda A, Barber L, Gray J, Lopez E, Woodling J, Norris D. Reproductive disruption in fish downstream from and estrogenic wastewater effluent. Environmental Science and Technology. 2008;42(9) 3407-14.

[32]  Campbell C, Borglin S, Green F, Grayson A, Wozei E, Stringfellow W. Biologically directed environmental monitoring, fate, and transport of estrogenic endocrine disrupting compounds in water: A review. Chemosphere. 2006;65 1265-80.

[33]  Islinger M, Pawlowski S, Hollert H, Volki A, Brumbeck T. Measurement of vitellogenin-mRNA expression in primary cultures of rainbow trout hepatocytes in a non-radioactive dot blot/RNAse protection-assay. Science of the Total Environment. 1999;233(1-3) 109-22.

[34]  Desbrow C, Routledge E, Brighty G, Sumpter J, Waldock M. Identification of estrogenic chemicals in STW effluent. 1. Chemicals fractionation and in vitro biological screening. Environmental Science and Technology. 1998;32(11) 1549-58.

[35]  Routeledge E, Sheahan D, Desbrow C, Brighty G, Waldock M, Sumpter J. Identification of estrogenic chemicals in STW effluent. 2. In vivo responses in trout and roach. Environmental Science and Technology. 1998;32(11) 1559-65.

[36]  Shelby M, Newbold R, Tully D, Chae K, Davis V. Assessing environmental chemicals for estrogenicity using a combination of in vitro and in vivo assays. Environmental Health Perspectives. 1996;104(12) 1296-300.

[37]  Snyder S, Villeneuve D, Snyder E, Giesy J. Identification and quantification of estrogen receptor agonists in wastewater effluents. Environmental Science and Technology. 2001;35(18) 3620-5.

[38]  Soto A, Sonneschein C, Chaung K, Fernandez M, Olea N, Serrano F. The E-SCREEN assay as a tool to identify estrogens: an update on estrogenic environmental pollutants. Environmental Health Perspectives. 1995;103(7) 113-22.

[39]  Thorpe K, Hutchinson T, Hetheridge M, Scholze M, Sumpter J. Assessing the biological potency of binary mixtures of environmental estrogens using vitellogenin induction in juvenile rainbow trout (Oncorhyncus mykiss). Environmental Science and Technology. 2001;35(12) 2476-81.

[40]  Sole M, Lopez De Alda M, Castillo M, Porte C, Ladegaard-Pedersen K, Barcelo D. Estrogenicity determination in sewage treatment plants and surface waters from the Catalonian area (NE Spain). Environmental Science and Technology. 2000;34 5076-83.

[41]  Ferrer I, Thurman E. Liquid chromatography/time-of-flight/mass spectrometry (LC/TOF/MS) for the analysis of emerging contaminants. Trends in Analytical Chemistry. 2003;22(10) 750-6.

[42]  Richardson S. Environmental mass spectrometry: emerging contaminants and current issues. Analytical Chemistry. 2008;80(12) 4373-402.

[43]  Richardson S. Water analysis: emerging contaminants and current issues. Analytical Chemistry. 2009;81(12) 4645-77.

[44] Routeledge E, Sumpter J. Estrogenic activity of surfactants and some of their degradation products using a recombinant yeast screen. Environmental Toxicology and Chemistry. 1996;15 241-8.

[45] Sanseverino J, Gupta R, Layton A, Patterson S, Ripp S, Saidak L, et al. Use of Saccharomyces cerevisiae BLYES expressing bacterial bioluminescence for rapid, sensitive detection of estrogenic compounds. Applied and Environmental Microbiology. 2005;71 4455-60.

[46] Bradley P, Barber L, Chapelle F, Gray J, Kolpin D, McMahon P. Biodegradation of 17B-Estradiol, estrone and testosterone in stream sediments. Environmental Science and Technology. 2009;43(6) 1902-10.

[47] Vazquez-Duhalt R, Marquez-rocha F, Ponce E, Licea A, Viana M. Nonylphenol, an integrated vision of a pollutant, scientific review. Applied Ecology and Environmental Research. 2005;4 1-25.

[48] Ying G, Williams B, Kookana R. Environmental fate of alkylphenols and alkylphenol ethoxylates - a review. Environment International. 2002;28 215-26.

[49] Ames B, Gold L. Paracelsus to parascience: the environmental cancer distraction. Mutation Research. 2000;447 3-13.

[50] Hotchkiss A, Rider C, Blystone C, Wilson V, Hartig P, Ankley G, et al. Fifteen years after "Wingspread"-environmental endocrine disrupters and human and wildlife health: where are today and where we need to go. Toxicological Sciences. 2008;105(2) 235-59.

[51] Crews D, Willingham E, Skipper J. Endocrine disruptors: present issues future directions. The Quarterly Review of Biology. 2000;75(3) 243-60.

[52] Vandenberg L, Maffini M, Sonnenshcein C, Rubin B, Soto A. Bispenol-A and the great divide: a review of controversies in the field in endocrin disruption. Endocrine Reviews. 2009;30(1) 75-95.

[53] Welshons W, Thayer K, Judy B, Taylor J, Curran E, vom Saal F. Large effects from small exposures: I. mechanisms for endocrine-disrupting chemicals with estrogenic activity. Environmental Health Perspectives. 2003;111 994-1006.

[54] heehan D, Vom Saal F. Low dose effects of hormones: a challenge for risk assessment. Risk Policy Report. 1997;4 31-9.

[55] Vandenberg L, Wadia P, Schaeberle C, Rubin B, Sonnenshcein C, Soto A. The mammary gland response to estradiol: monotonic at the cellular level, non-monotonic at the tissue-level of organization? Journal of Steroid Biochemistry and Molecular Biology. 2006;101 263-74.

[56] Allner B, Wegener G, Knacker T, Stahlschmidt-Allner P. Electrophoretic determination of estrogen-protein in fish exposed to synthetic and naturally occurring chemicals. Science of the Total Environment. 1999;233(21-31).

[57] Folmar L, Hemmer M, Hemmer R, Bowman C, Kroll K, Denslow N. Comparative estrogenicity of estradiol, ethynyl estradiol and diethylstilbestrol in an in vivo male sheepshead minnow (Cypinodon variegatus), vitellogenin bioassay. Aquatic Toxicology. 2000;49(1-2) 77-88.

[58] Jobling S, Nolan M, Tyler C, Brighty G, Sumpter J. Widespread sexual disruption in wild fish. Environmental Science and Technology. 1998;32(17) 2498-506.

[59] Kramer K, Miles-Richarson S, Pieren S, Giesy J. Reproductive impairment and induction of alkaline-labile phosphate, a biomarker of estrogen exposure, in fathead minnows (Pimephales promelas) exposed to waterborne 17β-estradiol. Aquatic Toxicology. 1998;40(4) 335-60.

[60] Nimrod A, Benson W. Reproduction and development of Japanes medaka following an early life stage exposure to xenoestrogens. Aquatic Toxicology. 1998;44(1-2) 141-56.

[61] Thorpe K, Hutchinson T, Hetheridge M, Sumpter J. Development of an in vivo screening assay for estrogenic chemicals using juvenile rainbow trout (Oncohychus mykiss). Environmental Toxicology and Chemistry. 2000;19(11) 2812-20.

[62] Thorpe K, Gros-Sorokin M, Johnson I, Brighty G. An assessment of the model of concentration addition for predicting the estrogenicity of chemical mixtures from wastewater treatment works effluents. Environmental Health Perspectives. 2006;114(S-1) 90-7.

[63] Bradley P, Barber L, Kolpin D, McMahon P, Chapelle F. Potential for 4-n-nonylphenol biodegradation in stream sediments. Environmental Toxicology and Chemistry. 2008;27(2) 260-5.

[64] Holz s. There is no "away" Pharmaceuticals, personal care products, and endocrine-disrupting substances: Emerging contaminants detected in water. Toronto, Ontario, Canada: Canadian Institute for Environmental Law and Policy; 2006. Available from: http://www.cielap.org/pdf/NoAway.pdf.

[65] Bradley P, Chapelle F, Landmeyer J. Effect of redox conditions on MTBE biodegradation in surface water sediments. Environmental Science and Technology. 2001;35(23) 4643-7.

[66] Bradley P, Landmeyer J, Chapelle F. TBA biodegradation in surface-water sediments under aerobic and anaerobic conditions. Environmental Science and Technology. 2002;36(19) 4087-90.

[67] Jurgens M, Holthaus K, Johnson A, Smith J, Hetheridge M, Williams R. The potential for estradiol and ethinylestradiol degradation in English rivers. . Environmental Toxicology and Chemistry. 2002;21(3) 480-8.

[68] Layton A, Gregory B, Seward J, Schultz T, Sayler G. Mineralization of steroidal hormones by biosolids in wastewater treatment systems in Tennessee U.S.A. . Environmental Science and Technology. 2000;34(18) 3925-31.

[69] Matsuoka S, Kikuchi M, Kimura S, Kurokawa Y, Kawai S. Determination of estrogenic substances in the water of Muko River using in vitro assays, and the degradation of natural estrogens by aquatic bacteria. Journal of Health Sciences. 2005;51(2) 178-84.

[70] Writer J, Barber L, Ryan J, Bradley P. Biodegradation and attenuation of steroidal hormones and alkylphenols by stream biofilms and sediments. Environmental Science and Technology. 2011;45 4370-6.

[71] Bradley P, Chapelle F, Landmeyer J. Methyl t-butyl ether mineralization in surface-water sediment microcosms under denitrifying conditions. Applied and Environmental Microbiology. 2001;67(4) 1975-8.

# Chapter 3

# Watershed-Scale Hydrological Modeling Methods and Applications

Prem B. Parajuli and Ying Ouyang

Additional information is available at the end of the chapter

## 1. Introduction

Pollution of surface water with harmful chemicals and eutrophication of rivers and lakes with excess nutrients are serious environmental concerns. The U.S. Environmental Protection Agency (USEPA) estimated that 53% of the 27% assessed rivers and streams miles and 69% of the 45% assessed lakes, ponds, and reservoirs acreage in the nation are impaired (USEPA, 2010). In Mississippi, 57% of the 5% assessed rivers and streams miles are impaired (USEPA, 2010). These impairment estimates may increase when assessments of more water bodies are performed and water quality criteria are improved. The most common water pollution concerns in U.S. rivers and streams are sediment, nutrients (Phosphorus and Nitrogen) and pathogens. Hydrological processes can significantly impact on the transport of water quality pollutants.

Non-point source pollution from agricultural, forest, and urban lands can contribute to water quality degradation. Total Maximum Daily Loads (TMDLs) are developed by states to improve water quality. The TMDL requires identifying and quantifying pollutant contributions from each source to devise source-specific pollutant reduction strategies to meet applicable water quality standards. Commonly, water quality assessment at the watershed scale is accomplished using two techniques: (a) watershed monitoring and (b) watershed modeling. Watershed models provide a tool for linking pollutants to the receiving streams. Models provide quick and cost-effective assessment of water quality conditions, as they can simulate hydrologic processes, which are affected by several factors including climate change, soils, and agricultural management practices. However, methods used to develop a model for watersheds can significantly impact in the model outputs. Here several hydrological and water quality models are described. Case studies of two commonly used models with calibration and validation are provided with current and future climate change scenarios. This book chapter briefly reviews currently available hydrologic and water quality models, and presents model application case studies, to provide a foundation for further model development and watershed assessment studies.

## 2. Review of water quality models

Several useful hydrologic and water quality models are available today, each with diverse capabilities for watershed assessment. Many of these models are relevant to water quality goal assessment and implementation. Modeling of hydrology, sediment and nutrients has developed substantially, but advances have not always been consistent with the needs of the water quality goals program. Comprehensive education and training with model applications and case studies are needed for users to understand the potentials, limitations, and suitable applications of a model. Review of several hydrological models (e.g. SWAT, An-nAGNPS, HSPF, SPARROW, GLEAMS, WEPP, EFDC etc.) including models description and application within the U.S. or other countries are discussed.

### 2.1. SWAT model

The SWAT model is developed and supported by the USDA/ARS. It is a physically based watershed-scale continuous time-scale model, which operates on a daily time step. The SWAT model can simulate runoff, sediment, nutrients, pesticide, and bacteria transport from agricultural watersheds (Arnold et al., 1998). The SWAT model delineates a watershed, and sub-divides that watershed in to sub-basins. In each sub-basin, the model creates several hydrologic response units (HRUs) based on specific land cover, soil, and topographic conditions. Model simulations that are performed at the HRU levels are summarized for the sub-basins. Water is routed from HRUs to associated reaches in the SWAT model. SWAT first deposits estimated pollutants within the stream channel system then transport them to the outlet of the watershed. The HRUs provide opportunity to include processes for possible spatial and temporal variations in model input parameters. The hydrologic module of the model quantifies a soil water balance at each time step during the simulation period based on daily precipitation inputs.

The SWAT model distinguishes the effects of weather, surface runoff, evapo-transpiration, crop growth, nutrient loading, water routing, and the long-term effects of varying agricultural management practices (Neitsch et al., 2005). In the hydrologic module of the model, the surface runoff is estimated separately for each sub-basin and routed to quantify the total surface runoff for the watershed. Runoff volume is commonly estimated from daily rainfall using modified SCS-CN method. The Modified Universal Soil Loss Equation (MUSLE) is used to predict sediment yield from the watershed. The SWAT model has been extensively applied for simulating stream flow, sediment yield, and nutrient modeling (Gosain et al., 2005; Vache et al., 2002; Varanou et al., 2002). The model needs several data inputs to represent watershed conditions which include: digital elevation model (DEM), land use land cover, soils, climate data. The SWAT model is an advancement of the Simulator for Water

Resources in Rural Basins (SWRRB) and Routing Outputs to Outlet (ROTO) models. The SWAT model development was influenced by other models like CREAMS (Knisel, 1980), GLEAMS (Leonard et al., 1987), and EPIC (Williams et al., 1984; Neitsch et al., 2002).

The SWAT model has been recently applied to assess watershed conditions of the U.S. (Gassman et al., 2007; Parajuli et al., 2008; 2009; Parajuli 2010a; 2011; 2012; Chaubey et al., 2010) and internationally such as Ethiopia (Betrie et al., 2011); Kenya and northwest Tanzania (Dessu and Melesse, 2012); Bulgaria and Greece (Boskidis et al., 2012); and Australia (Githui et al., 2012).

## 2.2. AnnAGNPS

The AnnAGNPS model is a product of the USDA Agriculture Research Service (USDA-ARS) and the USDA Natural Resources Conservation Service (USDA-NRCS) to evaluate non-point source pollution from agriculture watersheds. Similar to the SWAT model, it is a physically based continuous and daily time step model used to simulate surface runoff, sediment, and nutrient yields (Cronshey and Theurer, 1998; Bingner and Theurer, 2003). The AnnAGNPS is considered an enhanced modification to the single event based Agricultural Non-Point Source (AGNPS) model (Young et al., 1989), as it retains many features of AGNPS (Yuan et al., 2001). Unlike AGNPS, the AnnAGNPS delineates watershed, sub-divides the watershed into small drainage areas with homogenous land use, soils, etc. The sub-areas are integrated and simulated surface runoff and pollutant loads through rivers and streams within the sub-areas and watershed, which is enhanced from the AGNPS.

The AnnAGNPS model utilizes and incorporates components or sub-components from several other models such as; Revised Universal Soil Loss Equation (RUSLE) model (Renard et al., 1997); Chemicals, Runoff, and Erosion from Agricultural Management Systems (CREAMS) model (Knisel, 1980); Groundwater Loading Effects on Agricultural Management Systems (GLEAMS) model (Leonard et al., 1987); and Erosion Productivity Impact Calculator (EPIC) Model (Sharpley and Williams, 1990). The AnnAGNPS model represents small watershed areas using a cell-based approach, with land and soil property characterization similar to SWAT model HRUs. Daily soil moisture contents are calculated using the Curve Number (CN) method, which help to quantify surface and subsurface flows. The AnnAGNPS model uses the RUSLE to estimate sediment yields.

Refereed AnnAGNPS model based evaluations have been applied predominantly to watersheds located in the U.S. (Yuan et al., 2011; 2002; Zuercher et al., 2011; Polyakov et al., 2007). However, the model also has been applied in other countries such as Mediterranian (Licciardello et al., 2011; 2007); Australia (Baginska et al., 2003), and China (Hua et al., 2012).

## 2.3. WEPP

The Water Erosion Prediction Project (WEPP) model is a product of USDA. The WEPP model is a process-based, distributed parameter, single storm and continuous based model used to predict surface flow and sediment yields from the hill slopes and small watersheds. WEPP allows simulation of the effects of crop, crop rotation, contour farming, and strip cropping. The WEPP model components includes weather generation, snow accumulation and melt, irrigation, infiltration, overland flow process, water balance, plant growth, residue management, soil disturbance by tillage, and erosion processes. The WEPP model considers sheet and rill erosion processes to predict erosion. The WEPP model incorporates modified water balance and percolation components from the SWRRB model (Williams and Nicks, 1985). The WEPP model utilizes and incorporates components or sub-components from several other models such as; EPIC (Williams et al., 1984); and CREAMS model (Knisel, 1980). The WEPP model has undergone continuous development since 1992 (1992-1995 with DOS version; 1997-2000 with window interface; 1999-2009 with Geo-WEPP ArcView/ArcGIS extensions; and 2001-present with web-browser interface; Flanagan et al., 2007; Foltz et al., 2011).

Refereed WEPP-model-based evaluations exist predominantly for agricultural fields or small watersheds located in the U.S. (Dun et al., 2010; Flanagan et al., 2007; Foltz et al., 2011). However, the WEPP has been applied in other countries such as China (Zhang et al., 2008).

## 2.4. GLEAMS

Groundwater Loading Effects of Agricultural Management Systems (GLEAMS) is a daily time-step, continuous, field-scale hydrological and pollutant transport mathematical model (Leonard et al., 1987). The GLEAMS model can simulate surface runoff, percolation, nutrient and pesticide leaching, erosion and sedimentation. The GLEAMS model requires several daily climate data including mean daily air temperature, daily rainfall, mean monthly maximum and minimum temperatures, wind speed, solar radiation and dew-point temperature data. The soil input parameters in the model can be obtained from the State Soil Geographic Database (STATSGO) or Soil Survey Geographic Database (SSURGO) soil data. Previous studies described the ability of GLEAMS model to predict nitrate transport process from the agricultural areas (Shirmohammadi et al., 1998; Bakhsh et al., 2000; Chinkuyu and Kanwar, 2001).

Refereed GLEAMS model applications have been published predominantly for field scale studies in the U.S. (Bakhsh et al., 2000; Chinkuyu et al., 2004). However, GLEAMS also has been applied in a few other countries, such as China (Zhang et al., 2008).

## 2.5. HSPF model

The hydrological simulation program—FORTRAN (HSPF) is a product of U.S. Environmental Protection Agency (US-EPA), which is a comprehensive model used for modeling processes related to water quantity and quality in watersheds of various sizes and complexities (Bicknell et al. 2001). It simulates both the land area of watersheds and the water bodies. The HSPF model uses input data including hourly history of rainfall, temperature and solar radiation; land surface characteristics/land use conditions; and land management practices to predict parameters at watershed scales. The results of model simulations are based on a time history of the quantity and quality of runoff from an urban, forest or agricultural watershed, which include surface runoff, sediment load, nutrients and pesticide concentrations. The

HSPF model can simulate three sediment types (sand, silt, and clay) in addition to organic chemicals and alternative products. A detailed description of HSPF model can be found in Bicknell et al. (2001).

There have been hundreds of applications of HSPF around the world (Bicknell et al., 2001; Akter and Babel, 2012; Ouyang et al., 2012; Rolle et al., 2012). Examples include applications in a large watershed at the Chesapeake Bay, in a small watershed near Watkinsville, GA, with the experimental plots of a few hectares and in other areas such as Seattle, WA, Patuxent River, MD., and Truckee-Carson Basins, NV. Details are available at: (http://water.usgs.gov/cgi-bin/man_wrdapp?hspf).

### 2.6. SPARROW

The SPAtially-Referenced Regression On Watershed attributes (SPARROW) model is a watershed modeling tool for comparing water-quality data collected at a network of monitoring stations to characterize watersheds containing the stations (Smith et al., 1997; Schwarz et al., 2008). The SPARROW model has a nonlinear regression equation depicting the non-conservative transport of contaminants from the point and diffuse sources on land surfaces to streams and rivers. The SPARROW predicts contaminant flux, concentration, and yield in streams. It has been used to evaluate alternative hypotheses about important contaminant sources and watershed properties that control contaminant load and transport over large spatial scales. The SPARROW can be used to explain spatial patterns of stream water quality in relation to human activities and natural processes.

Numerous applications of SPARROW have been performed to assess water quality in watersheds in recent years. Brown (2011) investigated nutrient sources and transport in the Missouri River Basin with SPARROW. Saad et al. (2011) applied SPARROW to estimate nutrient load and to improve water quality monitoring design using a multi-agency dataset. Alam and Goodall (2012) examined the effects of hydrologic and nitrogen source changes on nitrogen yield in the contiguous United States with SPARROW.

### 2.7. EFDC

The Environmental Fluid Dynamics Code (EFDC) is a multifunctional surface water modeling system, which includes hydrodynamic, sediment-contaminant, and eutrophication components (Hamrick, 1996) and is available to the public through US-EPA website available at: http://www.epa.gov/ceampubl/swater/efdc/index.html. The EFDC can be used to simulate aquatic systems in multiple dimensions with the stretched or sigma vertical coordinates and the Cartesian (or curvilinear), and orthogonal horizontal coordinates to represent the physical characteristics of a water body. A dynamically-coupled transport process for turbulent kinetic energy, turbulent length scale, salinity and temperature are included in the EFDC model. The EFDC allows for drying and wetting in shallow water bodies by a mass conservation scheme.

Refereed EFDC-model-based evaluations exist predominately for stream ecosystems. Examples include a three-dimensional hydrodynamic model of the Chicago River, Illinois (Sinha et al., 2012); the effect of interacting downstream branches on saltwater intrusion in the Modaomen Estuary, China (Gong et al., 2012); and comparison of two hydrodynamic models of Weeks Bay, Alabama (Alarcon et al., 2012).

## 2.8. SWMM

The US-EPA's Storm Water Management Model (SWMM) was initially developed in 1971, and has been significantly upgraded (http://www.epa.gov/nrmrl/wswrd/wq/models / swmm/index.htm). The SWMM model is a widely used model for planning, analysis and design related to storm water runoff, sewers, and other drainage systems in urban areas. SWMM can simulate single storm-events or provide continuous prediction of surface-runoff quantity and quality from urban areas. In addition to predicting surface-runoff quantity and quality, the model can also predict flow rate, flow depth, and water quality in each pipe and channel.

There have been numerous applications of SWMM in the literature recently. Blumensaat et al. (2012) investigated sewer transport with SWMM under minimum data requirements. Cantone and Schmidt (2011) applied SWMM to improve understanding of the hydrologic response of highly urbanized watershed catchments like the Illinois Urban areas. Talei and Chua (2012) estimated the influence of lag-time on storm event-based hydrologic impacts (e.g. rainfall, surface-runoff) using the SWMM model and a data-driven approach.

# 3. Methods to develop a model

Appropriate methods are needed to develop a model, utilize different data sources (e.g. digital elevation, soil, land use, weather etc.), and develop methods to quantify pollutants source loads in the model. As examples, the methods development process is described here for two commonly used models (i.e., SWAT and HSPF).

## 3.1. SWAT model

The SWAT model utilizes digital elevation model (DEM), soils, land cover, and weather data such as precipitation, temperature, wind speed, solar radiation, and relative humidity. SWAT delineates watershed boundary and topographic characteristics of the watershed using National Elevation Dataset called digital elevation model (DEM) data, which are available in the grid form with different resolutions (e.g. 30m x 30m grid; 10m x 10m grid) generally collected by U.S. Geological survey (USGS, 1999) or other sources. The 30m grid data are commonly used in the large scale watershed modeling work. However, small watershed or field scale modeling may benefit from using of 10m x 10m resolution DEM data. Model defines land use inputs in the model are described using distributed land cover data (USDA-NASS, 2010) or other land use data. The time-specific land-cover data (e.g. 1992, 2001 and 2006) for the U.S. and Puerto Rico can be downloaded from the National Land Cover Database (NLCD), a publicly available data source. The distributed land cover data with land use classifications can provide essential model input for the watershed assessment. Currently, land-use data layers are available in geographic information systems (GIS) format, which is applicable for the watershed modeling.

The SWAT model also requires distributed detail soils data, which is available from either State Soil Geographic (STATSGO) database or Soil Survey Geographic (SSURGO) databases (USDA, 2005). The SSURGO database is the most detailed data source currently available in the U.S. as it provides more soil polygons per unit area. The DEM, soils, and landuse geographic data layers should be all projected in one projection system (e.g. Universal Transverse Mercator-UTM 1983, zone 16).

Most of the watershed or field scale models (e.g. SWAT, WEPP) have embedded weather stations and climate generators. However, more field-specific climate inputs (e.g. rainfall; daily minimum, maximum and mean temperatures; solar radiation; relative humidity, and wind speed) can be allowed in the model for the watershed assessment. Weather data such as daily rainfall and ambient temperature can be downloaded from the National Climatic Data Center (NCDC, 2012). Other field-specific model input parameters such as irrigation (e.g. auto or manual irrigation), fertilizer application (application rates, fertilizer type), crop rotation (e.g. corn after soybean), tillage (e.g. conventional, reduced, no-tillage), planting and harvesting dates can be defined (Parajuli, 2010b).

### 3.2. HSPF model

The major procedures in water quality modeling with HSPF are the construction of a conceptual model, mathematical description of the conceptual model, preparation of input data such as time series parameter values, calibration and validation of the model, and application of the model for field conditions. Time series input data can be supplied into the HSPF model by using a stand-alone program or the Watershed Data Management program (WDM) provided in BASINS (Better assessment science integrating point and nonpoint sources). BASINS is a multipurpose environmental analysis system model, which can be utilized by regional, state, and local agencies for conducting water quality based studies. The BASINS system incorporates an open source geographic information system (GIS) (i.e., MapWindow), the national watershed and meteorological data, and the state-of-the-art environmental models such as HSPF, Pollutant Loading Application (PLOAD), and SWAT into one convenient package (USEPA, 2010).

Normally, the development of a HSPF model starts with a watershed delineation process, which includes the setup of digital elevation model (DEM) data in the ArcInfo grid format, generation of stream networks in shape format, and designation of watershed inlets or outlets using the watershed delineation tool built in the BASINS. The HSPF also needs land use and soil data to determine the area and the hydrologic parameters of each land use pattern in the model, which can be done with the land use and soil classification tool in the BASINS. The HSPF is a lumped parameter model with a modular structure. The PERLAND modular represents the pervious land segments over which a considerable amount of water infiltrates into the ground. The IMPLND modular denotes the Impervious land segments over which infiltration is negligible such as paved urban surfaces. Processes involving water bodies like streams and lakes are represented with the RCHRES module. These modules have many components dealing with hydrological and water quality processes. Detailed information about the structure and functioning of these modules can be found in elsewhere (Donigian and Crawford 1976; Donigian et al. 1984; Bicknel et al. 1993; Chen et al. 1998).

# 4. Model application

Two watersheds in Mississippi (Upper Pearl River and Yazoo River Basin) were selected for modeling case studies using two hydrologic and water quality models (SWAT and HSPF). Models were calibrated and validated using USGS observed streamflow data for the current conditions and models were applied to predict future climate change scenarios impact on hydrology. Case studies demonstrated how future climate change scenarios impact stream-flow from the watersheds.

## 4.1. SWAT model

The main objective of this case study was to quantify the potential impact of future climate change scenarios on hydrologic characteristics such as monthly average streamflow with in the Upper Pearl River Watershed (UPRW) using the SWAT model. The specific objectives were to: (1) develop a site-specific SWAT model for the UPRW based on watershed charac-teristics, climatic, and hydrological conditions; (2) calibrate and validate model using USGS observed stream flow data; and (3) develop future climate change scenarios and quantify their impacts on stream flows.

### 4.1.1. Study area and model development

The SWAT model was developed and applied in the UPRW (7,588 km²), which is located in Mississippi (Fig. 1). The UPRW covers ten counties (Choctaw, Attala, Winston, Leake, Ne-sobha, Kemper, Madison, Rankin, Scott and Newton) in Mississippi with predominant land uses of woodland (72%), grassland (20%), urban land (6%) and others (2%).

To develop the SWAT model, this case study utilized national elevation data, which is also called DEM data of 30m x 30m grids to delineate watershed boundary. The STATSGO was used to create distributed soil data input in the model. The land cover data was created us-ing the cropland data layer in the model. The climate data (e.g. daily precipitation, tempera-ture) were used from several weather stations within or near the watershed as maintained by the National Climatic Data Center. The SWAT model allows several potential evapo-transpiration estimation method alternatives (e.g. Penman-Monteith, Hargreaves, Priestley-Taylor). This case study utilized the Penman-Monteith method to estimate PET, which requires daily rainfall, maximum and minimum temperatures, relative-humidity, solar radi-ation, and wind speed data. The additional data needed to simulate the SWAT model using Penman-Monteith PET method were generated by the SWAT model.

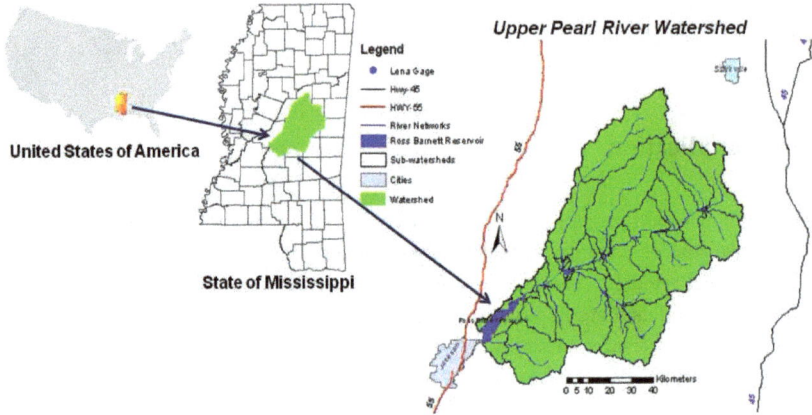

**Figure 1.** Location map of the watershed showing sub-watersheds and others

### 4.1.2. Model calibration and validation

The SWAT model predicted monthly streamflow values were compared separately for model calibration and validation periods using three common parameters (coefficient of determination – $R^2$; Nash–Sutcliffe efficiency index – E; and root mean square error - RMSE). The monthly model performances were ranked excellent for $R^2$ or E values > 0.90, very good for values between 0.75–0.89, good for values between 0.50–0.74, fair for values between 0.25–0.49, poor for values between 0–0.24, and unsatisfactory for values < 0 (Moriasi et al., 2007; Parajuli et al., 2008, 2009). The RMSE performance has no suggested values to rank, however the smaller the RMSE the better the performance of the model (Moriasi et al., 2007), and a value of zero for RMSE represents perfect simulation of the measured data.

The SWAT model was calibrated (from January 1998 to December 2003) and validated (from January 2004 to December 2009) using field observed monthly streamflow data from the Lena USGS gage station (USGS 02483500) within the UPRW. Model calibration and validation parameters were adopted from previous study (Parajuli, 2010a). Model simulated results showed good to very good performances for the monthly streamflow prediction both during model calibration ($R^2$ = 0.75, E = 0.70) and validation ($R^2$ = 0.73, E = 0.51) periods (Fig. 2). The SWAT model predicted monthly streamflow ($m^3$ $s^{-1}$) estimated very similar RMSE values (<2% difference) during model calibration (RMSE = 51.7 $m^3$ $s^{-1}$) and validation (RMSE = 50.7 $m^3$ $s^{-1}$) periods. This case study results were in close agreement with several previous studies that used the SWAT model (Gassman et al., 2007; Moriasi et al., 2007; Parajuli et al., 2009; Parajuli 2010a; Nejadhashemi et al., 2011; Sheshukov et al., 2011).

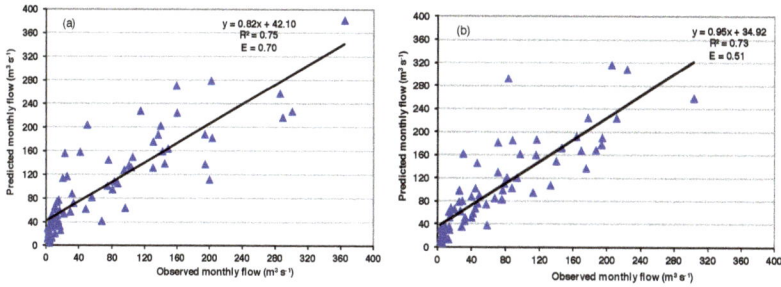

**Figure 2.** Monthly observed vs. predicted streamflow during (a) calibration and (b) validation periods

### 4.1.3. Future climate scenarios

The calibrated and validated SWAT model for the UPRW was simulated for an additional 30 years (January 2010 to December 2040) to provide fourteen future climate change scenarios (Table 1). The average streamflow value from the calibrated and validated model was considered as baseline scenario. The future climate change scenarios represented percentage change in the precipitation, temperature and $CO_2$ concentration values as described in the Table 1. The $CO_2$ values were adjusted from a baseline value of 330 ppmv (part per million by volume), which is a default value provided in the SWAT model. Two other $CO_2$ values (495 and 660) were tested in the model considering 50% and 100% increase from the model default value. Percentage changes in the precipitation were simulated for ±20% from the baseline value. Similarly, the model temperature factor was adjusted using +1 and +2 degrees in Celsius from the baseline. The fourteen future climate change scenarios were developed using interaction of three $CO_2$, three precipitation, and three temperature adjustment values.

The SWAT model results for fourteen scenarios (from Sc1 to Sc14 for Lena gage station) predicted an average maximum monthly stream flow decrease of 57% and average maximum monthly flow increase of 74% from the base simulation (Figure 3). Precipitation increase always had the greatest impact on monthly streamflow from the watershed. A twenty percent increase in precipitation resulted into the greatest impact in the future streamflow prediction. However, increases in $CO_2$ and temperature accelerated the magnitude of streamflow process.

Scenario 13 with the highest increase in the precipitation (+20%), $CO_2$ (660 ppmv), and temperature (+2 degree Celsius) had about 74% greater impact on streamflow prediction than the baseline condition (Fig. 3). Other scenarios that had high impact on streamflow prediction were Sc1, Sc4, Sc7, and Sc10. The increase in the temperature had medium impact on streamflow process as shown by Sc3, Sc6, Sc9, and Sc12. However, Sc12 had the greatest impact among medium scenarios as it predicted about 10% greater cumulative monthly streamflow than the baseline condition. Scenarios Sc2, Sc5, Sc8, Sc11, and Sc14 had lower cumulative monthly streamflow than the baseline condition, as they all had decreased precipitation (-20%). However, Sc14 had the greatest effect on stream flow among all low condition scenarios, due to the highest temperature (+2 degree Celsius) and $CO_2$ values (660 ppmv).

| $CO_2$ (ppmv) | Precip. (%) | Temp. (adj. °C) | Scenarios | Effect |
|---|---|---|---|---|
| 330 | 0 | 0 | Base | No |
| 330 | +20 | 0 | Sc1 | High |
| 330 | -20 | 0 | Sc2 | Low |
| 330 | 0 | +1 | Sc3 | Medium |
| 330 | +20 | +1 | Sc4 | High |
| 330 | -20 | +1 | Sc5 | Low |
| 330 | 0 | +2 | Sc6 | Medium |
| 330 | +20 | +2 | Sc7 | High |
| 330 | -20 | +2 | Sc8 | Low |
| 495 | 0 | +2 | Sc9 | Medium |
| 495 | +20 | +2 | Sc10 | High |
| 495 | -20 | +2 | Sc11 | Low |
| 660 | 0 | +2 | Sc12 | Medium |
| 660 | +20 | +2 | Sc13 | High |
| 660 | -20 | +2 | Sc14 | Low |

$CO_2$ = carbon dioxide, Precip. = precipitation, Temp. = temperature, Sc = scenario

**Table 1.** Simulated climate change parameters scenarios and effect

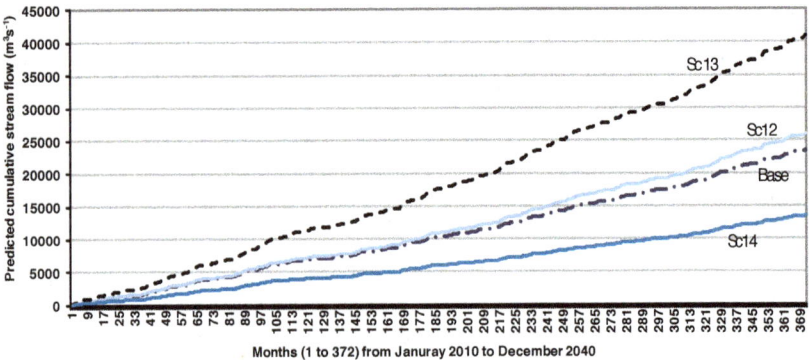

**Figure 3.** Model predicted cumulative monthly streamflow during thirty years period (2010-2040) showing greater than base condition and lower than base condition scenarios.

## 4.2. HSPF model

The goal of this case study was to estimate the potential impact of future climate change upon hydrologic characteristics such as river discharge, surface evaporation, and water out-flow in the YRB (Yazoo River Basin) using the HSPF model. The specific objectives were to: (1) develop a site-specific model for the YRB based on watershed, meteorological, and hydrological conditions; (2) calibrate the resulting model using existing field data and/or computational data; and (3) create simulation scenarios to project the potential impact of future climate changes upon hydrologic characteristics in the YRB.

### 4.2.1. Study area and model development

The YRB is the largest river basin in Mississippi, USA and has a total drainage area of 34,600 km$^2$ (Fig. 4). This basin is separated into two distinct topographic regions, one is the Bluff Hills (about 16600 km$^2$) and the other is the Mississippi Alluvial Delta (Guedon and Thomas, 2004; MDEQ, 2008; Shields et al., 2008). The Bluff Hills region is a hilly and upland area where streams originate from lush oak and hickory forests and pastures dominate the rural landscape. The Delta Region, on the other hand, is a flat and lowland area characterized by slow streamflow and an extensive system of oxbow lakes.

Data collection for the YRB (HUC 8030208) includes watershed descriptions, meteorological, and hydrologic data. Several agencies are active in the data collection efforts. Most of the data used in this study such as land use, soil type, topography, precipitation, and discharge are from National Hydrography Dataset, U.S. Geologic Survey National Water Information System, and 2001 National Land Cover Data.

Four future climate change scenario data, namely the HADCM3B2, CSIROMK35A1B, CSIROMK2A1B, and MIROC32A1B, were used in this case study. HADCM3, CSIROMK35, CSIROMK2, and MIROC32 are names of climate general circulation models (GCM). The B2 and A1B at the end of the names of the climate change scenarios are the Intergovernmental Panel on Climate Change (IPCC) emission scenarios under which the GCMs were run to produce the individual climate projection. The HADCM3B2 scenario data was obtained from the Hadley Centre for Climate Prediction and Research, United Kingdom. The CSIROMK35A1B and CSIROMK2A1B scenarios data were obtained from the Australian Commonwealth Scientific and Research Organization Atmospheric Research, and the MIROC32A1B scenario data was obtained from the Center for Climate System Research, University of Tokyo National Institute for Environmental Studies and Frontier Research Center for Global Change. More detail information about these climate scenarios are available at: http://www-pcmdi.llnl.gov/ipcc/model_documentation/ipcc_model_documentation.php. These four scenarios data involve monthly air temperature and precipitation for a period from 2000 to 2050, which were generated by GCMs and the Center for Climate System Research National Institute for Environmental Studies and Frontier Research Center for Global Change (University of Tokyo). These data were scaled to the 8-digit HUC watersheds for different regions. For the YRB watershed, the 8-digit HUC was 08030208. A descriptive statistics for these four scenarios data showed the amount of precipitation from high to low order as: CSIROMK35A1B > HADCM3B2 > CSIROMK2A1B > MIROC32A1B, whereas the magnitude of air temperature from high to low order as: MIROC32A1B > CSIROMK35A1B > CSIR-OMK32A1B > HADCM3B2.

The HSPF model for this case study was developed using the PERLND, IMPLND, and RCHRES modules that are available in HSPF. The PWATER section of the PERLND module is a major component that simulates the water budget, including surface flow, inter-flow and groundwater behavior. The HYDR section of the RCHRES module simulates the hydraulic behavior of the stream.

**Figure 4.** Location of modeled area in the Yazoo River Basin, Mississippi.

## 4.2.2. Model calibration and validation

Model calibration involves adjusting input parameters within a reasonable range to obtain a best fitness between field observations and model predictions. Model validation is a process of validating the calibrated model by comparing the field observations against the model predictions without changing any input parameter values. Table 2 shows a comparison of the observed and predicted annual water outflow volume. The annual differences in errors between the observed and predicted water outflow volumes were about 6% and were, therefore, acceptable (Bicknell et al., 2001). With prediction = 0.97*observation and R2 = 0.98 and E = 0.96, we determined that an excellent agreement was obtained between the field observations and model predictions during the model calibration process.

Comparison of annual water outflow between the observations and predictions for a time period from January 1, 2005 to December 31, 2010 during the model validation process was given in Table 2. The regression equation predictions = 0.97*observation and $R^2$ = 0.99 and E = 0.97 verified the excellent agreement between the model predictions and the field observations during the model validation process.

| Year | Simulated Outflow $(m^3)$ | Observed Outflow $(m^3)$ | Percent Different |
|---|---|---|---|
| Model Calibration | | | |
| 2000 | 1.77E+09 | 1.75E+09 | 0.88 |
| 2001 | 2.05E+09 | 1.92E+09 | 6.34 |
| 2002 | 1.93E+09 | 1.93E+09 | -0.37 |
| 2003 | 1.15E+09 | 1.16E+09 | -0.34 |
| 2004 | 2.00E+09 | 1.90E+09 | 5.58 |
| Total | 8.90E+09 | 8.66E+09 | 2.68 |
| Model Validation | | | |
| 2005 | 1.32E+09 | 1.30E+09 | 1.64 |
| 2006 | 1.33E+09 | 1.35E+09 | -2.10 |
| 2007 | 1.20E+09 | 1.19E+09 | 1.13 |
| 2008 | 9.71E+08 | 9.47E+08 | 2.54 |
| 2009 | 1.96E+09 | 1.82E+09 | 7.40 |
| Total | 6.78E+09 | 6.62E+09 | 2.50 |

**Table 2.** Comparison of the simulated and observed annual water outflow volumes during model calibration and validation.

*4.2.3. Past and future climate change*

Comparison of mean annual water yields between the past 10 years (2001-2011) and the future 40 years (2011-2050) for the four climate projections indicates that water yields will continue to decline (Table 3). The percent change in mean annual water yield varied from 29.47% for the CSIROMK35A1B projection to 18.51% for the MIROC32A1B projection, with four climate projections indicating continuing declines out to 2050. The same decline trends were observed for maximum annual water yields (Table 3). The declines in mean and maximum annual water yields occurred primarily due to the projected precipitation decrease. Mixed results were found for the mean annual evaporative loss (Table 3). The CSIROMK2B2 projection indicated a long-term increase while the other three projections indicated a long-term decrease in evaporative losses. Further work is thus necessary to better determine how evaporative losses will respond in the future.

Changes of monthly minimum, mean, and maximum in water discharges and yields for the four climate projections during the 40-year simulation period (2011-2050) are given in Figs. 5 and 6. The monthly minimum, mean, and maximum water discharges and yields varied among the four climate projections and changed from year to year within each projection. In general, the MIROC32A1B projection had highest monthly minimum, mean, and maximum water discharges and yields in most of the years during the 40-year simulation, which occurred because the MIROC32A1B projection had highest annual precipitation during the same simulation period (Table 3).

| Scenario | Precipitation (cm) | | | Evaporative Loss (m³) | | | Water Yield (m³) | | |
|---|---|---|---|---|---|---|---|---|---|
| | Past 10 Years (2001 to 2010) | Future 40 Years (2011 to 2050) | % Change | Past 10 Years (2001 to 2010) | Future 10 Years (2011 to 2050) | % Change | Past 10 Years (2001 to 2050) | Future 10 Years (2011 to 2050) | % Change |
| | Annual Mean | | | | | | | | |
| HADCM3B2 | 0.017 | 0.015 | -10.23 | 92.80 | 84.40 | -9.95 | 40000.00 | 32282.00 | -23.91 |
| MIROC32A1B | 0.016 | 0.015 | -10.01 | 99.90 | 86.89 | -14.97 | 37000.00 | 31222.00 | -18.51 |
| CSIROMK35A1B | 0.019 | 0.017 | -12.69 | 125.00 | 100.96 | -23.81 | 53300.00 | 41169.00 | -29.47 |
| CSIROMK2B2 | 0.016 | 0.016 | 0.10 | 96.20 | 104.62 | 8.05 | 34800.00 | 32787.00 | -6.14 |
| | Annual Maximum | | | | | | | | |
| HADCM3B2 | 1.052 | 0.754 | -39.45 | 484.00 | 330.25 | -46.56 | 2160000.00 | 129412.00 | -1569.09 |
| MIROC32A1B | 1.161 | 0.842 | -37.79 | 618.00 | 372.10 | -66.08 | 2150000.00 | 148377.00 | -1349.01 |
| CSIROMK35A1B | 1.346 | 1.088 | -23.73 | 580.00 | 438.07 | -32.40 | 2160000.00 | 205767.00 | -949.73 |
| CSIROMK2B2 | 0.991 | 0.749 | -32.33 | 488.00 | 371.37 | -31.41 | 2160000.00 | 130590.00 | -1554.03 |

**Table 3.** Comparison of the sum and mean values for precipitation, evaporative loss, and water yield between the past and future 10 years.

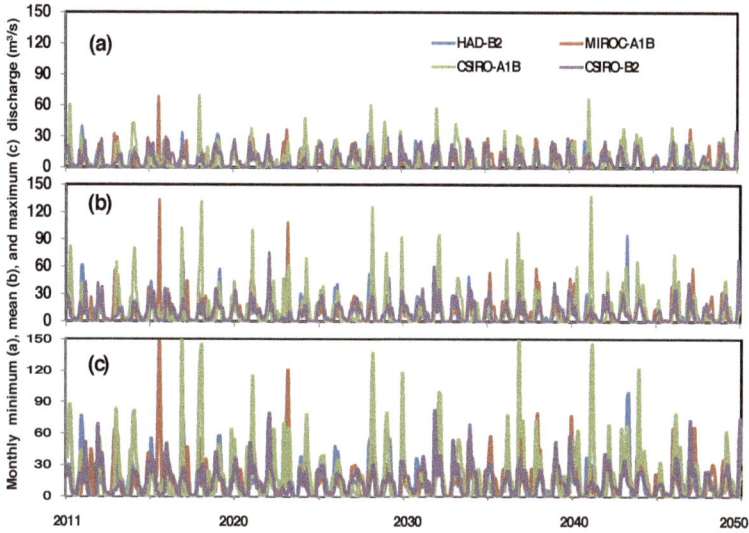

**Figure 5.** Simulated monthly minimum (a), mean (b), and maximum (c) discharge for the four simulation scenarios.

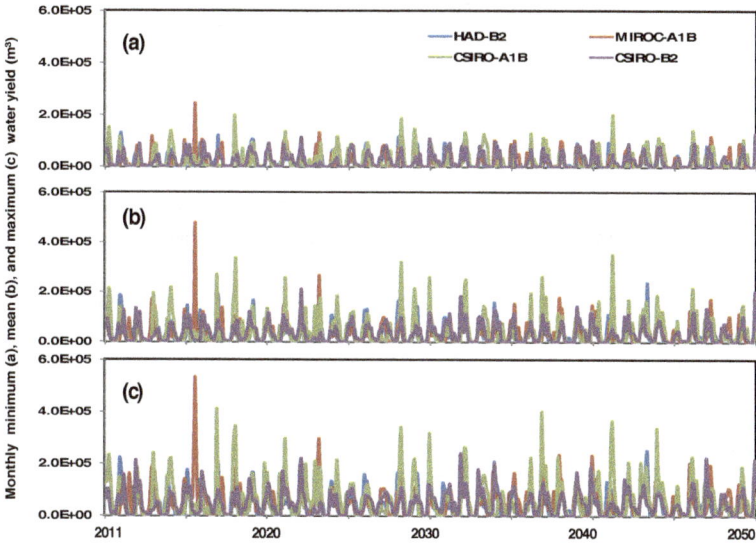

**Figure 6.** Simulated monthly minimum (a), mean (b), and maximum (c) water outflow volume for the four simulation scenarios.

## 5. Conclusions

Two models (SWAT and HSPF) commonly used in hydrological and water quality studies were applied here in two large scale watersheds (UPRW and YRB) in the state of Mississippi. Models were calibrated and validated using USGS observed streamflow data. The long-term hydrological impacts due to future climate change scenarios were assessed using the SWAT and HSPF models.

For one case study, simulated mean monthly streamflow results for the calibrated and validated SWAT model provided good to very good fits ($R^2$ and E values from 0.75 to 0.51) to USGS monthly observed streamflow data. Fourteen future climate change scenarios were developed using interaction of precipitation, $CO_2$, and temperature adjustment values in the SWAT model. The scenario with the highest increase in the precipitation (+20%), $CO_2$ (660 ppmv), and temperature (+2 degree Celsius) had about the greatest (> 74%) impact on streamflow simulation when compare with the baseline condition. Interaction of temperature adjustment and $CO_2$ factors had a medium and low impact respectively during thirty year's model simulation period in this study.

Another case study examined the impact of climate change on future water discharge, evaporation, and yield in the YRB using the BASINS-HSPF model. The model was calibrated using observed data from a five-year (2001 to 2004), and validated using observed data from another five-year (2005 to 2010). Excellent agreements were obtained between the model predictions and the field observations for model calibration and validation.

Four future climate scenarios (or projections) - CSIROMK35A1B, HADCM3B2, CSIROMK2B2, and MIROC32A1B were used to investigate water discharge, evaporative loss, and water outflow responses to predicted precipitation and air temperature changes over a 50-year period from 2001 to 2050. Comparison of simulation results between the past 10 years (2001-2010) and the future 40 years (2011-2050) shows that the mean and maximum annual water yields declined due to the projected precipitation decrease. In general, the MIROC32A1B projection had the highest monthly minimum, mean, and maximum water discharges and yields in most of the years during the 40-year simulation period (2011-2050). This projection had the highest projected annual precipitation. Results suggest that the projected precipitation had profound impacts upon water discharge and yield in the YRB.

Spatial data used in the models may have potential sources of errors. For example, the DEM data are used to delineate watershed boundary are available in different resolutions. Similarly, use of land use, soils and weather data may have some spatial errors, which can influence the hydrologic and climate change impact. However, these results will only have relative influence in model simulated results. This book chapter provided review of several watershed and water quality models and two case studies to evaluate future climate change impact on hydrology using two models.

# Acknowledgement

The part of the case study research presented in this book chapter is based on work supported by the Special Research Initiatives (SRI) and Mississippi Agricultural and Forestry Experiment Station (MAFES) at Mississippi State University.

# Author details

Prem B. Parajuli[1*] and Ying Ouyang[2]

*Address all correspondence to: pparajuli@abe.msstate.edu

1 Department of Agricultural and Biological Engineering at Mississippi State University, Mississippi State, USA

2 USDA-Forest Service Center for Bottomland Hardwoods Research, Mississippi State, USA

# References

[1] Akter A., and Babel M. S. 2012. Hydrological modeling of the Mun River basin in Thailand. Journal of Hydrology, 452-453: 232-246.

[2] Alam M. J., and Goodall J. L. 2012. Toward disentangling the effect of hydrologic and nitrogen source changes from 1992 to 2001 on incremental nitrogen yield in the contiguous United States. Water Resources Research, 48 (4), W04506, doi: 10.1029/2011WR010967 (in press).

[3] Alarcon V. J., McAnally W. H.,and Pathak S. 2012. Comparison of two hydrodynamic models of Weeks Bay, Alabama. Computational science and its applications – ICCSA, Lecture notes in Computer Science, 7334: 589-598, doi: 10.1007/978-3-642-31075-1_44.

[4] Arnold J. G., Srinivasan R., Muttiah R. S., and Williams J. R. 1998. Large area hydrologic modeling and assessment, Part I: model development. Journal of American Water Resources Association, 34(1): 73-89.

[5] Baginska B., Milne-Home W., and Cornish P. S. 2003. Modelling nutrient transport in Currency Creek, NSW with AnnAGNPS and PEST. Environmental Modelling & Software, 18: 801–808.

[6] Bakhsh A., R. S. Kanwar D. B. Jaynes T. S. Colvin and L. R. Ahuja. 2000. Prediction of NO3–N losses with subsurface drainage water from manured and UAN–fertilized plots using GLEAMS. Trans. ASAE, 43 (1): 69–77.

[7] Betrie G. D., Y. A. Mohamed A. van Griensven and R. Srinivasan. 2011. Sediment management modelling in the Blue Nile Basin using SWAT model. Hydrology and Earth System Sciences, 15: 807–818.

[8] Bicknell B. R., Imhoff J. C., Kittle J. L., Donigian A. S., and Johanson R. C. 1993. Hydrological Simulation Program – FORTRAN (HSPF): Users Manual for Release 10. EPA-600/R-93/174, U.S. EPA, Athens, GA, 30605

[9] Bicknell, B. R., Imhoff, J. C., Kittle, Jr., J. L., Jobes, T. H., Donigian, Jr., A. S., 2001. Hydrological Simulation Program – Fortran, HSPF, Version 12, User's Manual. National Exposure Research Laboratory, Office of Research and Development, U.S. Environmental Protection Agency, Athens, GA.

[10] Bingner R. L., and Theurer F. D. 2003. AnnAGNPS technical processes documentation, Version 3.2. USDA-ARS, National Sedimentation Laboratory: Oxford, MS.

[11] Blumensaat F., Wolfram M., and Krebs P. 2012. Sewer model development under minimum data requirements. Environmental Earth Sciences, 65:1427-1437.

[12] Boskidis, I., G. D. Gikas, G. K. Sylalos and V. A. Tsihrintzis. 2012. Water Resources Management, 26(10): 3023-3051.

[13] Brown, J. B. 2011. Application of the SPARROW watershed model to describe nutrient sources and transport in the Missouri River Basin: U.S. Geological Survey Fact Sheet, 3104, 4 p.

[14] Cantone, J., and Schmidt, A 2011. Improved understanding and prediction of the hydrologic response of highly urbanized catchments through development of the Illinois Urban Hydrologic Model. Water Resources Research, 47: W08538.

[15] Chaubey, I., L. Chiang, M. W. Gitau, and S. Mohamed. 2010. Journal of soil and water conservation, 65 (6): 424-437.

[16] Chen Y. D., Carsel R. F., Mccutcheon S. C., and Nutter W. L. 1998. Stream temperature simulation of forested riparian areas: I. Watershed model development. ASCE - Journal of Environ Engineering, 124:304–315

[17] Chinkuyu, A. J., T. Meixner, T. Gish, and C. Daughtry. 2004. The importance of seepage zones in predicting soil moisture content and surface runoff using GLEAMS and RZWQM. Trans of the ASAE, 47(2): 427–438.

[18] Chinkuyu, A. J., and R. S. Kanwar. 2001. Predicting soil nitrate–nitrogen losses from incorporated poultry manure using GLEAMS model. Trans. ASAE, 44 (6): 1643–1650.

[19] Cronshey R. G., and Theurer F. G. 1998. AnnAGNPS-non point pollutant loading model. In Proceedings First Federal Interagency Hydrologic Modelling Conference. Las Vegas, NV.

[20] Dessu, S. B. and A. M. Melesse. 2012. Modelling the rainfall-runoff process of the Mara River basin using the Soil and Water Assessment Tool. Hydrological Processes, DOI: 10.1002/hyp.9205.

[21] Donigian A. S. Jr., and Crawford N. H. 1976. Modeling Pesticides and Nutrients on Agricultural Lands. Environmental Research Laboratory, Athens, GA. EPA 600/2-7-76-043, 317 p

[22] Donogian A. S. Jr., Imhoff J. C., Bicknell B. R., Kittle J. I. 1984. Application guide for hydrological simulation program-FORTRAN (HSPF), EPA, Athens, GA. EPA-600/3-84-065.

[23] Dun S., J. Q. Wu, D. K. McCool, J. R. Frankenberger, and D. C. Flanagan. 2010. Improving frost-simulation sub-routines of the Water Erosion Prediction Project (WEPP) model. Trans. of the ASABE, 53(5): 1399-1411.

[24] Flanagan D. C., J. E. Gilley, and T. G. Franti. 2007. Water Erosion Prediction Project (WEPP): development history, model capabilities, and future enhancements. Trans. of the ASABE, 50(5): 1603-1612.

[25] Foltz R. B., W. J. Elliot, and N. S. Wagenbrenner. 2011. Soil erosion model predictions using parent material/soil texture-based parameters compared to using site-specific parameters. Trans. of the ASABE, 54(4): 1347-1356.

[26] Gassman P. W., Reyes M. R., Green C. H., and Arnold, J. G. 2007. The Soil and Water Assessment Tool: historical development, applications, and future research directions. Transactions of the ASABE, 50(4): 1211–1250.

[27] Githui F., B. Selle and T. Thayalakumaran. 2012. Recharge estimation using remotely sensed evapotranspiration in an irrigated catchment in southeast Australia. Hydrological Processes, 26: 1379-1389.

[28] Gong W., Wang Y., and Jia J. 2012. The effect of interacting downstream branches on saltwater intrusion in the Modaomen Estuary, China. Journal of Asian Earth Sciences, 45: 223-238.

[29] Gosain A. K., Rao S., Srinivasan R., and Reddy N. G. 2005. Return-flow assessment for irrigation command in the Palleru River basin using SWAT model. Hydrological Processes, 19: 673–682.

[30] Guedon, N. B., and Thomas, J. V. 2004. State of Mississippi Water Quality Assessment, section 305(b) Report 62. Mississippi Department of Environmental Quality, Jackson, MS.

[31] Hamrick, J. M. 1996. A User's Manual for the Environmental Fluid Dynamics Computer Code (EFDC), The College of William and Mary, Virginia Institute for Marine Sciences, Special Report 331.

[32] Hua L., Xiubin H., Yongping Y., and Hongwei N. 2012. Assessment of Runoff and Sediment Yields Using the AnnAGNPS Model in a Three-Gorge Watershed of China. International Journal of Environmental Research and Public Health, 9:1887-1907.

[33] Knisel W. G. 1980. CREAMS, a field scale model for chemicals, runoff and erosion from agricultural management systems. USDA Conservation Research Rept. No. 26, U.S. Department of Agriculture: Washington D. C.

[34] Leonard R. A., W. G. Knisel, and D. A. Still. 1987. GLEAMS: Groundwater loading effects of agricultural management systems. Trans. of the ASAE, 30:1403-1418.

[35] Licciardello F., D. A. Zema, S. M. Zimbone, and R. L. Bingner. 2007. Runoff and soil erosion evaluation by the AnnAGNPS model in a small Mediterranean watershed. Trans. of the ASAE, 50, 1585-1593.

[36] Licciardello F., D. A. Zema, S. M. Zimbone, and R. L. Bingner. 2011. Runoff and soil erosion evaluation by the AnnAGNPS model in a small Mediterranean watershed. Trans. of the ASABE, 50(5): 1585-1593.

[37] Mississippi Department of Environmental Quality (MDEQ), 2008. Sediment TMDL for the Yalobusha River Yazoo River Basin. PO Box 10385, Jackson, MS 39289-0385.

[38] Mississippi Department of Environmental Quality (MDEQ). 2010. Total Daily Maximum Daily Load Program. Available at: http://www.deq.state.ms.us/MDEQ.nsf/page/TWB_Total_Maximum_Daily_Load_Section?OpenDocument. Accessed on March 15, 2012.

[39] Moriasi D. N., Arnold, J. G., Van Liew, M. W., Bingner, R. L., Harmel, R. D., and Veith, T. L. 2007. Model evaluation guidelines for systematic quantification of accuracy in watershed simulations. Transactions of the ASABE, 50(3): 885-900.

[40] National Climatic Data Center (NCDC). 2012. Locate weather observation station record. Available at: http://www.ncdc.noaa.gov/oa/climate/stationlocator.html. Accessed on April 14, 2012.

[41] Neitsch S. L., Arnold, J. G., Kiniry, J. R., Williams, J. R., and King K. W. 2002. Soil and water assessment tool (SWAT), theoretical documentation, Blackland research center, grassland, soil and water research laboratory, agricultural research service: Temple, TX.

[42] Neitsch S. L., Arnold J. G., Kiniry J. R., and Williams J. R. 2005. Soil and water assessment tool (SWAT), theoretical documentation, Blackland research center, grassland, soil and water research laboratory, agricultural research service: Temple, TX.

[43] Nejadhashemi A. P., Woznicki S. A. and Douglas-Mankin K. R., 2011. Comparison of four models (STEPL, PLOAD, L-THIA, AND SWAT) in simulating sediment, nitrogen, and phosphorus loads and pollutant source areas. Trans. of the ASABE, 54(3): 875-890.

[44] Ouyang Y., J. Higman and J. Hatten. 2012. Estimation of dynamic load of mercury in a river with BASINS-HSPF model. Journal of Soils and Sediments, 12(2): 207-216.

[45] Parajuli P. B., Mankin K. R. and Barnes P. L. 2008. Applicability of Targeting Vegetative Filter Strips to Abate Fecal Bacteria and Sediment Yield using SWAT. Agricultural Water Management, 95 (10): 1189-1200.

[46] Parajuli P. B., Mankin K. R., and Barnes P. L. 2009. Source specific fecal bacteria modeling using soil and water assessment tool model. Bioresource Technology, 100 (2): 953-963.

[47] Parajuli P. B. 2010a. Assessing sensitivity of hydrologic responses to climate change from forested watershed in Mississippi. Hydrological Processes, 24 (26): 3785-3797.

[48] Parajuli P. B. 2010b. Methods for Modeling Livestock and Human Sources of Nutrients at Watershed Scale. MAFES Research Report, 24(8): 1-8.

[49] Parajuli P. B. 2011. Effects of Spatial Heterogeneity on Hydrologic Responses at Watershed Scale. Journal of Environmental Hydrology, 19(18): 1-18.

[50] Parajuli P. B. 2012. Evaluation of Spatial Variability on Hydrology and Nutrient Source Loads at Watershed Scale using a Modeling Approach. Hydrology Research, doi:10.2166/nh.2012.013.

[51] Polyakov V., Fares A., Kubo D., Jacobi J., and Smith C. 2007. Evaluation of a nonpoint source pollution model, AnnAGNPS, in a tropical watershed. Environmental Modelling & Software, 22(11): 1617–1627.

[52] Renard K. G., Foster G. R., Weesies G. A., McCool D. K., and Yoder D. C. 1997. Predicting Soil Erosion by Water: A Guide to Conservation Planning with the Revised Universal Soil Loss Equation (RUSLE). U.S. Department of Agriculture, Agriculture Handbook No. 703.

[53] Rolle K., Gitau M. W., Chen G., and Chauhan A. 2012. Assessing fecal coliform fate and transport in a coastal watershed using HSPF. Water Science and Technology, 66:1096-1102.

[54] Saad D. A., Schwarz G. E., Robertson D. M., and Booth N. L. 2011. A Multi-Agency Nutrient Dataset Used to Estimate Loads, Improve Monitoring Design, and Calibrate Regional Nutrient SPARROW Models. Journal of the American Water Resources Association, 47: 933-949.

[55] Schwarz, G. E. 2008. A Preliminary SPARROW model of suspended sediment for the conterminous United States, U.S. Geological Survey Open-File Report 2008–1205, 8 p.

[56] Sharpley A. N. and J. R. Williams. 1990. EPIC—Erosion/Productivity Impact Calculator: 1. Model Documentation. U.S. Department of Agriculture Technical Bulletin No. 1768.

[57] Sheshukov A. Y., Siebenmorgen C. B., and Douglas-Mankin K. R. 2011. Seasonal and annual impacts of climate change on watershed response using an ensemble of global climate models. Trans. of the ASABE, 54(6): 2209-2218.

[58] Shields F. D. Jr., Cooper C. M., Testa III, S., and Ursic, M. E. 2008. Nutrient Transport in the Yazoo River Basin, Research Report 60. U.S. Dept of Agriculture Agricultural Research Service National Sedimentation Laboratory: Oxford. http://www.ars.usda.gov /SP2UserFiles/person/5120/NSLReport60.pdf.

[59] Shirmohammadi A., B. Ulen, L. F. Bergstrom, and W. G. Knisel. 1998. Simulation of nitrogen and phosphorus leaching in a structured soil using GLEAMS and a new sub-model. Trans. of the ASAE, 41(2): 353-360.

[60] Sinha S., Liu X., and Garcia, M. H. 2012. Three-dimensional hydrodynamic modeling of the Chicago River, IL. Environmental Fluid Mechanics , Pages 1-24.

[61] Smith R. A., Schwarz G. E., and Alexander R. B. 1997. Regional interpretation of water-quality monitoring data. Water Resources Research, 33 (12): 2781-2798.

[62] Talei A., and Chua L. H. C. 2012. Influence of lag time on event-based rainfall-runoff modeling using the data driven approach. Journal of Hydrology, 438-439: 223-233.

[63] U.S. Department of Agriculture (USDA). 2005. Soil data mart. Natural Resources Conservation Service. Available at: http://soildatamart.nrcs.usda.gov/Default.aspx. Accessed on May 23, 2012.

[64] U.S. Department of Agriculture, National Agricultural Statistics Service (USDA/NASS). 2010. The cropland data layer. Available at: http://www.nass.usda.gov/research/Cropland/SARS1a.htm. Accessed on April 29, 2012.

[65] U.S. Environmental Protection Agency (US/EPA). 2010. National summary of impaired waters and TMDL information. U.S. Environmental Protection Agency: Washington, D.C. Available at: http://iaspub.epa.gov/waters10/attains_nation_cy.control?p_report_type=T. Accessed on June 28, 2012.

[66] U.S. Geological Society (USGS). 1999. National elevation dataset. Available at: http://seamless.usgs.gov/index.php. Accessed April 29, 2012.

[67] Vache K. B., Eilers J. M., and Santelmann M. V. 2002. Water quality modeling of alternative agricultural scenarios in the U.S. Corn Belt. Journal of American Water Resources Association, 38(3): 773-787.

[68] Varanou E., Gkouvatsou E., Baltas E., and Mimikou M., 2002. Quantity and quality integrated catchment modeling under climate change with use of soil and water assessment tool model. ASCE Journal of Hydrological Engineering, 7(3): 228-244.

[69] Vellidis G., P. Barnes, D. D. Bosch, and A. M. Cathey. 2012. Mathematical simulation tools for developing dissolved oxygen TMDLs. Trans of the ASABE, 49(4): 1003–1022.

[70] Williams, J. R., C. A. Jones and P. T. Dyke. 1984. A modeling approach to determining the relationship between erosion and soil productivity. Trans. ASAE 27(1): 129-144.

[71] Williams J. R. and A. D. Nicks. 1985. SWRRB, a simulator for water resources in rural basins: an overview. In: D.G. DeCoursey (editor), Proc. of the Natural Resources Modeling Symp., Pingree Park, CO. USDA-ARS, ARS-30, pp. 17-22.

[72] Yuan Y., Bingner R. L., and Rebich R. A. 2001. Evaluation of AnnAGNPS on Mississippi Delta MSEA watersheds. Trans. of the ASAE, 44(5): 1183–1190.

[73] Yuan Y, Dabney S, and Bingner R. L. 2002. Cost/benefit analysis of agricultural BMPs for sediment reduction in the Mississippi Delta. Journal of Soil and Water Conservation, 57(5): 259–267.

[74] Yuan Y., Locke M. A., and Bingner R. L. 2008. Annualized Agricultural Non-Point Source model application for Mississippi Delta Beasley Lake watershed conservation practices assessment. Journal of Soil and Water Conservation, 63(6): 542-551.

[75] Yuan Y., R. L. Bingner M. A. Locke F. D. Theurer, and J. Stafford. 2011. Assessment of Subsurface Drainage Management Practices to Reduce Nitrogen Loadings Using AnnAGNPS. Applied Engineering in Agriculture, 27(3): 335-344.

[76] Young R. A., Onstead C. A., Bosch D. D., and Anderson W. P. 1989. AGNPS: a nonpoint source pollution model for evaluating agricultural watersheds. Journal of Soil Water Conservation, 44(2):168–173.

[77] Zhang G. H., B. Y. Liu and X. C. Zhang. 2008. Applicability of WEPP sediment transport equation to steep slopes. Trans of the ASABE, 51(5): 1675-1681.

[78] Zuercher B. W., D. C. Flanagan, and G. C. Heathman. 2011. Evaluation of the AnnAGNPS model for atrazine prediction in Northeast Indiana. Trans of the ASABE, 54(3): 811-825.

# Contaminant Hydrology: Groundwater

# Chapter 4

# Occurrence and Mobility
# of Mercury in Groundwater

Julia L. Barringer, Zoltan Szabo and Pamela A. Reilly

Additional information is available at the end of the chapter

## 1. Introduction

### 1.1. Forms, toxicity, and health effects

Mercury (Hg) has long been identified as an element that is injurious, even lethal, to living organisms. Exposure to its inorganic form, mainly from elemental Hg (Hg(0)) vapor (Fitzgerald & Lamborg, 2007) can cause damage to respiratory, neural, and renal systems (Hutton, 1987; USEPA, 2012; WHO, 2012). The organic form, methylmercury ($CH_3Hg^+$; MeHg), is substantially more toxic than the inorganic form (Fitzgerald & Lamborg, 2007). Methylmercury attacks the nervous system and exposure can prove lethal, as demonstrated by well-known incidents such as those in 1956 in Minimata, Japan (Harada, 1995), and 1971 in rural Iraq (Bakir et al., 1973), where, in the former, industrial release of MeHg into coastal waters severely tainted the fish caught and eaten by the local population, and in the latter, grain seed treated with an organic mercurial fungicide was not planted, but eaten in bread instead. Resultant deaths are not known with certainty but have been estimated at about 100 and 500, respectively (Hutton, 1987). Absent such lethal accidents, human exposure to MeHg comes mainly from ingestion of piscivorous fish in which MeHg has accumulated, with potential fetal damage ascribed to high fish diets during their mothers' pregnancies (USEPA, 2001). Lesser human exposure occurs through ingestion of drinking water (USEPA, 2001), where concentrations of total Hg (THg; inorganic plus organic forms) typically are in the low nanograms-per-liter range[1], particularly from many groundwater sources, and concentrations at the microgram-per-liter level are rare.

---

1 Because many studies report Hg concentrations in units of nanograms per liter, results reported for aqueous samples in other units herein will be converted to nanograms per liter. Contents of solid materials will be reported in megagrams, kilograms and (or) milligrams per kilogram.

## 1.2. Standards for mercury in water, and other regulations

For drinking water the World Health Organization (WHO) guideline for THg is 1,000 ng/L, a level (in some instances given as 0.001 mg/L or 1.0 µg/L) adopted by the European Union countries, the United Kingdom, Canada, and India (WHO, 1993; NIEA, 2011; Environment Canada, 2010; Srivastava, 2003). These standards are lower than the maximum contaminant level (MCL) adopted by the U.S. Environmental Protection Agency (USEPA), which is 2,000 ng/L (USEPA, 2001a). The USEPA has also adopted a reference dose (RfD) for MeHg in drinking water of 0.1 µg/kg bw/day (or 100 ng/kg bw/day, where bw = body weight) (USEPA 2001b).

Measures, such as regulations to restrict commerce in Hg, have been taken to reduce the amount of elemental Hg available globally. The amount mined in 2010, for example, was estimated at 2042 megagrams (Mg; or 2250 tons)(Brooks, 2010). In 2007, the European Union passed a ban on the export of Hg(0) and in 2008 this was expanded to include various Hg compounds. The USEPA, in 2008, passed a ban on the export of elemental Hg from the United States (USA), to take effect in 2013 (USEPA 2009). Because the USA is ranked as one of the world's top exporters of Hg (~417 Mg in 2010 (Brooks, 2010)), implementation of the act will remove a substantial amount from the global market, as it is exported to foreign countries where, among other uses it is employed in small-scale gold (artisanal) mining (USEPA, 2012). China, the world leader in Hg production in 2010, plans to lower production of some metals, including Hg, by 2015 (Brooks, 2010).

## 1.3. Previous studies of mercury in the environment

Given the potentially severe health effects of MeHg, the need to understand the production of MeHg and its bioaccumulation in the food web, as well as the role of atmospheric deposition in supplying Hg to soils and surface water, has led to numerous investigations of Hg inputs and transformations. Hence, the majority of studies and reviews of Hg in the environment focus on atmospheric inputs (e.g., Engstrom et al., 2007; Glass et al., 1991; Pacyna et al., 2006; Schuster et al., 2002) and the fate and transport of Hg in soils, sediments and surface-water settings (e.g., Bradley et al., 2011; Driscoll et al., 1994; Grigal., 2002; Porvari et al., 2003; Rudd, 1995; Shanley et al., 2002; Schuster et al., 2008; Selvendiran et al., 2008; Skyllberg, 2008; Ullrich et al., 2001; Yu et al., 2012). Many of these studies have their emphasis on the production, mobility, and bioaccumulation of MeHg. Far fewer studies have been performed and published on Hg in groundwater. Some of those have shown that Hg concentrations are low, in the nanograms-per-liter range (e.g. Krabbenhoft & Babiarz, 1992). However, other studies have shown that regional levels of Hg in groundwater can be high, in the micrograms-per-liter range (e.g. Barringer & Szabo, 2006), and some show that microgram-per-liter levels of Hg are being found in groundwaters previously not tested (e.g. Khatiwada et al., 2002).

As discussed below, not all contamination with Hg at the land surface causes contamination of groundwater. It is critical to know the complex biogeochemistry of Hg in order to understand how existing important research applies to Hg fate and transport in groundwater systems.

## 2. Sources and fluxes of mercury

### 2.1. Mercury in natural materials

Mercury is one of the least abundant elements in crustal rocks. Concentrations in rocks of the upper continental crust typically range from 0.01 to about 2 mg/kg, although concentrations higher by more than two orders of magnitude are reported from igneous and sedimentary rocks of Crimea and the Donets Basin of Russia (Fleischer, 1970). A reasonable estimate of the average concentration of THg is 0.08 mg/kg (Fairbridge, 1972). The most abundant Hg mineral is cinnabar (HgS), which can be found in association with metacinnabar (β HgS—a metastable sulfide). Other Hg minerals, mainly sulfosalts that can also contain arsenic, are not common, and, when found, are generally associated with ore deposits.

Volcanic emissions have been estimated by some researchers to constitute the main natural source of Hg to the atmosphere where Hg is primarily present as gaseous elemental Hg (Martin et al., 2012). Kilauea, in Hawaii, was estimated to produce about 260,000 kg/yr (Siegel & Siegel, 1987). In contrast, Ferrara et al. (2000) suggest that volcanic emissions are not the main natural source of THg to the atmosphere, based on reported estimates of emissions from Mediterranean volcanoes: Vulcano, about 1 to 6 kg/yr; Stromboli, about 7 to 80 kg/yr; and Etna, about 60 to 500 kg/yr.

Volcanic emissions can cause local contamination of soils and surface waters, but, as shown at Mt. Etna in Sicily, concentrations of Hg in local groundwater were typically < 10 ng/L (Martin et al., 2012), indicating that Hg had not been easily mobilized from the land surface. Nevertheless, some of the Hg deposited to soils by volcanic emissions volatilizes and can be re-emitted to the atmosphere (Fig. 1). Further, evasion of Hg(0) from the ocean surface adds substantially to atmospheric Hg (Mason et al., 1994). Characterization of atmospheric THg along the central USA Gulf Coast has shown inputs from sea spray increases atmospheric deposition of THg in coastal regions (Engle et al., 2008).

Mineralization associated with igneous activity has produced volcanic-hosted massive sulfide deposits, in which the ore mineral cinnabar forms (some mined since the third century B.C. (Navarro, 2008)). Such major geologic sources of Hg are found in Spain, Slovenia, China, and the western USA, in California. Volcanically derived sedimentary deposits containing other sulfides also can contain substantial Hg (Navarro 2008).

Associated with igneous activity, circulating geothermal fluids can contribute Hg to groundwater, surface water, and to solids that precipitate around mineral springs, geysers and fumaroles. In Russia, Yudovich and Ketris (2005) report that Hg contents in condensate from fumarole gases ranged up to 0.11 mg/kg at the Kamchatka Peninsula, and even higher (up to 0.40 mg/kg) at fumaroles in Japan and Guatemala. In Yellowstone National Park in the western USA, Hg is present at concentrations of 12 to 640 ng/L in waters of hot springs and geysers (Ball et al., 2006), although Hg minerals were not recognized in precipitates there (White et al., 1970). Saline waters at Sulfur Bank and Wilbur Springs in the California Coast Range contained about 1,500 ng/L of Hg, and cinnabar and metacinnabar precipitate at Sulfur Bank (White et al., 1970).

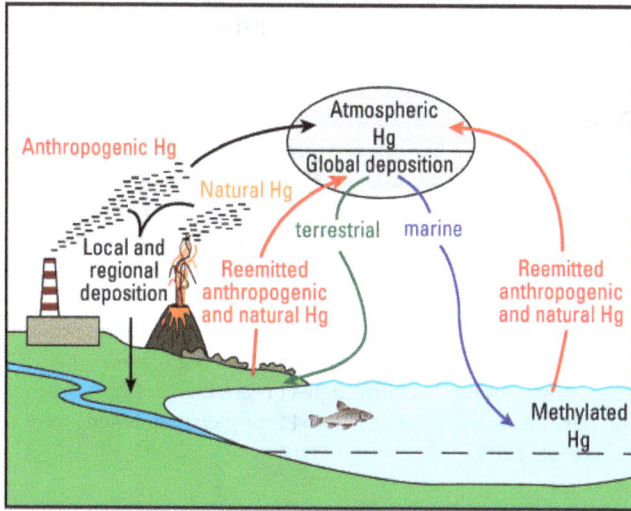

**Figure 1.** Sources and biogeochemical cycling of mercury between land surface, surface water, and the atmosphere (from Tewalt et al., 2001).

Fleischer (1970) compiled literature values for Hg in crustal rocks, worldwide. Average values for mafic igneous rocks ranged from 0.001 to 0.240 mg/kg, whereas for silicic igneous rock, averages ranged from 0.005 to 0.190 mg/kg, except for igneous rocks from the Crimea and Donets Basin, where the average Hg contents ranged from 0.250 to 17.6 mg/kg. Average Hg contents of sandstones and limestones ranged from 0.018 to 5.70 mg/kg, and in shales from 0.05 to 2.3 mg/kg, with the higher contents typically found in the Russian shales. In metamorphic rocks, contents ranged from 0.060 to 2.50 mg/kg.

Many of the widely used sources of coal contain high concentrations of Hg; these include from China (Pirrone et al., 2010), Ukraine (Kolker et al., 2009), and Texas, USA (Tewalt et al., 2001). Coal burning is a major source of Hg release into the atmosphere (Wang et al., 2004); but potential effects on groundwater are mostly unknown. The Hg content of coals ranges from 0.01 to 1.85 mg/kg, with the highest content found in some Chinese coals (Pirrone et al., 2010 and references therein). Much of the Hg in coals is in associated pyrites, with contents that can be 10 mg/kg or greater (Yudovich & Ketris, 2005). Direct effects from coal seams on groundwater may not be obvious, however. Cravotta (2008) did not find detectable Hg concentrations in samples of waters discharging from abandoned coal mines in Pennsylvania, USA.

## 2.2. Mercury in atmospheric deposition and its effects

Emissions of Hg to the atmosphere have increased greatly since the beginning of the industrial era, although estimates of the increase in atmospheric levels vary widely—from about 3 to 24 times that of the pre-industrial period (Wang et al., 2004, and references therein). Recently, it has been estimated that total global emissions account for about 6500 megagrams (Mg) of Hg released annually to the environment. The uncertainty for current estimates is about a factor of two, according to Lohman et al., (2008). Nevertheless, as Gustin et al. (2008) point out, estimates of natural emissions vary widely (volcanic emissions are not constant, for example). Gustin et al. (2008) also note that the range in estimates of anthropogenic releases is small, relative to the range of estimates for natural sources. Coal burning and combustion of other fossil fuels were estimated to constitute about 60 % of the annual amount contributed to the atmosphere from anthropogenic sources (Swain et al., 2007). Total Hg contents of ice-core samples from the Upper Fremont Glacier in Wyoming, USA, when integrated over the past 270 years indicated that anthropogenic emissions accounted for 52%, volcanic events contributed 6%, and background sources supplied 42% (Schuster et al., 2002), and most of the anthropogenic contributions have occurred since about 1850. A current estimate of total annual emissions, both natural and anthropogenic, is about 7,300 Mg (Pirrone et al., 2010) (Table 1), higher than earlier estimates, and also with a higher uncertainty. In terms of regional emissions, those from Asia have increased from 38% to 64% of global emissions over the period of 1990-2007 (Pirrone et al., 2010).

Global production (mining and processing) of Hg has been reduced since the mid 20th century, although present-day production may still contribute about one third of emitted Hg from anthropogenic sources (Hylander & Meili, 2003). Amounts of Hg in atmospheric deposition are declining in some parts of the USA as there have been efforts to curb industrial and power-plant emissions. For example, in a study of Hg loading to Minnesota (USA) lakes, Engstrom et al. (2007) found that inputs, mainly from atmospheric deposition and subsequent soil erosion, peaked in the 1970s, and have declined substantially in recent years. Additionally, in the upper Great Lakes (Superior & Huron) in the USA and Canada, a decline in Hg in fish tissue is noted, although not in the lower Great Lakes (Erie & Ontario) (Bhavsar et al., 2010).

Atmospherically deposited Hg has affected mainly surface water and the organisms that live in water bodies. Hg deposited as Hg(0) may be oxidized to Hg(II), then transformed to MeHg by bacterial activity at and below the sediment/water interface or in algal mats. Low-trophic-level organisms (invertebrates) take up both THg and MeHg (Fig. 1). The concentration of MeHg in organisms increases with each step up the food chain—a process known as biomagnification (Alpers & Hunerlach, 2000; USGS, 2000). During the late 20th century, emissions from coal-fired plants in the Midwest of the USA, which are mainly in vapor (Hg(0)) form (Lindberg, 1987), deposited Hg on surface-water bodies. These emissions and other industrial emanations have resulted in fish consumption advisories because of high levels in the tissues of edible fish (Brooks, 2002). Advisories for non-commercial fish, as of 2006, now extend to freshwaters in 48 USA States, of which 23 are statewide advisories. In addition, 13 States have coastal and estuarine advisories (USEPA, 2010).

Results of two studies suggest that Hg from atmospheric deposition contributes to elevated Hg concentrations in groundwater. Bradley et al. (2012, p. 7507) found that strong hydraulic gradients toward a stream indicated deep groundwater discharge was the primary source of filtered Hg (FTHg) to a Coastal Plain stream, USA. Additionally, higher concentrations

| Source | Mercury[e] (Mg/yr) | Year for estimate |
|---|---|---|
| *Natural* | | |
| Oceans | 2,680 | 2008 |
| Lakes | 96 | 2008 |
| Forests | 342 | 2008 |
| Tundra/grasslands[b] | 448 | 2008 |
| Desert/non vegetated areas[c] | 546 | 2008 |
| Volcanoes and geothermal areas | 90 | 2008 |
| *Anthropogenic* | | |
| Non-ferrous metal production | 310 | 2003-06 |
| Pig iron and steel production | 43 | 2003-06 |
| Cement production | 236 | 2003-06 |
| Caustic soda production (chlor alkali process) | 163 | 2003-06 |
| Mercury production | 50 | 2003-06 |
| Gold production | 400 | 2003-06 |
| Waste disposal | 187 | 2003-06 |
| Stationary (fossil fuel) combustion | 810 | 2003-06 |
| Coal-bed fires | 32 | 2003-06 |
| Vinyl chloride monomer production | 24 | 2003-06 |
| Other | 65 | 2003-06 |
| Biomass burning[d] | 675 | 2008 |
| Agricultural areas[d] | 128 | 2008 |

[a] the mean uncertainty associated with these estimates is + 25% (Pirrone et al., 2010)

[b] Includes savannah, prairie, chaparral.

[c] Includes metalliferous areas.

[d] Amounts may need reconfiguring from Pirrone et al.(2010) category of natural sources, p. 5953, because some burning is caused by humans, agricultural inputs of Hg are anthropogenic, and evasion is amplified by soil disturbance.

[e] Values have been rounded to the nearest whole number.

**Table 1.** Estimates[a] of global mercury emissions from natural and anthropogenic sources, and year(s) for which estimate is made. [Mg, megagrams; Data from Pirrone et al., 2010 and references therein]

of FTHg in deeper wells near the stream (compared to shallower wells) and a lack of geologic Hg deposits in soils and sediments were "consistent with atmospheric Hg deposition on the terrestrial landscape as the distal source of FTHg to groundwater." Barringer and Szabo (2006) contend that Hg deposited from the atmosphere to soils of the Coastal Plain of southern New Jersey, USA, may be mobilized, along with pesticide residues, to groundwater where Hg concentrations in domestic well water are found to exceed the State and USEPA MCL of 2,000 ng/L.

### 2.3. Effects of mercury from manufacturing and agriculture on soils, surface water and groundwater

Mercury in consumer products and their manufacture constitutes a source of Hg to aquatic systems. Mercury is used in dry-cell batteries, fluorescent light bulbs and thermostats, and, in the mid-to late 20th century, as a fungicide in paints and in wood preservatives; Hg is also used in dental amalgams (Barringer et al., 1997; Brooks, 2000, 2002; USEPA, 2006). Use of Hg in dentistry has consumed large amounts of the metal (about 63,500 kg in Europe), resulting in emissions as well as constituting a substantial portion of the Hg in municipal waste streams (Hylander et al., 2006; Metro, 1991). Several studies, summarized by Morrison (1981), indicated that subsurface disposal of municipal sewage sludge in the USA resulted in several metals, including Hg, leaching to groundwater. Since the mid 20th century, industrial use of Hg has declined in the USA and elsewhere, however. There are now ongoing attempts in some countries, such as the USA and the United Kingdom, to reduce the amount of Hg used in products; particularly those that are then discarded (Bradley & Journey, 2012).

Groundwater contamination with THg from both inactive and active industrial sources is found in many countries where, in general, the contamination is relatively local. Chlor-alkali plants that once produced chlorine gas and sodium hydroxide using Hg-cell technology have been shown to contribute Hg to surface water, soils and groundwater—up to 22,900,000 ng/L in groundwater at a facility in Sydney, Australia (Orica, 2012). There are numerous such facilities; other examples are chlor-alkali plants in New York State, USA, and in Kazakhstan, which have also produced Hg contamination of surrounding soils and waters (Gbondo-Tugbawa & Driscoll, 1998; Ullrich et al., 2007). The use of Hg-cell technology is being voluntarily phased out in most countries; however, wastes from chlor alkali plants still remain at some locations (Hylander & Meili, 2003).

In some instances, Hg contamination of groundwater is more diffuse, perhaps coming from multiple (potentially small) point sources. Suspected industrial and wastewater discharges have introduced Hg contamination to the surficial alluvial aquifer in urban Madras, India (Somasundaram et al., 1993), where, along with other metals, Hg concentrations are reported to range from 1,000 to 18,000 ng/L in water from wells near the River Cooum in the Madras urban area, mainly exceeding the Indian drinking-water standard of 1,000 ng/L.

In some instances, Hg released from industrial operations results in contamination of soils at the site, but the soil characteristics are such that the Hg is sequestered or attenuated, and does not leach to groundwater, as found, for example, at a site in Trondheim, Norway (Saether et al., 1997). In other instances, Hg contamination of groundwater is noted locally near industrial sites. At a site in southern Germany where "kyanizing" (mercuric chloride ($HgCl_2$) used to preserve wood from decay) was performed, groundwater in a 1.3-km-long plume was found to be contaminated with Hg at concentrations that reached 230,000 ng/L (Bollen et al., 2008). $HgCl_2$ is extremely soluble in water, ($7.4 \times 10^{10}$ ng/L; O'Neil, 2006), and thus easily mobilized from inputs at the land surface to the water table.

Mercury has been used in munitions and for other purposes on military bases. This use has, in some cases, led to contamination of soils at the bases (Bricka et al., 1994); with the potential for further contamination of surface and groundwaters. Mercury contamination of soils, surface water and groundwater is found at a former military base in southern New Jersey, USA. The Hg content of soils ranged up to 555 mg/kg, and concentrations in some streamwater and groundwater samples exceeded 2,000 ng/L; determining whether the Hg derives solely from military activities is a part of ongoing investigations there (Barringer et al., 2012).

From the early-to-late 20th century, Hg was used in agricultural pesticides, but use of Hg compounds decreased substantially after 1970 in the USA (Murphy & Aucott, 1999). Mercurial compounds also were once used on golf courses in the USA, being most heavily applied in northern States to control snow mold. On a golf course in New Hampshire, USA, Estes et al. (1973) estimated that annual fungicide applications contributed 2.1 kg of Hg per hectare. In Australia, sugar-cane setts were treated with a fungicide containing methoxymethylmercury chloride before planting; concentrations of Hg from 30 to 670 ng/L were found in groundwater underlying cane cropland, but were interpreted as being within the range of naturally occurring Hg concentrations in area groundwater (Brodie et al., 1984). It is apparent that not all applications of Hg to soils result in groundwater contamination, but there are instances where Hg can be mobilized from soils to groundwater.

In the early to mid-20th century, phenylmercuric acetate was used in orchards; calomel and $HgCl_2$ were used on row crops; these are known to be used on sandy agricultural soils of the Coastal Plain of southern New Jersey, USA (Murphy & Aucott, 1999). Inputs of Hg would have been high if the federally recommended application rates of 3.4 kg/hectare for highly soluble $HgCl_2$ were followed (Barringer et al., 1997 and references therein). Mobilization of Hg from applications of mercurial compounds, enhanced by subsequent disturbance from residential development of the land, may be the cause of elevated Hg concentrations on groundwater downgradient from such sources. Water from wells completed in the quartz sand aquifer beneath the former agricultural land contains Hg at concentrations exceeding 2,000 ng/L (Barringer et al., 2005; Barringer & Szabo, 2006).

In addition to pesticides, Hg also could be introduced to soils via fertilizers. A commercial 20-20-20 fertilizer solution prepared according to manufacturer's directions contained 280,000 ng/L of Hg (Barringer & MacLeod, 2001). Mercury was measured in several common fertilizers, with the highest concentration (5.1 mg/kg) found in calcium superphosphate, and a lower concentration (1.2 mg/kg) in 15-5-5 Nitrogen-Phosphorus-Potassium (NPK) fertilizer (Zhao & Wang, 2010). Some States in the USA are acting to regulate the amount of metals permissible in fertilizers and other agricultural chemicals (e.g. ODA, 2002).

## 2.4. Effects of human activities on naturally occurring mercury

Seawater intrusion along coasts typically occurs because withdrawals of freshwater resources reduce the freshwater hydraulic head, allowing seawater to enter aquifers on land. In southern Tuscany, Italy, the high chloride concentrations brought in by seawater intrusion may have been responsible for mobilizing Hg in the geologic materials to groundwater, perhaps as a chloride complex (Grassi & Netti, 2000; Protano et al., 2000). In Sardinia, dewatering of a lead-zinc (Pb-Zn) mine resulted in intrusion of deeper saline water that may have been responsible for increased Hg concentrations in shallow groundwater (Cidu et al., 2001). Experiments support these interpretations; Behra (1986) showed that elevated chloride (Cl⁻) concentrations mobilized Hg(II) in columns packed with quartz sand.

Contamination of soils, surface water and groundwater also arises from mining of cinnabar deposits. Mining of other metal ores has also resulted in Hg contamination, such as at the McLaughlin gold-mercury deposit in California (Sherlock, 2005), because of trace amounts of cinnabar or occurrence of Hg with other sulfide minerals. As a result of early 20th century copper mining activities along the Alaskan coast, USA, dissolved concentrations of Hg up to 4,100 ng/L were found in pore waters of sediments affected by mining waste (Koski et al., 2008). In a Canadian gold and silver mine, accessory cinnabar in waste piles oxidized and leached to groundwater, resulting in concentrations that ranged up to 150,000 ng/L (Foucher et al., 2012). Not all mining activities and mining waste disposal procedures result in extremely high levels of Hg in groundwaters, however; concentrations in adit water samples from an abandoned mercury mine in Turkey were still high relative to most waters, but ranged from 250 to 274 ng/L (Gemici, 2008).

The process of amalgamation of silver and gold with Hg was developed in the 16th century and used on an industrial scale into the 19th century in Central and South Americas, primarily to extract silver, and into the 20th century in North America for gold extraction (Nriagu, 1994). Elemental Hg was lost to the environment during the process; Hg losses during gold and silver extraction in the Americas are estimated to be about 240,000 Mg (Nriagu, 1994). Similar extraction activities have taken place in the Philippines, Indonesia, Thailand, Vietnam, Tanzania, and China (Lacerda, 1997), though the scale is not as great as in the Americas. Although some of this mining-activity-related Hg has volatilized, adding to the atmospheric Hg burden, much of it apparently still remains in the areas where metal processing took place. A few of these mining/extraction sites have been studied. In Tanzania and Zimbabwe, Hg sorption to iron-rich lateritic soils appears to have prevented groundwater contamination from Hg in tailings (van Straaten, 2000). Extreme surface-water loads have been documented from many of these mines including the Sierra Nevada gold mines in the USA (Domagalski, 1998), the Idrija Hg mine in Slovenia (Covelli et al., 2007), as well as Hg mining in Spain (Navarro, 2008) among others, with Hg sorption to iron-hydroxide-rich stream deposits a likely mechanism for Hg attenuation in some cases (Rytuba, 2000). Use of Hg in small-scale (artisanal) gold mining operations remains a substantial concern even today (Bradley & Journey, 2012).

# 3. Biogeochemistry of mercury: Field and experimental studies in soil and surface-water environments

### 3.1. Impact of mercury reactivity on transport

Oxidation-reduction, precipitation-dissolution, aqueous complexation, and adsorption-desorption reactions will strongly influence the fate and transport of Hg in groundwater, and in the environment, generally. At the land surface, Hg participates in photochemical reactions (see the review of Zhang, 2006), but these reactions are not relevant to groundwater. Biogeochemical reactions in soils are of great importance to the fate and transport of Hg, however. Characteristics of soils, which include pH, carbon content, mineralogy, drainage properties, slope, and texture, all play a role in Hg retention or mobility and whether THg inputs to the land surface reach the water table. Concentrations of THg typically are higher in organic soil horizons than in the deeper mineral horizons because Hg typically is closely associated with organic matter (Amirbahman & Fernandez, 2012), and Andersson (1979) reports sorption to iron oxides (typical of some temperate-climate subsoils, and also tropical soils) at pH > 5.5. The reactions described below can occur in soils, in the surface-water environment, and, apparently, in groundwater as well.

## 3.2. Oxidation-reduction and sorption reactions

The three stable oxidations states of Hg in low-temperature aquatic systems are Hg(II), Hg(I) and Hg(0). The mercurous (Hg(I)) species is stable over a more limited range of conditions in sulfidic aqueous systems than it is when sulfur is absent (Hem, 1970). The Hg species vary in their solubility, complexation, adsorption (Stumm and Morgan, 1995) and their availability for microbial processes. Therefore, oxidation-reduction (redox) reactions will have a profound influence on Hg concentrations and mobility in groundwater. Both abiotic and biotic (primarily microbial) processes can drive Hg redox transformations.

Iron geochemistry is intimately associated with that of Hg. Anaerobic column experiments showed transport of Hg(II) retarded by sorption (as a Hg-Cl complex) to pyrite ($FeS_2$) (Bower, et al., 2008), and Hg(II)) has been shown to sorb to iron oxides at pH > 5.5 (Andersson, 1979). Given a positive association of Hg with iron (Fe) in iron-hydroxide-rich sub-soils in the New Jersey Coastal Plain, USA, sorption of Hg to Fe hydroxides appears to be a mechanism for attenuating Hg (Barringer & Szabo, 2006). The same mechanism appears present at some mining sites (Rytuba, 2000; van Staaten, 2000), although formation of aqueous and solid-phase sulfides controls Hg(II) concentrations in tailings-contaminated sediments from California, USA, mines (Rytuba et al., 2005). Fe(II) hydroxides can be reductively dissolved by sulfide, resulting in the release of sorbed Hg. Experiments showed that, in the presence of sulfide ($S^{2-}$), Fe (III) was reduced and concentrations of dissolved Hg increased (Slowey & Brown, 2007). It appears that these and the experiments of Bower et al. (2008) were not done in the dark, however. Consequently, applicability of results to a groundwater setting is not clear.

Field examples also demonstrate that oxygen-depleted conditions caused by septic-system-effluent releases led to reductive dissolution of Fe hydroxides, resulting in release of sorbed Hg(II) (Barringer & MacLeod, 2001). Further, the Fe(II) generated in such a reaction may adsorb to minerals where it can then reduce Hg(II) to Hg(0) (Charlet et al., 2002). Recent experiments show that Fe(II) in minerals also can reduce Hg(II) to Hg(0). For example, in sealed, dark bottles, magnetite was found to reduce Hg(II) to Hg(0) within minutes (Wiatrowski et al., 2009; Yee et al., 2010). Mercury (Hg(II)) was also rapidly reduced in anoxic solutions by Fe(II) under varying pH conditions, with aqueous $Fe(OH)^+$ being the species that best described the electron transfer that occurred in the experiments (Amirbahman et al., 2012). Metals other than iron, such as tin, are also known to reduce Hg(II) to Hg(0) (e.g., Biester et al., 2000).

Natural organic matter has been shown to abiotically reduce Hg(II) to Hg(0) (Allard & Arsenie, 1991). Experiments under dark anoxic conditions by Gu et al. (2011) showed that dissolved organic matter (DOM) reduced Hg(II) to Hg(0) when low concentrations of DOM were present. At higher DOM concentrations, however, complexation with Hg inhibited Hg reduction reactions.

Microbially mediated redox reactions involving Hg also have been demonstrated. Hg(II) was reduced to gaseous Hg(0) by a *Pseudomonas* strain (Baldi et al., 1993). A newly isolated *mer*A-carrying *Bradyrhizobium* bacterium recently was found by Wang et al. (2012) to also reduce Hg(II) to Hg(0). The recent work by Wang et al. (2012) shows that Hg inhibits denitrification in groundwater in direct proportion to the concentration of Hg.

### 3.3. Organic and inorganic complexation

Mercury (Hg(II)) can be present not only as $Hg^{2+}$, but as $Hg(OH)_2$, $HgCl_2$ and other minor $OH^-$ and $Cl^-$ complexes, and in complexes with various organic anions, depending on pH, Eh, chloride concentrations and presence of DOM. Were DOM to be low or absent, then Hg could be present as hydroxide or chloride complexes in fresh waters (Reimers & Krenkel, 1974; Stumm & Morgan, 1995); at low to moderate pH and moderate to high chloride concentrations, chloride complexes would be most likely (Ravichandran, 2004). In the presence of dissolved sulfide, mercury-sulfide species may form (Benoit et al., 2001).

Mercury tends to form strong complexes with $S^{2-}$ and, in DOM, Hg(II) binds preferentially to sulfur-containing functional groups such as thiols (Gabriel and Williamson, 2004; Ravichandran, 2004; Reimers & Krenkel, 1974). In anoxic environments, Hg can form complexes such as dissolved HgS, $HgS_2^{2-}$, $Hg(SH)_2$, $HgSH^+$, HgOHSH and HgClSH (Gabriel & Williamson, 2004). Although metals typically bind to acid sites (carboxyls, phenols, ammonium, alcohols, and thiols) in organic matter, Hg(II) binds preferentially with thiols and other reduced sulfur groups with which it forms strong covalent-like bonds. These sulfur-bearing groups are found in moderate abundance in organic matter in soils, in some surface water, and in wastewater (Hsu-Kim & Sedlack, 2003; Ravichandran, 2004). When the Hg/DOM ratio is high (> 10,000 ng Hg to 1 mg DOM), however, Hg also binds to the more abundant but less Hg-selective oxygen (ie., carboxyl) functional groups (Haitzer et al., 2002). Further, binding of DOM with Hg(II) is less strong at low pH than at high pH (Haitzer et al, 2003); this occurs because the extent of protonation of functional groups serving as Hg(II) ligands on DOM increases as pH decreases. Given the affinity of Hg(II) for thiol groups on DOM, it has been shown that DOM can dissolve cinnabar, inhibiting or preventing precipitation of metacinnabar and aggregation of HgS nanoparticles (Ravichandran et al., 1999; Reddy & Aiken, 2001; Slowey, 2010; Waples et al., 2005).

### 3.4. Microbial transformations

An important transformation of inorganic Hg involves its methylation to monomethyl- or dimethyl-mercury. Methylation of Hg(II) in soils and surface-water was found to be carried out under anoxic conditions by dissimilatory sulfate-reducing bacteria (DSRBs) (Gilmour et al., 1992). Dissimilatory iron-reducing bacteria (DIRBs) later were found to be able to methylate Hg(II) as well (e.g., Kerin et al., 2006). Populations of both DSRBs and DIRBs have been found to coexist in stream-bottom sediments where fine-grained sediments were "potential hot spots for both methylation and demethylation activities" (Yu et al., 2012).

Further, at low sulfate ($SO_4^{2-}$) concentrations, the methylating activity of SRBs is stimulated, but at high concentrations the methylating activity is inhibited because precipitated sulfides incorporate the Hg (Ullrich et al., 2001). Concentrations of $SO_4^{2-}$ between 0.2 and 0.5 mM (about 19 to 48 mg/L) appeared to be optimum for promoting Hg methylation in freshwater (Gilmour and Henry, 1991). Barkay et al., (1997) discovered that high concentrations of DOM and salinity inhibited Hg(II) methylation because the Hg was complexed into forms that were not bioavailable to the methylating bacteria. The Hg in aqueous HgS complexes, which form in the presence of dissolved sulfide, was found to be bioavailable to the methylating bacteria, however (Benoit et al., 1999). Recent research shows that, although DOM can inhibit Hg bioavailability by complexing the Hg, DOM can also prevent HgS nanoparticles from aggregating, and thus the nanoparticles are bioavailable (Graham et al., 2012).

Mercury demethylation has also been studied in stream and lake sediments (e.g., Achá et al, 2011; Hintelmann et al., 2000; Pak & Bartha, 1998; Steffan et al., 1988). In experiments using sediments from southern New Jersey lakes, USA, Pak and Bartha (1998) showed that demethylation of MeHg is carried out by sulfidogenic and methanogenic bacteria, which are obligate anaerobes. Although the Hg methylation process was inhibited by low pH (4.4) conditions (which are common in southern New Jersey surface waters and groundwaters), demethylation of MeHg did not appear to be similarly affected for the pH range 4.4 to 8; inhibition occurred at pH < 4.4 (Steffan et al., 1988).

### 3.5. Colloids and particles

Extensive sorption of Hg(II) can limit concentration and mobility in groundwater unless the Hg(II) binds to colloidal solids under conditions where the colloids are stable and mobile. Colloids (particles < 1 μm in one dimension (Kretchmaar & Schafer, 2005)) in waters provide transport for various contaminants and generally are sufficiently small so as to pass the 0.45 μm pore-size filters used by most researchers in collected filtered water samples. Because of their large surface area relative to their volume, small particles and colloids can provide many sorption sites for strongly sorbing contaminants whose mobility would otherwise be minimal through soils and aquifers. Such movement can be triggered by chemical or physical disturbance of soils and sediments. For example, Hg sorbed to particles was released to runoff from boreal forest soils following clear cutting and scarification (Porvari et al., 2003).

Colloids can be formed by clay minerals; oxides and hydroxides of iron, aluminum, and manganese; silica; humic and fulvic acids; carbonates; phosphates (Ryan & Gschwend, 1994), also bacteria; and viruses (Kretchmaar & Schafer, 2005). Colloids are common in surface waters, soil and sediment porewaters. Colloids are found in groundwater as well. Changes in pH and redox reactions can cause dissolution or precipitation reactions that can form or release colloidal material. Examples are precipitation of colloid-sized minerals such as iron hydroxides when Fe(II) is oxidized, and oxides and carbonates that would precipitate at high pH.

In a southern New Jersey, USA, aquifer with Hg contamination, a study found colloids were more abundant in anoxic groundwater of an undeveloped area than in oxic groundwater (Ryan & Gschwend, 1990). The greater abundance likely occurred because iron hydroxide cements that bound clays to quartz-grain surfaces were being dissolved, liberating both Fe(II) and clay particles to solution (Ryan & Gschwend, 1994). With fluctuating water tables, some of the Fe(II) could be re-oxidized, forming colloidal precipitates.

In groundwater, colloids are subject to forces exerted by pumping. Sequential sampling of domestic wells in the unconfined aquifer system of southern New Jersey found that particulate Hg concentrations were commonly higher in first-draw water samples than in samples collected later during well purging (Szabo et al., 2010).

## 4. Background concentrations of THg in groundwater and groundwater/ surface-water interactions

"Background" concentrations—that is—naturally occurring concentrations of THg (and MeHg) in groundwater probably depend upon ambient geochemical conditions, which would include pH that favors adsorption or desorption, presence and amount of Fe and DOM, and oxidizing or reducing conditions favorable to mineral precipitation or dissolution or microbial activities, including Hg(II) methylation. Absent known sources of contamination, background concentrations of THg in groundwater in several studies were found to be < 10 ng/L (Andren & Nriagu, 1979; Barringer & Szabo, 2006; Krabbenhoft & Babiarz, 1992; Kowalski et al., 2007). Total mercury concentrations in other groundwater studies, depending on filtration or lack thereof, ranged from <5 to 210 ng/L (Wiklander, 1969/1970; Reimann et al., 1999), with concentrations generally higher in unfiltered samples because of particulate material[2].

Relatively few studies have examined interactions between groundwater and surface water. In the Everglades swamp area in Florida, USA, groundwater pumping, dredging of canals, levee construction, and land subsidence have altered area hydrology (Harvey et al., 2002). In the surficial aquifer, THg, which is an element of concern because of severe MeHg impacts on Everglades biota, is recharged from surface water to groundwater, with higher concentrations (0.8 to 2.7 ng/L) tending to be in recharge from agricultural areas. Methylmercury (0.2 ng/L) was found only in groundwater recharged from agricultural areas and was not detected in groundwater elsewhere (Harvey et al. 2002). In a study of Hg inputs to Lake Superior, USA,

---

2 An issue with studies of low concentrations of Hg historically has been sample contamination. Studies since the early 1990s have used sampling protocols that typically obtain reliable samples; e.g. Krabbenhoft and Babiarz, (1992).

however, groundwater was found to be an important source of MeHg to the lake, with concentrations as high as 12 ng/L in a hyporheic-zone sample (Stoor et al., 2006).

At a lake in glacial outwash (Wisconsin, USA), groundwater (sampled by piezometers and dug wells) both discharges to the lake and receives recharge from lake waters. Mercury enters from atmospheric deposition to the lake, and, apparently, through soils to groundwater. Mercury concentrations in groundwater discharge (mean 12 ng/L) to the lake was higher than that of water from nearby wells (mean 2.8 ng/L), showing the importance of reactions near and at the sediment/water interface (Krabbenhoft & Babiarz, 1992). In a New Jersey, USA, Coastal Plain watershed, groundwater discharging to a major river contained concentrations of THg in urban areas (some mainly in particulate form) that were higher than those in forested wetlands areas. Concentrations of THg in unfiltered water were 36 and 177 ng/L in discharge to the river at two sites in an urban area (Barringer et al., 2010a) and were not representative of background concentrations for that aquifer, which typically are < 10 ng/L (Barringer & Szabo, 2006). Bradley et al. (2012) also found Hg in groundwater discharge to be an important input to a southern USA Coastal Plain stream, although the THg concentrations in the groundwater were an order of magnitude lower than the concentrations in urban groundwater discharge in the study of Barringer et al. (2010a).

# 5. Biogeochemical processes and mechanisms for mobilizing mercury in groundwater — Case studies

## 5.1. New Jersey Coastal Plain, USA

At more than 70 residential sites underlain by an areally extensive (7,770 km$^2$) unconfined non-calcareous quartz sand aquifer system in the Coastal Plain of New Jersey, USA, water from domestic wells has been found to contain Hg at concentrations that exceed the State MCL of 2,000 ng/L. In the same aquifer system, background levels of THg in the groundwater in neighboring forested areas and unaffected residential areas are <10 ng/L (Barringer et al., 1997; Barringer & Szabo, 2006). This highly permeable system is vulnerable to contamination, with a water-table depth commonly less than 7 m below land surface. Past agricultural use of mercurial pesticides and atmospheric deposition are two likely sources of THg to the aquifer. Currently, about 700 wells are known to have withdrawn water containing THg at concentrations ranging as high as 80,000 ng/L (Dr. Judith Louis, New Jersey Department of Environmental Protection, 2010, oral commun.); the New Jersey Department of Environmental Protection supplies treatment systems or connections to public supplies for the affected households. A small number of public-supply wells have also been affected in urban/suburban areas (Fischer et al., 2010). Groundwater contaminated with THg was found in a similar Coastal Plain aquifer in the neighboring State of Delaware (Koterba et al., 2006).

Redox reactions were found to be important in explaining the mobility of Hg in the New Jersey Coastal Plain aquifer system. Septic-system effluent discharges, which were ubiquitous in the unsewered residential areas with Hg-contaminated well water, apparently drove redox reactions that influenced THg mobility in the aquifer. Discharges with elevated concentrations of electron donors promoted reductive dissolution of Fe hydroxide coatings on subsoil and aquifer sediment grains and sorbed Hg (presumably as Hg(II)), was released to groundwater (Barringer & MacLeod, 2001; Barringer et al., 2006; Barringer & Szabo, 2006). Septage and leach-field effluent contained THg concentrations in the range <20 to 60 ng/L. Therefore the effluent is seen, not as a prominent source of Hg, but as a source of electron donors (or reduced solutes) capable of driving redox reactions, such as dissolution of Fe hydroxides and reduction of Hg(II) to Hg(0), that favor increased concentrations and mobility of Hg.

The THg in New Jersey groundwater is hypothesized to be mobilized from soils. The mobility of Hg in soils was tested during U.S. Geological Survey (USGS) studies by pulsed leaching of samples of disturbed subsoils with de-ionized water, with artificial road-salt solution and with 20-20-20 fertilizer solution. In all cases the Hg removed was particulate, and was associated with removal of particulate Fe, indicating that the most likely mode of transport for the THg from the mineral soil horizons was in association with Fe hydroxide particles (Reilly et al., 2012; Barringer et al., (2012)). In the Fe hydroxide-rich subsoils, the introduction of septic systems, and onset of reducing conditions in some instances provided the conditions for reductive dissolution of hydroxides and release of sorbed Hg(II) to the water table. From the water table, gradients toward numerous pumping domestic wells brought mobile Hg to levels in the aquifer where drinking-water supplies were impacted. The question then arises—in what form is the Hg mobile?

Although well-water samples were not analyzed for Hg(0) in the USGS studies in the New Jersey Coastal Plain, geochemical modeling results indicated that Hg in the groundwater could be reduced and that Hg(0) would be a likely species in the groundwater (Barringer & Szabo, 2006). The presence of Hg(0) in water from the Coastal Plain aquifer system has been shown by sampling at the military base mentioned previously, where reducing conditions are present in DOM-rich wetland areas (Barringer et al., 2012). As mentioned earlier, experimental studies have shown that reduction of Hg(II) to Hg(0) by sorbed Fe(II) is sufficiently fast to promote mobilization as Hg(0) (Amirbahman et al, 2012; Charlet et al., 2002). Because Hg(0) is slightly soluble in water (56,000 ng/L at 25° (O'Neil, 2006)), it is likely that, in parts of the aquifer system affected by effluent discharges, Hg(0) could form and be mobile as dissolved Hg within the aquifer.

Mercury in the New Jersey Coastal Plain aquifer also appears to be mobile in particulate and colloidal form. The differences between unfiltered Hg concentrations and filtered concentra-tions (passing 0.45- μm pore-size filters) is large for some groundwater samples; the magnitude of this difference is interpreted as the concentration of particulates greater than 0.45 μm in diameter (Barringer et al., 2006; Barringer & Szabo, 2006). The finding of Hg on particulates/ colloids indicates presence of THg, either as oxidized Hg(II) or as reduced Hg(0) (Bouffard & Amyot (2009) sorbed to the mobile particles. Currently (2012) in New Jersey, the composition of the particulates is not known at the various residential Hg contamination sites—they could be Fe hydroxides (Ryan & Gschwend, 1990), clay minerals (Ryan & Gschwend, 1994), organic material (Hurley et al., 1998), mixed organo-oxide colloids (Chadwick et al., 2006), or sulfide particles (Slowey et al., 2005), depending on the geochemical environment in which they form.

The presence of particulates indicates that the dissolved THg concentrations measured in the New Jersey Coastal Plain groundwater represent only some fraction of the total pool of THg present in parts of the aquifer system

## 5.2. Nepal

An incidence of Hg in groundwater in Nepal bears some resemblance to the New Jersey Coastal Plain contamination. In Nepal, public supply wells, deep private wells, and shallow dug wells, hand pumps and spouts completed in gravelly unconfined and confined aquifers were sampled. Twenty-three of the samples from 31 sampling sites contained Hg at concentrations that exceed the WHO Guideline of 1,000 ng/L; the highest concentration measured was about 300,000 ng/L in water from a dug well in an urban area. Concentrations of nitrogen species ($NH_3$ and $NO_3^-$) and dissolved organic carbon (DOC) were particularly high in several of the well water samples (ammonia ($NH_3$), up to 62 mg/L as N; nitrate ($NO_3^-$), 25 mg/L as N; and DOC, 63.6 mg/L), and reducing (anoxic) conditions were apparent in the confined aquifer (Khatiwada et al., 2002). No suggestions as to the source of the Hg were given, but the elevated nutrient and organic carbon levels are suggestive of anthropogenic inputs such as sewage that either include Hg, or that mobilize naturally occurring Hg, presumably under reducing conditions.

## 5.3. Cape Cod, Massachusetts, USA

The importance of redox conditions in Hg release and mobility also is demonstrated in recent research at the USGS research site downgradient from the Massachusetts Military Reservation on Western Cape Cod. Land disposal of treated wastewater to the unconsolidated sands and gravels of the shallow aquifer from the 1930s to December 1995 resulted in development of a plume of contaminated groundwater that has been investigated since the early 1980s (LeBlanc, 1984). In addition to standard monitoring wells, the area affected by the plume has been instrumented with numerous multi-level samplers (MLS) (LeBlanc et al, 1991). These MLS allow collection of point samples from distinct biogeochemical zones, which would otherwise be difficult owing to steep vertical gradients in groundwater chemistry in the plume. An extensive suboxic zone with elevated nitrate concentrations and denitrification, and an anoxic zone with dissimilatory iron reduction (DIR) have persisted for more than 15 years following cessation of inputs from the source (Repert et al. 2006; Savoie et al., 2012). A recent study that examined Hg fate and transport within the plume found that Hg(0) constituted > 50% of the Hg present in the DIR zone near the source. About 1-2 km downgradient from the source, the anoxic zone had essentially no dissolved iron, but concentrations of both ammonium ($NH_4^+$) and $NO_3^-$ were high. Methylmercury comprised nearly 100% of the dissolved THg present in some samples from this region of the plume. Under the original infiltration beds, THg concentrations ranged up to about 200 ng/L (1,000 pM), but concentrations were rapidly attenuated with distance downgradient from the beds. Concentrations of dissolved THg in the oxic, uncontaminated groundwater, where pH values ranged from 5.0 to 5.6, were about 0.2 ng/L. The study shows there are at least two distinct redox environments and two different microbial regimes operating within the plume. However, the distribution of dissolved THg suggests that downgradient from the source, dissolved THg has been mainly mobilized from the non-calcareous, quartzitic aquifer sediments rather than being transported long distances in the wastewater stream (Carl Lamborg, Woods Hole Oceanographic Institution & Douglas Kent, USGS, 2012, written commun.).

## 5.4. Maine, USA

Crystalline rocks can contain Hg that may become mobile. Mercury in water from domestic wells completed in fractured rock aquifers of the Waldoboro pluton complex in Maine, USA, exceeded the USEPA drinking water standard of 2,000 ng/L (Sidle, 1993). Most high Hg concentrations were found in water from granitoid rock aquifers that contain several joint sets as well as cataclasite-fault breccias zones and shear-mylonite zones. Mineral-rich pegmatite dikes follow one of the joint sets in the Waldoboro pluton. A few anomalously high concentrations occurred in water from wells completed in associated metamorphic rock aquifers (gneisses and amphibolites) of the Bucksport Formation.

The Hg content of the rocks varied substantially, from 0.005 to about 500 mg/kg, with a median content of 78 mg/kg. Mercury concentrations in well-water samples ranged from 40 to 6200 ng/L. Mercury in surficial sediments, which contained glacially-derived deposits, ranged up to about 0.100 mg/kg. The presumption is that the Hg derived from the geologic materials, and that the fracturing aided in transporting Hg-rich water through the aquifers. There were no other chemical data reported for the domestic wells, and therefore it is not clear whether natural weathering of the Hg-enriched rocks was the cause of the high THg levels measured in well water, or whether inputs of anthropogenic chemicals could have mobilized Hg from the geologic substrate.

## 5.5. Groundwater discharge to coastal areas

Largely within the last decade, researchers have turned to an examination of the impacts of fresh groundwater inputs on coastal waters. Groundwater that discharges to bays and estuaries has been found to contribute nutrients to coastal waters (e.g., Slomp & van Cappellen, 2004), and recent research has expanded to examine contributions of trace elements from this source. Within the past 5 years, at the volcanic island of Jeju, south of Korea, on the south coast of the English Channel at Caux, France, and along both east and west coasts of the USA, submarine groundwater discharge (SGD) of Hg and MeHg to estuaries and bays has been investigated (Black et al., 2009; Bone et al, 2007; Ganguli et al., 2012; Laurier et al., 2007; Lee et al., 2011).

Mass balance indicated that Hg in SGD at Jeju Island constituted 34 and 67% of the total Hg annually in waters of two bays, whereas atmospheric deposition contributed from 23 to 25% of the Hg. Methylmercury in SGD at the island also constituted the majority of MeHg in the coastal waters (Lee et al., 2011).

At Waquoit Bay in Massachusetts, USA, sampling of SGD showed that Hg concentrations in groundwater ranged from <0.64 ng/L to 52.4 ng/L, and that daily discharge of Hg was from 94 to 380 ng/m$^2$ (Bone et al., 2007). Further, total dissolved Hg and DOC were not correlated, a lack of relation that was also observed in some of the New Jersey, USA, Coastal Plain studies, particularly at THg concentrations much higher than those found in waters discharging to Waquoit Bay.

At Caux, France, the Hg concentrations in tissues of blue mussels (*Mytilus edulis*) were higher than elsewhere along the coast, prompting an investigation of Hg in SGD. Although the total Hg concentrations were not greatly elevated at Caux (a mean of 0.73 ng/L), they were higher than in the nearby industrialized Seine estuary (mean of 0.32 ng/L) and the dissolved Hg likely represented a source of bioavailable Hg to the mussel population. Dissolved MeHg concentrations at Caux constituted from 2.4 to 7.8% of the total Hg present in SGD (Laurier, et al., 2007).

On the west coast of the USA, north of San Francisco, SGD contributed from 0.24 ng/L to 5.7 ng/L to California coastal waters; the Hg concentrations were significantly correlated with $NH_4^+$ and silica ($SiO_2$) (Black et al., 2009). Farther south, at Malibu Lagoon, SGD also transported Hg and MeHg to coastal water. Mixing between groundwater and seawater was inferred. MeHg concentrations in seawater increased at low tide, as did filtered Hg concentrations; inputs from SGD were thought to change Hg partitioning and solubility in the seawater (Ganguli et al., 2012).

## 6. Conceptual model for processes influencing mercury fate and transport in groundwater

Overall, the studies of the Hg-contaminated groundwater in New Jersey and Cape Cod, in particular, have provided opportunities to investigate mechanisms for releasing Hg from the land surface to the water table, and to suggest further avenues to explore biogeochemical reactions that mobilize Hg from the subsurface and within aquifers. A conceptual model for Hg mobilization has been developed by New Jersey researchers (Reilly et al., 2011) and is shown in figure 2.

Mercury, either naturally occurring, of anthropogenic origin, or both, is released to the water table as Hg(II) from surface soils and subsoils by weathering or by inputs of anthropogenic chemicals such as road salt or fertilizers, or by subsurface inputs of septic-system effluent. Under oxidizing conditions, dissolved THg is mobile as a complex with DOM, and sorbed Hg(II) is mobile on Fe hydroxide particles. Effluent discharges provide electron donors and sorbed Hg(II) is released as Fe hydroxides reductively dissolve, and Hg(II) may be reduced either by DOM or by Fe(II). Where anoxic conditions are present, sulfate reduction is an important terminal electron accepting process, and methylation may take place. Additionally, sulfides may precipitate, removing Hg from the aqueous phase. Hg(0) may be re-oxidized to Hg(II) should groundwater become more enriched in oxygen farther down a flowpath.

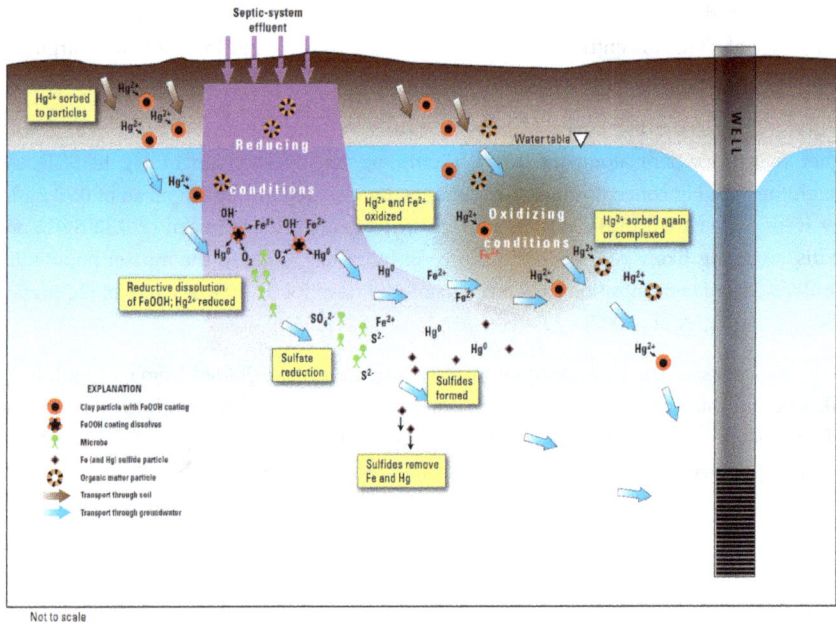

**Figure 2.** Mobilization of mercury from land surface to groundwater and biogeochemical transformations along flow paths in an unconsolidated sandy, acidic aquifer.

## 7. Conclusions

Mercury is relatively rare compared to most other elements, but owing to its toxicity at low concentrations, Hg is an important potential contaminant. There is a large reservoir of inorganic Hg in the environment—much of it derived from human activities, most associated with industrialization. Some of that Hg enters freshwater supplies where conditions may be conducive to methylation and thus the production of MeHg that readily bioaccumulates. Relatively less Hg is mobilized to groundwater than to surface water, in part because Hg can be attenuated by sorption to clays, iron oxides, and residual soil organic matter. Studies of the fate and transport of Hg in the subsurface are beginning to reveal how transport from land surface to groundwater might occur and how Hg remains mobile within aquifers.

In none of the above studies of Hg discharge to coastal waters have the sources of the Hg in SGD been identified, nor have the mechanisms for maintaining Hg mobility in groundwater discharging to the coasts been discerned. Given the recent discoveries of Hg and MeHg inputs from SGD, there clearly are avenues for further investigations into this phenomenon. Uptake into estuarine biota in the biodiverse and biomass rich estuarine and coastal waters is of key concern for these sensitive ecosystems and for human health, given the great potential for biomagnification. The importance of groundwater inputs thus cannot be ignored.

Studies in New Jersey and Cape Cod, USA, have investigated processes leading to Hg mobility in groundwater in settings that involve inputs from sewage to the subsurface. Given that sewage effluent contains materials that can fundamentally alter biogeochemical environments, mobilization of metals such as Hg, whatever their origin, may be an ever increasing process as humans continue to develop their surroundings. It is hoped that the research, past and ongoing, that is discussed herein will be of use to readers who seek to understand, to prevent, or to mitigate Hg contamination of groundwater supplies.

(Any use of trade, product, or firm names is for descriptive purposes only and does not imply endorsement by the U.S. Government.)

## Author details

Julia L. Barringer, Zoltan Szabo and Pamela A. Reilly

U.S. Geological Survey, USA

## References

[1] Achá, D., Hintelmann, H., & Yee, J. (2011). Importance of sulfate reducing bacteria in mercury methylation and demethylation in periphyton from Bolivian Amazon region. *Chemosphere, 82,* 911-916.

[2] Allard, B. & Arsenie, I. (1991). Abiotic reduction of mercury by humic substances in aquatic systems—an important process for the mercury cycle. *Water, Air, and Soil Pollution, 56,* 457-464.

[3] Alpers, C.N. & Hunerlach, M.P. (2000). Mercury contamination from historic gold mining in California. U.S. Geological Survey Fact Sheet FS-061-00.

[4] Amirbahman, A. & Fernandez, I.J. (2012). Chapter 7. The role of soils in storage and cycling of mercury. In. Bank, M.S. (ed.) Mercury in the environment: Pattern and Process. Berkely, University of California Press, 97-116.

[5] Amirbahman, A., Kent, D.B., & Curtis, G.P. (2012). Kinetics of abiotic mercury (II) reduction by iron (II). (Abstract), Program: BIOGEOMON, the 7th International Symposium on Ecosystem Behavior, July 15-20. 2012, Northport, Maine.

[6] Andersson, A. (1979). Mercury in soils. 79-112. In: Nriagu, J.O., (ed). The biogeochemistry of mercury in the environment. Amsterdam:, The Netherlands:Elsevier/ North Holland Biomedical Press.

[7]   Andren, A.W. & Nriago, J.O. (1979). The global cycle of mercury. 1-21. In. Nriagu, J.O., (ed.), The biogeochemistry of mercury in the environment. Amsterdam, The Netherlands: Elsevier/North Holland Biomedical Press.

[8]   Bakir, F., Damluji, S.F., Amin-Zaki, L., Murtada, M., Khalidi, A., al-Rawi, N.Y., Tikriti, S., Dahahir, H.I., Clarkson, T.W., Smith, J.C., & Doherty, R.A. (1973). Methylmercury poisoning in Iraq. *Science*, 181, 230-241.

[9]   Baldi, F., Parati, F., Semplici, F. & Tandoi, V. (1993). Biological removal of inorganic Hg(II) as gaseous elemental Hg(0) by continuous culture of a Hg-resistant *Pseudomonas-Putida* strain FB-1, *World Journal of Microbiology and Biotechnology*, 9, 275.

[10]  Ball, J.W., McCleskey, R.B., Nordstrom, D.K., & Holloway, J.M. (2006). Water-chemistry data for selected springs, geysers, and streams in Yellowstone National Park, Wyoming, 2003-2005. *U.S. Geological Survey Open-File Report 2006-1339*. 137 p.

[11]  Barkay, T., Gillman, M. & Turner, R.R. (1997). Effects of dissolved organic carbon and salinity on bioavailability of mercury. *Applied and Environmental Microbiology*, 63, 4267-4271.

[12]  Barringer, J.L., & MacLeod, C.L. (2001). Relation of mercury to other chemical constituents in ground water in the Kirkwood-Cohansey aquifer system, New Jersey Coastal Plain, and mechanisms for mobilization of mercury from sediments to ground water. *U.S. Geological Survey Water-Resources Investigations Report*, 00-4230.

[13]  Barringer,.J.L., MacLeod, C.L., & Gallagher, R.A. (1997). Mercury in ground water, soils, and sediments of the Kirkwood-Cohansey aquifer system in the New Jersey Coastal Plain. *U.S. Geological Survey Open-File Report 95-475*.

[14]  Barringer, J.L., Riskin, M.L., Szabo, Z., Reilly, P.A., Rosman, R., Bonin, J.L., Fischer, J.M. & Heckathorn, H.A. (2010a). Mercury & methylmercury dynamics in a Coastal Plain watershed, New Jersey, USA. *Water, Air, and Soil Pollution*, 212, 251-273.

[15]  Barringer, J.L., & Szabo, Z. (2006). Overview of investigations into mercury in ground water, soils, and septage, New Jersey Coastal Plain. *Water Air, and Soil Pollution*, 175, 193-221.

[16]  Barringer, J.L., Szabo, Z., Kauffmann, L.J., Barringer, T.H., Stackelberg, P.E. Ivahnenko, T., Rajagopalan, S. & Krabbenhoft, D.P. (2005). Mercury concentrations in water from an unconfined aquifer system, New Jersey Coastal Plain. *Science of the Total Environment*, 346, 169-183.

[17]  Barringer, J.L., Szabo, Z., Reilly, P.A. & Riskin, M.L. (2010b) Mobilization of mercury to a first-order Coastal Plain stream in New Jersey: Geological Society of America Abstracts, v. 42, no. 1, p. 154, Abstract 63-6.

[18]  Barringer, J.L., Szabo, Z., Schneider, D., Atkinson, W.D. & Gallagher, R.A. (2006). Mercury in ground water septage, leach-field effluent, and soils in residential areas, New Jersey Coastal Plain. *Science of the Total Environment*, 361, 144-162.

[19] Barringer, J.L., Szabo, Z. & Reilly, P.A. (2012). Mercury in waters, soils and sediments of the New Jersey Coastal Plain: A comparison of regional distribution and mobility to the mercury contamination at the William J. Hughes Technical Center, Atlantic County, New Jersey. *U.S. Geological Survey Scientific Investigations Report 2012-5115*

[20] Behra, P. (1986). Evidences for the existence of a retention phenomenon during the migration of a mercurial solution through a saturated porous medium. *Geoderma, 38,* 209-222.

[21] Benoit, J.M., Gilmour, C.C., Mason, R.P. & Heyes, A. (1999). Sulfide controls on mercury speciation and bioavailability to methylating bacteria in sediment pore waters. *Environmental Science and Technology*, 33, 951-957.

[22] Benoit, J.M., Mason, R.P., Gilmour, C.C. & Aiken, G.R. (2001). Constants for mercury binding by dissolved organic matter isolated from the Florida Everglades. *Geochimica et Cosmochimica Acta*, 65, 4445-4451.

[23] Bhavsar, S.P., Gewurtz, S.B., McGoldrick, D.J., Keir, M.J. & Backus, S.M., (2010). Changes in mercury levels in Great Lakes fish between 1970s and 2007. *Environmental Science and Technology*, 44, 3273-3279.

[24] Biester, H., Schihmacher, P. & Müller, G. (2000). Effectiveness of mossy tin filters to remove mercury from aqueous solution by Hg(II) reduction and Hg(0) amalgamation. *Water Research*, 34, 2031-2036.

[25] Black, F.J., Paytan, A., Knee, K.L., de Sieyes, N.R., Ganguli, P.M., Gray, E. & Flegal, R. (2009). Submarine groundwater discharge of total mercury and monomethylmercury to Central California coastal waters. *Environmental Science and Technology*, 43, 5652-5659.

[26] Bollen, A., Wenke, A. & Biester, H. (2008). Mercury speciation analyses in $HgCl_2$-contaminated soils and groundwater—implications for risk assessment and remediation strategies. *Water Research*, 42, 91-100.

[27] Bone, S.E., Charette, M.A., Lamborg, C.H. & Gonneea, M.E. (2007). Has submarine groundwater discharge been overlooked as a source of mercury to coastal waters? *Environmental Science and Technology*, 41, 3090-3095.

[28] Bouffard & Amyot (2009). Importance of elemental mercury in lake sediments. *Chemosphere*, 74, 1098-1103.

[29] Bower, J., Savage, K.S., Weinman, B., Barnett, W.P. & Harper, W.F. (2008). Immobilization of mercury by pyrite ($FeS_2$), *Environmental Pollution*, 156, 504-514.

[30] Bradley, P., Burns, D., Murray, K., Brigham, M., Button, D., Chasar, L., Marvin-DiPasquale, M., Lowery, M. & Journey, C.A. (2011). Spatial and seasonal variability of dissolved methylmercury in two stream basins in the eastern United States. *Environmental Science and Technology*, 45, 2048-2055.

[31] Bradley, P. & Journey, C. (2012). Hydrology and methylmercury availability in Coastal Plain streams. Chapter 8, 169-190. In: Water Resources Management and Modeling: Nayak, P., (ed.). InTech Open Access Publishing. Available at http:// www.intechopen.com/books/ water-resources-management-and-modeling.

[32] Bradley, P.M., Journey, C.A., Lowery, M.A., Brigham, M.E., Burns, D.A., Button, D.T., Chapelle, F.H., Luz, M.A., Marvin-DiPasquale, M.C. & Riva-Murray, K. (2012). Shallow groundwater mercury supply in a Coastal Plain stream. *Environmental Science and Technology*, 46, 7503-7511.

[33] Bricka, R.M., Williford, C.W. & Jones, L.W. (1994). Heavy metal soil contamination at U.S. Army installations: Proposed research and strategy for technology development. Technical Report IRRP-94-1, U.S. Army Corps of Engineers, Waterways Experiment Station, Vicksburg, MS.

[34] Brodie, J.E., Hicks, W.S., Richards, G.N. & Thomas, F.G. (1984). Residues related to agricultural chemicals in the groundwaters of the Burdekin River Delta, North Queensland. *Environmental Pollution (Series B)* 8, 187-215.

[35] Brooks S.C. & Southworth, G.R. (2011). History of mercury use and environmental contamination at the Oak Ridge Y-12 Plant. *Environmental Pollution*, 159, 219-228.

[36] Brooks, W.E., (2000). Mercury. U.S. Geological Survey 2000 Minerals Yearbook.

[37] Brooks, W.E., (2002). Mercury. U.S. Geological Survey 2002 Minerals Yearbook.

[38] Chadwick, J.P., Babiarz, C.L., Hurley, J.P. & Armstrong, D.E. (2006). Influences of iron, manganese, and dissolved organic carbon on the hypolimnetic cycling of amended mercury. *Science of the Total Environment*, 306, 177-188.

[39] Charlet, L., Bosbach, D. & Peretyashko, T. (2002). Natural attenuation of TCE, As, Hg linked to oxidation of Fe(II): an AFM study. *Chemical Geology*, 190, 303-319.

[40] Cidu, R., Biagini, C., Fanfani, L., La Ruffa, G. & Marras, I. (2001) Mine closure at Monteponi (Italy): effect of the cessation of dewatering on the quality of shallow groundwater. *Applied Geochemistry*, 16, 489-502.

[41] Covelli, S., Piani, R., Acquavita, A., Predonzani, S. & Frageneli, J., (2007). Transport and dispersion of particulate Hg associated with a river plume in coastal Northern Adriatic environments. *Marine Pollution Bulletin*, 55, 436-450.

[42] Cravotta, C.A., III. (2008). Dissolved metals and associated constituents in abandoned coal-mine discharges, Pennsylvania, USA, Part I: Constituent quantities and correlations. *Applied Geochemistry*, 23, 166-202.

[43] Davidson, B. & Fisher, R.S. (2005). Groundwater quality in Kentucky: Mercury. Information Circular 8, Series XII, Kentucky Geological Survey. Available at www.uky.edu/kgs.

[44] Domagalski, J.L. (1998). Occurrence and transport of total mercury and methyl mercury in the Sacramento River Basin, California. *J. Geochemical Exploration*, 64, 277-291.

[45] Driscoll, C.T., Yan, C., Schofield, C.L., Munson, R. & Holsapple, J. (1994). The mercury cycle and fish in the Adirondack lakes. *Environmental Science and Technology*, 28, 136A-143A.

[46] Engle, M. A., Tate, M. T., Krabbenhoft, D. P., Kolker, A., Olson, M. L., Edgerton, E. S., DeWild, J. F. & McPherson, A. K., 2008, Characterization and cycling of atmospheric mercury along the central U.S. Gulf Coast: *Applied Geochemistry*, v. 23, #3, p. 419-437, doi:1016/j.apgeochem.2007.12.024.

[47] Engstrom, D.R., Balogh, S.J. & Swain, E.B. (2007) History of mercury inputs to Minnesota lakes: Influences of watershed disturbance and localized atmospheric deposition. *Limnology and Oceanography*, 52, 2467-2483.

[48] Environment Canada (2010). Environment Canada Risk Management Strategy for Hg. Available at www.ec.gc.ac/doc/mercury-mercury/1241/index_e.htm#gotoC00

[49] Estes, G.O., Knoop, W.E. & Houghton, F.D., 10973. Soil-plant response to surface-applied mercury. *Journal of Environmental Quality*, 2, 451-452.

[50] Fairbridge, R.W. (1972). The encyclopedia of geochemistry, and environmental sciences. *Encyclopedia of Earth Sciences Series, vol. IV.,* New York: A. Van Nostrand Reinhold Co.

[51] Ferrara, R., Mazzolai, B., Lanzillotta, E., Nucaro, E. & Pirrone, N. (2000). Volcanoes as emission sources of atmospheric mercury in the Mediterranean basin. *Science of the Total Environment*, 259, 115-121.

[52] Fischer, J.M., Szabo, Z., Barringer, J.L. & Jacobsen, E. (2010). Mercury in a North Atlantic Coastal Plain aquifer system: A flowpath study: *Geological Society of America Abstracts, v. 42,* no. 1, p. 178, Abstract 78-4.

[53] Fitzgerald, W. & Lamborg, C.H. (2007). Geochemistry of mercury in the environment. 107-148. In: *Environmental Geochemistry, V. 9;* Lollar, B.S. (ed.), Oxford, Elsevier.

[54] Fleischer, M. (1970) Summary of the literature on the inorganic geochemistry of mercury. 6-13. In. Mercury in the Environment, *U.S. Geological Survey Professional Paper 713.*

[55] Foucher, D., Hintelmann, H., Al, T.A. & MacQuarrie, K.T. (2012) Mercury isotope fractionation in waters and sediments of the Murray Brook mine watershed (New Brunswick, Canada): Tracing mercury contamination and transformation. *Chemical Geology*, (in press).

[56] Gabriel, M.C. & Williamson, D.G. (2004). Principal biogeochemical factors affecting the speciation and transport of mercury through the terrestrial environment. *Environmental Geochemistry and Health*, 26, 421-434.

[57]  Ganguli, P.M., Conaway, C.H., Swarzenski, P.W., Izbicki, J.A. & Flegal, A.R. (2012) Mercury speciation and transport via submarine groundwater discharge at a southern California coastal lagoon system. *Environmental Science and Technology*, 46, 1480-1488.

[58]  Gbondo-Tugbawa, S. & Driscoll, C.T. (1998) Application of the regional mercury cycling model (RMCM) to predict the fate and remediation of mercury in Onondaga Lake, New York. *Water, Air, and Soil Pollution*, 105, 417-426.

[59]  Gemici, U. (2008) Evaluation of the water quality related to the acid mine drainage of an abandoned mercury mine (Alasehir, Turkey). *Environmental Monitoring and Assessment*, 147, 93-106.

[60]  Gilmour, C.C. & Henry, E.A. (1991) Mercury methylation in aquatic systems affected by acid deposition. *Environmental Pollution*, 71, 131-169.

[61]  Gilmour, C.C., Henry, E.A. and Mitchell, R. (1992). Sulfate stimulation of mercury methylation in freshwater sediments. *Environmental Science and Technology*, 26, 2281.

[62]  Glass, G.E., Sorensen, J.A., Schmidt, K.W., Rapp. G.R., Yap, D. & Fraser, D. (1991). Mercury deposition and sources for the upper Great Lakes region. *Water, Air, and Soil Pollution*, 56, 235-249.

[63]  Graham, A.M., Aiken, G.R. & Gilmour, C.G. (2012). Dissolved organic matter enhances microbial mercury methylation under sulfidic conditions. *Environmental Science and Technology*, 46, 2715-2723.

[64]  Gu, B., Bian, Y., Miller, C.L., Dong, W., Jiang, X. & Liang, L. (2011). Mercury reduction and complexation by natural organic matter in anoxic environments. *Proceedings of the National Academy of Sciences*, 108, 1479-1483.

[65]  Grassi, S. & Netti, R., (2000). Sea water intrusion and mercury pollution of some coastal aquifers in the province of Grosseto (Southern Tuscany, Italy). *Journal of Hydrology*, 237, 198-211.

[66]  Grigal, D.F. (2002). Inputs and outputs of mercury from terrestrial watersheds: A review. *Environmental Reviews*, 10, 1-39.

[67]  Gustin, M.S., Lindberg, S.E. & Weusberg, P.J. (2008). An update on the natural sources and sinks of atmospheric mercury. *Applied Geochemistry*, 23, 482-493.

[68]  Haitzer, M., Aiken, G.R. & Ryan, J.N. (2002). Binding of mercury (II) to dissolved organic matter: the role of mercury-to-DOM concentration ratio. *Environmental Science and Technology*, 36, 3564-3570.

[69]  Haitzer, M., Aiken, G.R. & Ryan, J.N. (2003). Binding of mercury (II) to aquatic humic substances: Influence of pH and source of humic substances. *Environmental Science and Technology*, 37, 2436-2441.

[70] Harada, M. (1995). Minimata disease: Methylmercury poisoning in Japan caused by environmental pollution. *Critical Reviews in Toxicology*, 25, 1-24.

[71] Harvey, J.W., Krupa, S.L., Gefvert, C., Mooney, R.H., Choi, J., King, S.A. & Giddings, J.B. (2002). Interactions between surface water and ground water and effects on mercury transport in the North-central Everglades. *U.S Geological Survey Water-Resources Investigations Report 02-4050.*

[72] Hem, J.D. (1970). Study and interpretation of the chemical characteristics of natural water. *U.S. Geological Survey Water-Supply Paper 2254.*

[73] Hintelmann, H., Keppel-Jones, K. & Evans, R.D. (2000). Constants of mercury methylation and demethylation rates in sediments and comparison of tracer and ambient mercury availability. *Environmental Toxicology and Chemistry*, 19, 2204-2211.

[74] Hsu-Kim, H. & Sedlak, D.L. (2005). Similarities between inorganic sulfide and the strong Hg(II)-complexing ligands in municipal wastewater effluent. *Environmental Science and Technology*, 39, 4035-4041.

[75] Hurley, J.P. Cowell, S.E., Shafer, M.M. & Hughes, P.E. (1998). Tributary loading of mercury to Lake Michigan: Importance of seasonal events and phase partitioning. *Science of the Total Environment 213*, 129-137.

[76] Hutton, M. (1987). Human health concerns of lead, mercury, cadmium and arsenic, Chapter 6. 53-68. In Hutchinson, T.C. and Meema, K.M. (eds.) Lead, Mercury, Cadmium and Arsenic in the Environment, SCOPE, London, John Wiley and Sons, Ltd.

[77] Hylander, L.D., Lindvall, A. & Gahnberg, L. (2006). High mercury emissions from dental clinics despite amalgam separators. *Science of the Total Environment*, 362, 74-84.

[78] Hylander L.D. & Meili, M. (2003). 500 years of mercury production: global annual inventory by region until 2000 and associated emissions. *The Science of the Total Environment*, 304, 13-27.

[79] Kerin, E.J., Gilmour, C.C., Roden, E., Suzuki, M.T., Coates, J.D. & Mason, R.P. (2006). Mercury methylation by dissimilatory iron-reducing bacteria. *Applied and Environmental Microbiology*, 72, 7919-7921.

[80] Khatiwada, N.R., Takizawa, S., Tran, T.V.N. & Inoue, M. (2002). Groundwater contamination assessment for sustainable water supply in Kathmandu Valley, Nepal. *Water Science and Technology*, 46, 147-154.

[81] Kolker, A., Panov, B. S., Panov,Y. B., Landa, E. R., Korchemagin, V. A., Conko, K. M. & Shendrik, T., 2009, Geochemistry of Donbas Coals and associated mine water in the vicinity of Donetsk, Ukraine: *International Journal of Coal Geology*, v. 79, p. 83-91.

[82] Koski, R.A., Munk, L., Foster, A.L., Shanks, W.C., III & Stillings, L.L. (2008). Sulfide oxidation and distribution of metals near abandoned copper mines in coastal environments, Prince William Sound, Alaska, USA. *Applied Geochemistry*, 23, 227-254

[83]  Koterba, M.T., Andres, A.S.,Vrabel, J., Crilley, D.M., Szabo, Z. & DeWild, J.T.. (2006). Occurrence and distribution of mercury in the surficial aquifer, Long Neck peninsula, Sussex County, Delaware, 2003-04. *U.S. Geological Survey Scientific Investigations Report 2006-5011.*

[84]  Kowalski, A., Siepak, M. & Boszke, L. (2007). Mercury contamination of surface and ground waters of Poland. *Polish Journal of Environmental Studies*, 16, 67-74.

[85]  Krabbenhoft, D.P. & Babiarz, C.L. (1992). The role of groundwater transport in aquatic mercury cycling. *Water Resources Research*, 28, 3119-3128.

[86]  Kretzschmar, R. & Schäfer, T. (2005). Metal retention and transport on colloidal particles in the environment. *Elements*, 1, 205-210.

[87]  Lacerda, L.D. (1997). Global mercury emissions from gold and silver mining. *Water, Air, and Soil Pollution*, 97, 209-221.

[88]  Laurier, F., J.G., Cossa, D., Beucherm C. & Brévière, E. (2007). The impact of groundwater discharges on mercury partitioning, speciation, and bioavailability to mussels in a coastal zone. *Marine Chemistry*, 104, 143-155.

[89]  LeBlanc, D.R. (1984). Sewage plume in a sand and gravel aquifer, Cape Cod, Massachusetts. *U.S. Geological Survey Water-Supply Paper* 2218.

[90]  LeBlanc, D.R., Garabedian, S.P., Hess, K.M., Gelhar, L.W., Quadri, R.D., Stollenwerk, K.G. & Wood, W.W. (1991). Large-scale natural-gradient tracer test in sand and gravel, Cape Cod, Massachusetts: 1. Experimental design and observed tracer movement. *Water Resources Research*, 27, 895-910.

[91]  Lee, Y-g., Moklesur Rahman, M.D., Kim, G. & Han, S. (2011) Mass balance of total mercury and monomethylmercury in coastal embayments of a volcanic island: significance of submarine groundwater discharge. *Environmental Science and Technology*, 45, 9891-9900.

[92]  Lindberg, S.E., (1987). Emission and deposition of atmospheric mercury vapor, Chapter 8, 89-106. In; Lead, Mercury, Cadmium and Arsenic in the Environment, Hutchinson, T.C. and Meema, K.M., (eds.) SCOPE, London, John Wiley and Sons, Ltd.

[93]  Lohman, K., Seigneur, C, Gustin, M. &Lindberg, S. (2008). Sensitivity of the global atmospheric cycle of mercury to emissions. *Applied Geochemistry*, 23, 454-466.

[94]  Martin, R.S., Witt, M.L.I., Sawyer, G.M., Thomas, H.E., Watt, S.F.L., Bagnato, E., Calabrese, S., Aiuppa, A., Delmelle, P., Pyle, D.M. &Mather, T.A. (2012). Bioindication of volcanic mercury (Hg) deposition around Mt. Etna (Sicily). *Chemical Geology*, 310-311, 12-22.

[95] Mason, R.P., Fitzgerald, W.F. & Morel, F.M.M. (1994). The biogeochemical cycling of elemental mercury: Anthropogenic influences. *Geochimica et Cosmochimica Acta,* 58, 3191-3198.

[96] Metro. (1991). Dental office waste stream characterization study (prepared by Cynthia Welland). *Seattle, WA, Municipality of Metropolitan Seattle,* September 1991.

[97] Morrison, R.D. (1981). Potential health impacts of subsurface sewage sludge disposal upon groundwater resources. In: Quality of groundwater; Proceedings of an International symposium, Noordwijkerhout, Netherlands, 23-27 March, 1981. Van Duijvenbooden, W., Glasbergen, P. & van Lelyveld, H. (eds.) Studies in Environmental Science, 17. Netherlands, Elsevier.

[98] Murphy, E.A. & Aucott, M. (1999). A methodology to assess the amounts of pesticidal mercury used historically in New Jersey. *Water, Air, and Soil Pollution,* 78, 61-72.

[99] Navarro, A. (2008). Review of characteristics of mercury speciation and mobility from areas of mercury mining in semi-arid environments. *Reviews of Environmental Science and Biotechnology,* 7, 287-306.

[100] NIEA. (2011). European and National Drinking Water Qualty Standards. Northern Ireland Environmental Agency (NIEA), October, 2011. Available at www.doeni.gov.uk/niea.

[101] Nriagu, J.O. (1994). Mercury pollution from the past mining of gold and silver in the Americas. *The Science of the Total Environment,* 149, 167-181.

[102] O'Neil, M.J., 2006, (Ed.) Merck Index—14th Edition: Whitehouse Station, NJ, Merck and Co., Inc.,. p.1017-1018.

[103] Orica (2012). Former chlor alkali plant mercury contamination—Investigation overview. Available from http://www.oricabotanytransformation.com/index?page=107andproject=103; accessed 7/12/12.

[104] Pacyna, E.G. Pacyna, J.M., Fudala, J., Strzelecka-Jastrzab, E., Hlawiczka, S., and Panasiuk, D. (2006). Mercury emissions to the atmosphere from anthropogenic sources in Europe in 2000 and their scenarios until 2020. *Science of the Total Environment,* 370, 147-156.

[105] Pak, K-R, and Bartha, R. (1998). Mercury methylation and demethylation in anoxic lake sediments and by strictly anaerobic bacteria. *Applied and Environmental Microbiology,* 64, 1013-1017.

[106] Pirrone, N., Cinnirella, S., Feng, X., Finkelman, R.B., Friedli, H.R. Leaqner, J., Nason, R., Mukherjee, A.B., Stracher, G.B., Streets, D.G. & Telmer, K. (2010). Global mercury emissions to the atmosphere from anthropogenic and natural sources. *Atmospheric Chemistry and Physics,* 10, 5951-5964.

[107] ODA (2002). Administrative rule developed to declare limits of certain metals in fertilizer, agricultural amendments, agricultural mineral and lime products distributed in Oregon. *Oregon Department of Agriculture; ODA White Paper 08/2002.*

[108] Porvari, P, Verta, M., Munthe, J. & Haapanen, M. (2003). Forestry practices increase mercury and methyl mercury output from boreal forest catchments. *Environmental Science and Technology*, 37, 2389-2393.

[109] Protano, G., Riccobono, F. & Sabatini, G. (2000). Does salt water intrusion constitute a mercury contamination risk for coastal fresh water aquifers? *Environmental Pollution*, 110, 451-458.

[110] Ravichandran, M. (2004). Interactions between mercury and dissolved organic matter —a review. *Chemosphere*, 55, 319-331.

[111] Ravichandran, M., Aiken, G.R., Ryan, J.N. & Reddy, M.M., (1999). Inhibition of precipitation and aggregation of metacinnabar (mercuric sulfide) by dissolved organic matter isolated from the Florida Everglades. *Environmental Science and Technology*, 33, 1418-1423.

[112] Reddy, M.M. & Aiken, G.R. (2001). Fulvic acid-sulfide ion competition for mercury ion binding in the Florida Everglades. *Water, Air, and Soil Pollution*, *132*, 89-104.

[113] Reilly, P.A., Barringer, J.L. & Szabo, Z. (2012). Mobility of mercury from sandy soils into acidic groundwater, New Jersey Coastal Plain: Geological Society of America Abstracts, v. 44, no. 2, p. 118, Abstract 53-2.

[114] Reimann, C., Siewers, U., Skarphagen, H. & Banks, D. (1999). Influence of filtration on concentrations of 62 elements analyzed on crystalline bedrock groundwater samples by ICP-MS. *Science of the Total Environment*, 234, 155-173.

[115] Reimers, R.S. & Krenkel, P.A. (1974). Kinetics of mercury adsorption and desorption in sediments. *Journal of the Water Pollution Control Federation*, 46, 3520365.

[116] Repert, D.A, Barber, L.B., Hess, K.M., Keefe, S.H., Kent, D.B., LeBlanc, D.R. & Smith, R.L. (2006). Long-term natural attenuation of carbon and nitrogen within a groundwater plume after removal of the treated wastewater source. *Environmental Science and Technology*, 40, 1154-1162.

[117] Rudd, J.W.M. (1995). Sources of methyl mercury to freshwater ecosystems—a review. *Water, Air, and Soil Pollution*, 80, 697-713.

[118] Ryan, J.N. & Gschwend, P.M. (1990). Colloid mobilization in two Atlantic Coastal Plain aquifers: field studies. *Water Resources Research*, 26, 307-322.

[119] Ryan, J.N. & Gschwend, P.M. (1994). Effect of solution chemistry on clay colloid release from an iron oxide-coated aquifer sand. *Environmental Science and Technology*, 28, 1717-1726.

[120] Rytuba, J.J. (2000). Mercury mine drainage and processes that control its environmental impact. *Science of the Total Environment* 260, 57-71.

[121] Rytuba, J.J., Ashley, R., Slowey, A.,J. Brown, G.E. & Foster, A. (2005). Mercury speciation and transport from historic placer gold and mercury mines, California. (Abstract). *Geological Society of America, Abstracts with Programs*, 37, 104.

[122] Saether, O.M., Storroe, G., Segar, D. & Krog, R. (1997). Contamination of soil and groundwater at a former industrial site, Trondheim, Norway. *Applied Geochemistry*, 12, 327-332.

[123] Savoie, J.G., LeBlanc, D.R., Fairchild, G.M., Smith, R.L., Kent, D.B., Barber, L.B., Repert, D.A., Hart, C.P., Keefe, S.H. & Parsons, L.A. (2012). Groundwater-quality data for a treated-wastewater plume near the Massachusetts Military Reservation, Ashumet Valley, Cape Cod, Massachusetts, 2006-08. *U.S. Geological Survey Data Series* 648, 11 p.

[124] Schuster, P.F., Krabbenhoft, D.P., Naftz, D.L., Cecil, L.D., Olson, M.L., DeWild, J.F., Susong, D.D., Green, J.R. & Abott, M.L. (2002). Atmospheric mercury deposition during the last 270 years: a glacial ice core record of natural and anthropogenic sources. *Environmental Science and Technology*, 36, 2303-2310.

[125] Schuster, P.F., Shanley, J.B., Marvin-diPasquale, M, Reddy, M.M., Aiken, G.R., Roth, D.A., Taylor, H.E, Krabbenhoft, D.P. & DeWild, J.F. (2008). Mercury and organic carbon dynamics during runoff episodes from a northeastern USA watershed. *Water, Air, and Soil Pollution*, 187, 89-108.

[126] Selvendiran, P., Driscoll, C.T., Bushey, J.T. & Montesdeoca, M.R. (2008). Wetland influence on mercury fate and transport in a temperate forested watershed. *Environmental Pollution*, 154, 46-55.

[127] Shanley, J.B., Schuster. P.F., Reddy, M.M., Roth, D.A., Taylor, H.E. & Aiken, G.R. (2002). Mercury on the move during snowmelt in Vermont. *EOS*, 83, 45-48.

[128] Sherlock, R.L. (2005). The relationship between the McLaughlin gold-mercury deposit and active hydrothermal systems in the Geysers-Clear Lake area, Coast Ranges, California. *Ore Geology Reviews*, 26, 349-382.

[129] Sidle, W.C. (1993). Naturally occurring mercury in a pristine environment? *Environmental Geology*, 21, 42-50.

[130] Siegel, B.Z. & Siegel, S.M. (1987). Hawaiian volcanoes and the biogeology of mercury. In: Decker, R.W., Wright, T.L. & Stauffer, P.H. (eds.) Volcanism in Hawaii. *U.S. Geological Survey Professional Paper* 1350, 1, 827-839.

[131] Skyllberg, U. (2008) Competition among thiols and inorganic sulfides and polysulfides for Hg and MeHg in wetland soils and sediments under suboxic conditions: illumination of controversies and implications for MeHg net production. *Journal of Geophysical Research: Biogeology*, 113, G00C03.

[132] Slomp, C.P. & van Cappellen, P. (2004). Nutrient inputs to the coastal ocean through submarine groundwater discharge. Controls and potential impact. *Journal of Hydrology*, 295, 64-86.

[133] Slowey, A.J. (2010). Rate of formation and dissolution of mercury sulfide nanoparticles: the dual role of natural organic matter. *Geochimica et Cosmochimica Acta*, 74, 4693-4708.

[134] Slowey, A.J. & Brown, G.E., Jr. (2007). Transformations of mercury, iron, and sulfur during reductive dissolution of iron hydroxides by sulfide. *Geochimica et Cosmochimica Acta*, 71, 877-894.

[135] Slowey, A.J., Johnson, S.B., Rytuba, J.J. & Brown, G.E., Jr. (2005). Role of organic acids in promoting colloidal transport of mercury from mine tailings. *Environmental Science and Technology*, 39, 7869-7874.

[136] Smith, R.L., Harvey, R.W. & LeBlanc, D.R. (1991). Importance of closely spaced vertical sampling in delineating chemical and microbiological gradients in groundwater studies. *Journal of Contaminant Hydrology*, 7, 285-300.

[137] Somasundaram, M.V., Ravindran, G. & Tellam, J.H. (1993). Ground-water pollution of the Madras urban aquifer, India. *Ground Water*, 31, 4-11.

[138] Srivastava, R.C. (2003). Guidance and awareness raising materials under new UNEP Mercury Programs (Indian Scenario). Center for Environmental Pollution Monitoring and Mitigation. Available at http://www.chem.unep.ch/mercury/2003-gov-sub/India-submission.pdf..

[139] Steffan, R.J., Korthals, E.T., and Winfrey, M.R. (1988). Effects of acidification on mercury methylation, demethylation, and volatilization in sediments from an acid-susceptible lake. *Applied and Environmental Microbiology*, 54, 2003-2009.

[140] Stoor, R.W., Hurley, J.P, Babiarz, C.L. & Armstrong, D.E. (2006). Subsurface sources of methyl mercury to Lake Superior from a wetland-forested watershed. *Science of the Total Environment.* 368, 99-110.

[141] Stumm, W. & Morgan, J.J. (1995). Aquatic chemistry, third ed. New York, John Wiley and Sons.

[142] Swain, E.B., Jakus, P.M., Rice, G., Lupi, F., Maxson, P.A., Pacyna, J.M., Penn, A., Spiegel, S.J. & Veiga, M.M. (2007). Socioeconomic consequences of mercury use and pollution. *Ambio*, 36, 45-61.

[143] Szabo, Z., Barringer, J.L., Jacobsen, E., Smith, N.P., Gallagher, R.A. & Sites, A. (2010) Variability of mercury concentrations in domestic well water, New Jersey Coastal Plain: Geological Society of America Abstracts, v. 42, no. 1, p. 178, Abstract 78-5.

[144] Tewalt, S.J., Bragg, L.J. & Finkelman, R.B., (2001). Mercury in U.S. coal – Abundance, distribution, and modes of occurrence. *U.S. Geological Survey Fact Sheet FS-095-01.*

[145] Ullrich, S.M., Ilyushchenko, M.A., Kamberov, I.M. & Tanton, T.W. (2007). Mercury contamination in the vicinity of a derelict chlor-alkali plant. Part I: Sediment and water contamination of Lake Balkyldak and the River Irtysh. *Science of the Total Environment*, 381, 1-16.

[146] Ullrich, S.M., Tanton, T.W. & Abdrashitova, S.A. (2001). Mercury in the aquatic environment: a review of factors affecting methylation. *Critical Reviews in Environmental Science and Technology*, 31, 241-293.

[147] USEPA (2001a). National primary drinking water standards. EPA 816-F-01-007

[148] USEPA (2001b). Water quality criterion for the protection of human health: Methylmercury. EPA-823-R-01-001.

[149] USEPA (2009). Report to Congress: Potential export of mercury compounds from the United States for conversion to elemental mercury. U.S. Environmental Protection Agency, Office of Pollution Prevention and Toxic Substances. Available at http://www.epa.gov/mercury/pdfs/mercury-rpt-to-congress.pdf.

[150] USEPA (2010). Mercury in fish: Data. America's Children and the Environment. Available at ttp://www.epa.gov/ace/emerging_issues/fish-table.html#tableei1

[151] USEPA (2012). Mercury: Laws and Regulations. Available from http://www.epa.gov/hg/regs.htm 2012, Accessed 8/13/2012.

[152] USGS, (2000). Mercury in the Environment. *U.S. Geological Survey Fact Sheet 146-00*. Available at http://www.usgs.gov/themes/factsheet/146-00/index.html.

[153] Van Straaten, P. (2000). Mercury contamination associated with small-scale gold mining in Tanzania and Zimbabwe. *The Science of the Total Environment*, 259, 105-113.

[154] Wang, Q, Kim, D., Dionysiou, D.D., Sorial, G.A. & Timberlake, D. (2004). Sources and remediation for mercury contamination in aquatic systems—a literature review. *Environmental Pollution*, 131, 323-336.

[155] Wang, Y., Wiatrowski, H.A., John, R., Lin, C-C, Young, L.Y., Kerkhof, L.J., Yee, N. & Barkay, T. (2012). Impact of mercury on denitrification and denitrifying microbial communities in nitrate enrichments of subsurface sediments. *Biodegradation*, in press, DOI 10.1007/s10532-012-9555-8.

[156] Waples J.C., Nagy,.L., Aiken, G.R. & Ryan, J.N. (2005). Dissolution of cinnabar (HgS) in the presence of organic matter. *Geochimica et Cosmochimica Acta*, 69, 1575-1588.

[157] White, D.E., Hinckle, M.E. & Barnes, I. (1970). Mercury contents of natural thermal and mineral fluids, 25-28. In. Mercury in the Environment. U.S. Geological Survey Professional Paper 713.

[158] WHO (1993). Guidelines for drinking water quality, Vol, 2, Health critera and other supporting nformation.

[159] WHO (2012). Mercury and health. Fact Sheet No. 361. Available from http://www.who.int/medicentre/factsheets/fs361/en/index.html; accessed 7/12/12.

[160] Wiatrowski, H.A., Das, S., Kukkadupu, R., Ilton, E.S., Barkay, T. & Yee, N. (2009). Reduction of Hg(II) to Hg(0) by magnetite. *Environmental Science and Technology*, 43, 5307-5313.

[161] Wiklander, L. (1969). The content of mercury in Swedish ground and river water. *Geoderma*, 3, 75-79.

[162] Yee, N., Barkay, T., Parikh, M., Lin, C-C., Wiatrowski, H. & Das, S. 2010. Biotic/abiotic pathways of Hg(II) reduction by dissimilatory iron reducing bacteria. Geological Society of America Abstracts, v. 42, no. 1, p. 178, Abstract 78-1.

[163] Yu, R-Q., Flanders, J.R., Mack, E.E., Turner, R., Mirza, M.B. & Barkay, T. (2012). Contribution of coexisting sulfate and iron reducing bacteria to methylmercury production in freshwater river sediments. *Environmental Science and Technology*, 46, 2691-2684.

[164] Yudovich. Ya E. & Ketris, M.P.(2005). Mercury in coal: a review. Part I. Geochemistry. *International Journal of Coal Geology*, 62, 107-134.

[165] Zhang, H. (2006). Photochemical redox reactions of mercury. 37-79. In: *Recent developments in mercury science. Structure and bonding*, 120. DOI: 10.1007/430_015.

[166] Zhao, X. & Wang, D. (2010). Mercury in some chemical fertilizers and the effect of calcium superphosphates on mercury uptake by corn seedlings (*Zea mays. L*). *Journal of Environmental Science*, 22, 1184-1188.

# Chapter 5

# Arsenic in Groundwater: A Summary of Sources and the Biogeochemical and Hydrogeologic Factors Affecting Arsenic Occurrence and Mobility

Julia L. Barringer and Pamela A. Reilly

Additional information is available at the end of the chapter

## 1. Introduction

### 1.1. World-wide occurrences of arsenic–contaminated groundwater – Forms and toxicity

Arsenic (As) is a metalloid element (atomic number 33) with one naturally occurring isotope of atomic mass 75, and four oxidation states (-3, 0, +3, and +5) (Smedley and Kinniburgh, 2002). In the aqueous environment, the +3 and +5 oxidation states are most prevalent, as the oxyanions arsenite ($H_3AsO_3$ or $H_2AsO_3^-$ at pH ~9-11) and arsenate ($H_2AsO_4^-$ and $HAsO_4^{2-}$ at pH ~4-10) (Smedley and Kinniburgh, 2002). In soils, arsine gases (containing $As^{3-}$) may be generated by fungi and other organisms (Woolson, 1977).

The different forms of As have different toxicities, with arsine gas being the most toxic form. Of the inorganic oxyanions, arsenite is considered more toxic than arsenate, and the organic (methylated) arsenic forms are considered least toxic (for a detailed discussion of toxicity issues, the reader is referred to Mandal and Suzuki (2002)). Arsenic is a global health concern due to its toxicity and the fact that it occurs at unhealthful levels in water supplies, particularly groundwater, in more than 70 countries (Ravenscroft et al., 2009) on six continents.

### 1.2. Health effects and standards

Despite its use in medicines for nearly 2,500 years (Mandal and Suzuki, 2002; Cullen, 2008) As has long been recognized as a toxic and often lethal substance. Chronic exposure to As can cause harm to the human cardiovascular, dermal, gastrointestinal, hepatic, neurological, pulmonary, renal and respiratory systems (ATSDR, 2000) and reproductive system (Mandal and Suzuki, 2002). Research on health effects is summarized and discussed by

Mandal and Suzuki (2002) and Ng et al. (2003). A compilation of their reviews is found in Table 1.

| System | Health effects |
| --- | --- |
| Cardiovascular | Heart attack, cardiac arrhythmias, thickening of blood vessels, loss of circulation leading to gangrene of extremities, hypertension |
| Dermal | Hyperpigmentation, abnormal skin thickening, narrowing of small arteries leading to numbness (Raynaud's Disease), squamous and basal-cell cancer |
| Gastrointestinal | Heartburn, nausea, abdominal pain |
| Hematological | Anemia, low white-blood-cell count (leucopenia) |
| Hepatic | Cirrhosis, fatty degeneration, abnormal cell growth (neoplasia) |
| Neurological | Brain malfunction, hallucinations, memory loss, seizures, coma, peripheral neuropathy |
| Pulmonary | Chronic cough, restrictive lung disease, cancer |
| Respiratory | Laryngitis, tracheal bronchitis, rhinitis, pharyngitis, shortness of breath, perforation of nasal septum |
| Renal | Hematuria, proteinuria, shock, dehydration, cortical necrosis, cancer of kidneys and bladder |
| Reproductive | Spontaneous abortions, still-births, congenital malformations of fetus, low birth weight |

**Table 1.** Summary of effects of chronic arsenic exposure on human health. (Data from Mandal and Suzuki, 2002, and Ng et al., 2003, and references therein.)

The carcinogenic properties of As were suspected as early as the late 19th Century (Smith et al., 2002). Arsenic is now widely recognized and regulated as a carcinogen (ATSDR, 2000; National Research Council, 1999; USEPA, 2001). Consequently, the occurrence of As in waters at concentrations that exceed existing standards for drinking-water supplies has become of increasing concern, leading to recommended or legislated decreases in concentrations of As in drinking water in many countries. In 1993, the World Health Organization provisionally recommended a decrease from 50 µg/L to 10 µg/L (WHO, 1993). The United States (USA) federal standard, the European Union (EU) Drinking Water Directive (98/83/EC), the New Zealand Drinking Water Standard, the Japanese standard, and recent laws in many Latin American countries (Argentina, Bolivia, Brazil, Chile, Colombia, Costa Rica, El Salvador, Guatemala, Honduras, Nicaragua, and Panama) now place 10 µg/L as the drinking water maximum contaminant level (MCL) (Bundschuh et al., 2012; Robinson et al., 2004; Rowland et al., 2011; Smedley and Kinniburgh, 2002; USEPA, 2001). Mexico has adopted 25 µg/L as a standard (Bundschuh et al., 2012), whereas Australia has instituted a standard of 7 µg/L (NHMRC, 1996) and the State of New Jersey in the USA adopted an As MCL of 5 µg/L in 2006 (NJDEP, 2009). Some developing countries (Bangladesh, for example) have maintained the earlier 50 µg/L MCL standard (Ng et al., 2003). Many instances of As concentrations in groundwater that far exceed standards have been reported throughout much of the world

(Smedley, 2008) and the number of countries in which groundwater is found to be contaminated by As has increased substantially over the past 80 years. This chapter presents a brief overview of the history of groundwater As contamination and summarizes information about the sources, occurrence and mobility of As in groundwater. A compilation of worldwide hazardous waste sites is beyond the scope of the chapter, and only a few examples will be presented. Information on As occurrence reported in previous important summaries and discussions (e.g., Bhattacharya et al., (eds) 2007; Smedley and Kinniburgh, 2002; Welch and Stollenwerk, (eds. ) 2003) is presented in addition to recent findings from the past decade.

## 1.3. Chronology of discoveries of geogenic arsenic contamination

Currently (2012), As contamination of groundwater resources has been identified in many parts of the world, although recognition of the widespread nature of the problem has been advanced only relatively recently. Despite localized inputs of As from human activities, much of the contamination of groundwater with As is shown to arise from geogenic sources and affected groundwater has been found in countries on nearly every continent or major land mass. To date, none has been reported for Greenland, and Antarctica (Figure 1).

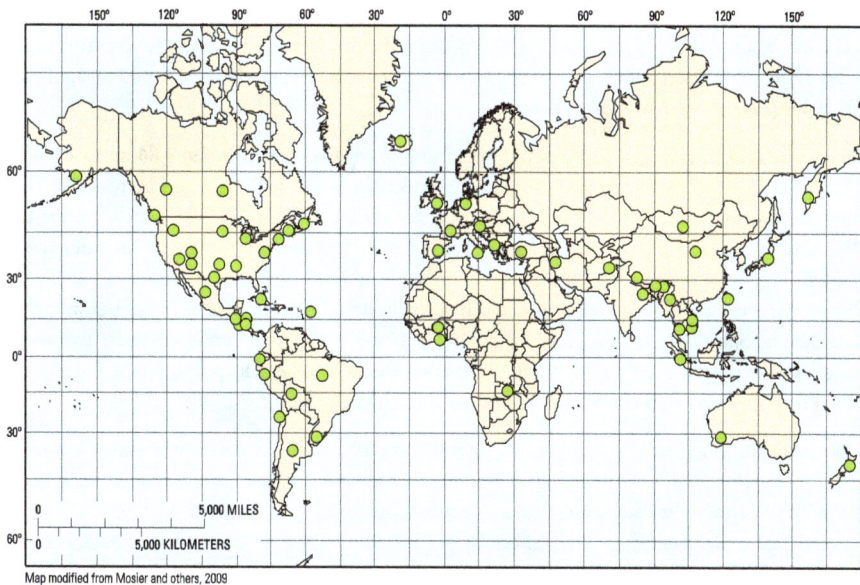

Map modified from Mosier and others, 2009

**Figure 1.** Countries, states, provinces, or areas mentioned in this chapter in which arsenic concentrations in groundwater, including geothermal waters and water contaminated by mining, exceed 10 micrograms per liter (the World Health Organization (WHO) recommended standard (WHO, 1993)). Arsenic concentrations at most locations shown exceed 50 micrograms per liter. Locations of dots are spatially generalized and do not indicate precise locations of arsenic-contaminated waters.

The discoveries of As contamination of groundwater have occurred over a span of nearly 100 years, the most recent within the last decade. Observations of health problems first led to the realization that As was being inadvertently ingested. Arsenic poisoning in humans in Argentina was recognized as early as 1913, and attributed to the drinking of groundwater (Bundschuh et al., 2012, and reference therein). A possible connection between skin cancers and drinking water was recognized in Taiwan during the 1930s (Chen et al., 1994). In the 1940s, As contamination of well water in the Pannonian Basin in Romania and adjacent Hungary was discovered (Gurzau and Pop, 2012; Mukherjee et al., 2006). Recognition of similar occurrences in other European countries such as southwestern England, Germany, Greece, and Spain followed, and groundwaters with As concentrations that exceed standards are now observed in more than 70 countries worldwide (Nordstrom, 2002; Ravenscroft et al., 2009).

During the mid-20th century, instances of As contamination of groundwater were reported for the western USA and Alaska (Welch et al., 1988; Mueller et al., 2001), but were not fully recognized in the states of Oklahoma, Texas, and Arkansas and in the Midwest and Northeastern parts of the country until the 1980s (Ayotte et al., 2003; Ayotte et al., 1999; Haque et al., 2008; Peters, 2008; Peters and Burkert, 2007; Scanlon et al., 2009; Sharif et al., 2008; Welch et al., 2000 and references therein). Most recently, groundwater containing As in excess of USA Federal and State MCLs was found in parts of the Atlantic Coastal Plain (Barringer et al., 2010; Drummond and Bolton, 2010; Haque et al., 2008; Mumford et al., 2012; Pearcy et al., 2011). In Canada, As-contaminated groundwaters in New Brunswick and Nova Scotia were noted in the 1970s (Bottomley, 1984), and instances in western Canada were noted in the 1960s and 1980s (Wang and Mulligan, 2006a).

In West Bengal, India, cases of arsenic poisoning were first noted in 1983-84, according to Rahman et al. (2005), although Mandal and Suzuki place 1978 as the time when arsenicosis (skin lesions) and groundwater contamination were first noticed in West Bengal. Since the 1980s, extensive sampling of well water in West Bengal has revealed levels of As that exceed 50 µg/L —concentrations in some samples exceeding 1000 µg/L (Rahman et al., 2005). In 1983-84, several patients treated for arsenicosis in West Bengal came from neighboring Bangladesh. Sampling in Bangladesh during the early 1990s of waters from tube wells (installed two decades earlier to provide what was thought to be safe, pathogen-free drinking water) revealed elevated concentrations of As (Smith et al., 2000).

Continued sampling throughout West Bengal showed concentrations ranging from <1 to 2,500 µg/L (Nordstrom, 2002). Although the population exposed to As-contaminated water in West Bengal was ultimately estimated to have been about 6,000,000, the exposed population in Bangladesh has been estimated to be about five times that number (Nordstrom, 2002), making the contamination in Bangladesh the arsenic-related public health problem of the greatest magnitude yet observed. Following the early discoveries in West Bengal and Bangladesh, well-water sampling in Vietnam, Cambodia, and Pakistan also identified contamination of groundwater with As (Agusa et al., 2006; Berg et al., 2001; Chanpiwat et al., 2011; Hoang et al., 2010; Luu et al., 2009; Nickson et al., 2005; Smedley, 2008; Sthiannopkao et al., 2008;), as well as in 1999, Nepal (Gurung et al., 2005), and in 2000, Myanmar (Tun, 2003).

Health problems attributed to As exposure were first noted within a district in a province of Thailand in 1987. Sampling indicated As at concentrations in groundwater that exceeded 5,000 µg/L with about 15,000 people thought to have been exposed (Nordstrom, 2002; Smedley, 2008). Sampling of well waters during the 1990s and later in northern parts of China, including Inner Mongolia, have shown groundwater As concentrations that range widely, from <1 to about 2,400 µg/L (Nordstrom, 2002; Smedley, 2008). In Japan, As levels were found to be high in geothermal waters and springs as early as the 1950s (Noguchi and Nakagawa, 1969), and As contamination of groundwater was noted in 1994. Exposure to As from industrial sources was noted as early as the 1950s (Mukherjee et al., 2006). In 1981 in Iran, chronic arsenic poisoning was noticed and subsequent well-water sampling revealed concentrations exceeding 1,000 µg/L (Mukherjee et al., 2006).

Arsenic contamination of both surface water and groundwater also is found in many Latin American countries, but the full extent of the problem is not yet clear (Bundschuh et al., 2012). Groundwater concentrations in Argentina are reported to range as high as about 15,000 µg/L (Bundschuh et al., and references therein). High levels of As were found in waters in Chile and Mexico in the 1950s and 60s (Bundschuh et al., 2012; Rosas et al., 1999). In northern Chile, As concentrations in groundwater ranged from 20 to 5,000 µg/L, and there is strong evidence from hair, skin, bones and funerary preparations (clay, paint) of Chinchorro culture mummies that the population there was exposed to high levels of As more than 7,000 years ago (Bundschuh et al., 2012), and, presumably, has been ever since. Concentrations of As in groundwater in Mexico generally are not reported to reach the higher levels found in Argentina except for a geothermal area in Michoacán, where 24,000 µg/L are reported (Bundschuh et al., 2012, and references therein). In mined areas of Mexico, leaching from tailings piles have contributed As to groundwater (Carillo-Chávez et al., 2000; Méndez and Armienta, 2003). Arsenic-contaminated surface water and groundwater subsequently was recognized in parts of Peru in the 1970s (Bundschuh et al., 2010). Although As contamination of waters in Argentina, Chile, Mexico and Peru was known from the early and mid-20th century, surface water and (or) groundwaters (and geothermal waters) containing geogenic As at contaminant levels were discovered in alluvial, metasedimentary, volcanic, and metavolcanic aquifers only since the late 1990s in Bolivia, Brazil, Columbia, Cuba, Ecuador, El Salvador, Guatemala, Honduras, Nicaragua, and Uruguay (Bundschuh et al., 2012).

In Africa, the occurrence of high concentrations of As in groundwater from wells in a village in Burkina Faso was first noticed in the 1970s, but more extensive sampling did not occur until several decades later (Smedley, 2008). Arsenic-related skin diseases were noted in the region, and, although many wells yielded water with As concentrations < 10 µg/L, concentrations as high as 1,600 µg/L were found (Smedley et al., 2007). Effects of mining on soils and waters in other African countries (Ghana and Zambia) have been studied only recently (e.g., Bowell et al., 1994; Nakayama et al., 2011; Smedley and Kinniburgh (2002); As contamination of groundwater has been reported in Ghana. Arsenic contamination of well water was discovered during the drought in Perth, Australia in 2002 (Appleyard et al., 2006).

## 2. Sources of arsenic

### 2.1. Anthropogenic sources

Sources of As that arise from human activities include mining and processing of ores and manufacturing using As-bearing sulfides. Smelters in numerous countries, including Canada, Chile, Italy, South Africa, the USA, and the former USSR have processed metal ores (mainly copper, but also zinc, gold, and tin) that contain As. The smelting process, both recent and ancient, has released As to the air and soils both locally and globally (Matschullat, 2000). Elevated As levels in precipitation and in soils surrounding smelters have frequently been documented (e.g., Ball et al., 1983; Beaulieu and Savage, 2005; Carpenter et al., 1978). Disposal of mining wastes has caused As contamination of groundwater in numerous places, including in Thailand, Ghana, and Turkey (Gunduz et al., 2010; Smedley and Kinniburgh, 2002; Smedley, 2008). In southwestern England, mineral deposits and mineral processing (tin, copper, with accessory As minerals) are recognized sources of As to soils and groundwater (Brunt et al., 2004; Camm et al., 2004; Palumbo-Roe et al., 2007). In southeastern Europe (Serbia, Bosnia, Poland), and in Spain, mining activities left a legacy of arsenic that has contaminated soils and waters (Dangic, 2007; Gomez et al., 2006; Karczewska et al., 2007). Groundwater contamination in mined areas is also found in parts of the western USA and Canada (Moncur et al., 2005; Welch et al., 2000).

Arsenic compounds have been used in the manufacture of numerous products. Arsenic has been used in glass production and by the wood-preservation industry. The latter industry has been, through the end of the 20th century, the most active user of such compounds in the USA (Welch et al., 2000), but the industry's voluntary reduction of the use of chromated copper arsenate (CCA) since 2003 has resulted in a more limited availability of CCA treated wood and products (Brooks, 2008). Contamination of soils and surface-water bodies has resulted from use of CCA-treated wood (Khan et al., 2006; Rice et al., 2002).

The use of inorganic arsenical pesticides has waned in recent years owing to bans in the 1980s and 90s (Welch et al., 2000), but, in the past, manufacture and use of arsenical pesticides were important contributors of As to the environment (e.g., Barringer et al., 1998; Barringer et al., 2001; USEPA, 2011). Inorganic arsenicals have been used on a variety of crops (citrus, cotton, tobacco, and potatoes) and on fruit trees; (Walsh and Keeney, 1975; Welch et al., 2000). The use of lead-arsenate pesticides in orchards has prompted concern that additions of phosphate fertilizers could displace arsenate sorbed to soil particles, mobilizing As to groundwater (e.g. Davenport and Peryea, 1991). Manufacture of pesticides has been responsible for As contamination of soil, surface water, and groundwater. Examples of contamination caused by former pesticide manufacture are found in India and in the USA (Barringer et al., 1998; Mukherjee et al., 2006; USEPA, 2011). Soil contamination by As may also occur in areas where soils are amended with chicken and swine manure. Such fertilizers can contain As due to use of chicken and swine feeds containing the growth additive Roxarsone, an organic arsenical (4-hydroxy-3-nitrophenylarsonic acid) (Hileman, 2007; O'Day, 2006).

Arsenic in Groundwater: A Summary of Sources and the Biogeochemical and
Hydrogeologic Factors Affecting Arsenic Occurrence and Mobility

121

There have been military uses of As, and attempts to recycle As-bearing materials. Arsenic was used in chemical warfare agents, most recently in the first half of the 20$^{th}$ century (Krüger et al., 2007). In the 1930s, concern over disposal of stockpiles of As compounds led to experiments with incorporating As in cement which was then used to coat pilings and other wooden structures (van Siclen and Gerry, 1936). One ounce of white arsenic was added to 12 pounds of sand and 3 pounds of cement, water was added and the slurry was applied to wood pilings by air gun. To the authors' knowledge, no studies of the effects of this practice on the environment have been published. Arsenic currently is used in various electronic devices, and improper disposal and lack of care in recycling these materials also can add As to the environment (Brooks, 2008).

Anthropogenic sources of As can affect the quality of surface water through groundwater discharge and runoff (Hemond, 1995; Martin and Pederson 2002). In the case of pesticides, the effect can be through direct applications to water bodies for control of nuisance vegetation (Kobayashi and Lee, 1978; Tanner and Clayton, 1990; Durant et al., 2004). Although groundwater contamination does exist at various sites affected by agricultural, industrial or military releases, for example (e.g., Hemond, 1995; Krüger et al. 2007; USEPA, 2011), contamination introduced at the land surface does not always move to groundwater. Owing to the affinity of As for soil constituents such as metal oxides and hydroxides (mainly iron (Fe), aluminum (Al), and manganese (Mn)) and clays (Goldberg and Glaubig, 1988; Manning and Goldberg, 1996), the As can be attenuated in the intervening soils by sorption to Fe hydroxides or clays, or by precipitation reactions, such as formation of As- or Fe-sulfides in anoxic soils (e.g., Brunt et al., 2004; Cancès et al, 2008).

## 2.2. Geologic sources

### 2.2.1. Arsenic minerals

For most known areally extensive instances of As contamination of groundwater, the sources of the As have been shown to be geogenic (Smedley and Kinniburgh, 2002). A summary table of the worldwide occurrence of As in groundwater (Nordstrom, 2002) indicates that mining of arsenic and metal ores and natural geologic sources of As dominate the environmental conditions listed for inputs of As to groundwater.

There are about 24 As-bearing minerals that are commonly found in hydrothermal veins, ore deposits, and rocks. Most primary As minerals are sulfides, of which arsenopyrite is the most common (Ehrlich and Newman, 2009). Secondary minerals tend to be less common arsenates and oxides. WHO (2001) provides a list of these minerals, which also can be found tabulated in Mandal and Suzuki (2002, p. 203). Arsenic in crustal rocks also has an affinity for, and is associated with, pyrite or Fe hydroxides and oxides (Nordstrom, 2002) for which chemical formulas are $FeS_2$, $FeOOH$, $Fe_2O_3$, and $Fe_3O_4$, respectively. The As content of crustal rocks varies widely; Smedley and Kinniburgh, 2002, p. 531) compile and tabulate those results. Concentrations of As in water associated with crustal rocks are described below.

*2.2.2. Geothermal activity, volcanic and plutonic rocks, and mineralized zones*

Geothermally active zones occur along plate boundaries, in tectonic rift areas such those in East Africa and at seafloor spreading centers, such as in Iceland, and at "hot spots" where mantle-derived plumes ascend, such as in Hawaii and Yellowstone National Park, USA. Arsenic is one of a suite of incompatible elements (these do not fit easily into the lattices of common rock-forming minerals), which include antimony (Sb), boron (B), fluoride (F), lithium (Li), mercury (Hg), selenium (Se) and thallium (Tl). Together with hydrogen sulfide, these elements are ubiquitous in high-temperature geothermal settings (Webster and Nordstrom, 2003). Concentrations of arsenic are high mainly in geothermal waters that leach continental rocks; geothermal waters in basaltic rocks, such as in Iceland, contain lesser amounts of As. Arsenic in hot geothermal fluids was shown to be derived mainly from leaching of host rocks at Yellowstone National Park, in Wyoming, USA (Stauffer and Thompson, 1984), rather than derived from magmas. The As in the hot fluid is present as As(III) in arsenious acid ($H_3AsO_3$); in low-sulfide fluids, the arsenite in the arsenious acid is oxidized to arsenate as the rising fluid mixes with cold oxygenated groundwater or encounters the atmosphere. In high sulfide solutions, As may be present as thioarsenate complexes (Webster and Nordstrom, 2003; Planer-Friedrich et al., 2007).

In the western USA, there are As inputs to groundwater and surface water from geothermal fluids in and near Yellowstone National Park (e.g., Ball et al., 1998, 2002; Nimick et al., 1998) and in other western mineralized areas (Welch et al., 1988). Groundwater associated with volcanics (tuffs and rhyolites) in California contain As at concentrations ranging up to 48,000 µg/L, with As-bearing sulfide minerals as the main source of As (Welch et al., 1988, and references therein). Geothermal waters on Dominica in the Lesser Antilles also contain concentrations of As >50 µg/L (McCarthy et al., 2005).

In general, because arsenic is an incompatible element, it accumulates in differentiated magmas, and is commonly found at higher concentrations in volcanic rocks of intermediate (andesites) to felsic (rhyolites) composition than in mafic (basaltic) rocks—as shown for the western USA (Welch et al., 1988; Welch et al., 2000). In Maine and New Hampshire, USA, where As-contaminated groundwater is present, pegmatites, granites and metamorphic rocks (granofels) were found to have substantial As contents—up to 60, 46 and 39 mg/kg, respectively (Peters, 2008; Peters et al., 1999; Peters and Blum, 2003). Weathering of pegmatite veins in Connecticut, USA, was thought to contribute As to groundwater (Brown and Chute, 2002). Although the As content of mafic rocks can be relatively low, fractured ultramafic rocks in Vermont, USA, contribute up to 327 µg/L of As to groundwater (Ryan et al., 2011).

Leached from surrounding rocks, As in hot springs from geothermal fields in New Zealand is found at concentrations that range to 4,800 µg/L (Brown and Simmons, 2003). Acidic (pH 1.2) geothermal springs in Japan contained As at 2,600 µg/L (Noguchi and Nakagawa, 1969). Arsenic in these springs precipitated out as As sulfides and lead (Pb) As sulfides in surface-water sediments. The As contents of the sediments ranged from about 5 to 56 wt. % (Noguchi and Nakagawa, 1969).

In Latin America, volcanic and geothermal activity along the Pacific tectonic plate boundaries produces As-rich waters and gases in springs and fumaroles. In sodium-chloride (Na-Cl)-rich waters, As concentrations can reach about 50,000 µg/L at the El Tatio geothermal field in Chile (Lopez et al, 2012). Arsenic concentrations in waters of geothermal fields in Mexico vary; 250 to 73,600 µg/L are reported, where the host rocks through which these waters have risen include sandstones and shales, lava flows and pyroclastics, and metamorphosed carbonate rocks, basalts and hornblende andesites (Lopez et al., 2012).

In the coastal volcanic areas of Central and South American countries, rocks are mainly andesitic or rhyolitic in composition. Arsenic concentrations in the geothermal springs throughout these areas vary widely—concentrations of several thousand micrograms per liter are reported, but none are reported as high as the highest concentration in a Mexican geothermal field (see above). Where springs and fumaroles discharge water and gases to a lake in the Bolivian Altiplano, however, As concentrations in the lake water are reported as high as 4,600,000 µg/L, the As apparently being contributed by oxidation of sulfide deposits (Lopez et al., 2012).

Dissolution of volcanic glasses in ash layers and leaching of loess-type deposits in the Chaco-Pampean plain of Argentina have resulted in groundwater As concentrations that range from <10 to 5,300 µg/L (Nicolli, et al., 2012) where the potentially affected rural population numbers several million people. Mining of various metals (gold, copper, silver) in Latin America has played an important role in mobilizing As from the geologic materials and exacerbating contamination of groundwater resources (McClintock et al., 2012).

In Europe, groundwaters containing As at concentrations that exceed 50 µg/L are found in geothermal fields of the Massif Central in France, and in Greece, (Brunt et al., 2004; Karydakis et al., 2005). Iceland, where As concentrations in groundwater can exceed 10 µg/L (Arnórsson, 2003) sits astride the Mid-Atlantic Ridge, and is subject to outpourings of basaltic lava that typically contain less As than do more silicic lavas (Onishi and Sandell (1955; Baur and Onishi, 1969; Ure and Berrow, 1982). In geothermal systems of northern and northeastern Spain, the As concentrations are high and deposit As-rich minerals (Navarro et al., 2011); the Caldes de Malavella field in northern Spain contributes substantial As to groundwater—50-80 µg/L in springs, and from <1 to 200 µg/L in groundwater (Piqué et al., 2010). Groundwaters (including brines) contained As concentrations ranging from 1.6 to 6,900 µg/L in the Phlegraean Fields in southern Italy (Aiuppa et al., 2006). Quaternary volcanic rocks with hydrothermal activity on the island of Ischia (offshore from Naples, Italy) impart As to groundwater at concentrations that range up to 3,800 µg/L (Aiuppa, et al., 2006; Daniele, 2004).

In the Mid-East, in northwestern Iran, As concentrations in thermal waters and hot springs are as high as 3,500 and 890 µg/L, respectively, in the area of Mt. Sabalan, a stratavolcano (Haeri et al, 2011). In western Anatolia, in Turkey, natural leaching, aided by pumping and discharge of waste geothermal fluids from an active geothermal system, has mobilized As from metamorphic, igneous and sedimentary rocks to groundwater in a shallow alluvial aquifer, where highest As concentrations in groundwater and geothermal waters were 561 and 594 µg/L, respectively (Gunduz and Simsek (2008).

Groundwater in early Proterozoic silicic volcanics and granites of the Chhattisgarh Basin of India contains As at concentrations that exceed 10 µg/L. The As is perhaps emplaced there by hydrothermal fluids (Acharyya, 2002).

### 2.2.3. Sedimentary and meta-sedimentary bedrock

Arsenic is found in coals, with the content of some coals from southwestern China being highest—826 to 2,578 mg/kg is reported (Nriagu et al., 2007) and up to 32,000 mg/kg is listed by Wang et al., (2006). In Germany, the As content of bituminous shales ranges from 100 to 900 mg/kg (Smedley and Kinniburgh, 2002). The As contents of American coals are reported to range as high as 2,200 mg/kg (Wang et al., 2006), but the mean concentration for more than 7000 samples is about 24 mg/kg (Kolker et al., 2006). Pyrite is the main source of As in coals with high As content, whereas in lower As coals, the As tends to be associated with the organic material (Yudovich and Ketris, 2005). In Pennsylvania, USA, As concentrations in water discharging from abandoned anthracite mines ranged from <0.03 to 15 µg/L and from abandoned bituminous mines, from 0.10 to 64 µg/L, with 10% of samples exceeding the USEPA MCL of 10 µg/L (Cravotta, 2008).

In Wisconsin, USA, As concentrations of water in sandstone and dolomite aquifers were as high as 100 µg/L. Oxidation of pyrite hosted by these formations was the likely source of the As, the transport of which was, in some instances, retarded by its association with Fe oxy-hydroxides (Burkel and Stoll, 1999; Thornburg and Sahai, 2004). In the adjacent State of Michigan, USA, As concentrations in groundwater reached 220 µg/L in another sandstone aquifer (Haack and Rachol, 2000). In Australia, a combination of increased water withdrawals during development and declining recharge due to drought caused oxidation of pyrite in sedimentary aquifers, resulting in As contamination of well water (Appleyard et al., 2006). In England, groundwater from a sandstone aquifer contained As at concentrations that spanned 10 to 50 µg/L; the As content of the sandstone ranged from 5 to 15 mg/kg. Desorption at pH of about 8 appeared to be the mechanism for As release to groundwater (Kinniburgh et al., 2006). Water from wells completed in a Mesozoic Era sandstone in northern Bavaria also contained As at concentrations from 10 to 150 µg/L (Heinrichs and Udluft, 1999), although the mineralogy contributing the As was not identified.

In the Piedmont of Pennsylvania and New Jersey, USA, groundwater in Mesozoic age aquifers of red and black shale, mudstone, and siltstone contains elevated levels of As—domestic well waters from Pennsylvania contained up to 65 µg/L (Peters and Burkert, 2007), whereas in New Jersey the highest concentration measured recently was 215 µg/L (Serfes, 2005). Pyrites in the reduced black shales and mudstones are a major source of As, with measured As contents of 3,000 and 40,000 mg/kg in some samples (Serfes, 2005). Arsenic (as arsenate) is also released from the red shales by desorption as pH rises above 6.5 (Serfes, 2005). These Piedmont rocks also contain hornfels along contacts with diabase intrusions, some of which contain mineralization by copper (Cu), As, and uranium (U). Typical As-bearing minerals in the Piedmont rocks include arsenopyrite (FeAsS), cobaltite ((Ni, Co, Fe)AsS), alloclasite ((Co, Fe)AsS), gersdorffite (NiAsS), erythrite ($Co_3(AsO_4)_2.8H_2O$), and safflorite ((Co, Fe, Ni)As$_2$) (Senior and Sloto, 2006).

In Taiwan, groundwater from artesian wells completed in black shales, muds and fine sands is contaminated with As. In northern Bavaria, mineralized sandstone yields As-contaminated groundwater (Smedley and Kinniburgh, 2002).

Carbonate rocks typically contain low concentrations of As (Baur and Onishi, 1969; Smedley and Kinniburgh, 2002), although some limestones may contain As-bearing pyrite (e.g. Price and Pichler, 2006). Because arsenate can substitute for phosphate in minerals, phosphorite deposits can contain substantial amounts of As — up to about 400 mg/kg is reported (Smedley and Kinniburgh, 2002). Barringer et al. (2011) report 19.5 to 56.6 mg/kg of As in phosphorite deposits in the Coastal Plain of New Jersey, USA.

The As content of meta-sedimentary rocks varies widely, with the contents of gneisses and quartzites generally < 10 mg/kg, and higher As contents for slates and phyllites — up to 143 mg/kg (Boyle and Jonasson, 1973). In the New England states of northeastern USA, ground-water contamination with As was found to be prevalent in water from wells completed in formations containing metapelite rocks (schists, phyllites, slates), particularly those rocks adjacent to intrusive bodies. In New England, meta-shales contain As-bearing minerals pyrrhotite, cobaltite, and arsenopyrite, and supergene minerals include orpiment (AsS) and loellingite ($FeAs_2$)(Foley et al., 2002). Mineralization in a Proterozoic marble in New Jersey, USA, has resulted in zinc ores and, in addition to arsenopyrite and loellingite, a variety of uncommon As-bearing minerals. Past mining activities have contributed to the release of As from the bedrock to shallow groundwater; concentrations ranged from 2.02 to 22.0 µg/L in water discharging to the area's major river, the Wallkill River (Barringer et al., 2007). In mineralized meta-sedimentary rocks containing sulfide minerals in Fairbanks, Alaska, USA, groundwater contains As at concentrations that range from <3 µg/L to 1,670 µg/L (Mueller et al., 2001).

*2.2.4. Alluvial and coastal plain unconsolidated sediments*

The sediments shed from the Himalayas have formed the extensive alluvial plain and delta through which the Ganges, Brahmaputra and Meghna Rivers flow and which form aquifers in India (West Bengal Delta Region) and adjacent Bangladesh. To the east, the deltas of the Mekong and Red Rivers form aquifers in Vietnam and adjacent Cambodia and Laos. Arsenic released from these sediments has caused the most widespread contamination in the world, with populations of many millions affected by drinking As-laden well water. The largest number of people (about 35 million) exposed to As contamination is in Bangladesh, and about 6 million in neighboring West Bengal, India (Nordstrom, 2002). The As contents of these young sediments are not extremely high — about 1 to 15 mg/kg, but vary with depth (Smedley, 2008). The concentrations of As in groundwater range, however, from undetectable to several thousand µg/L, with a survey of about 3,200 wells in Bangladesh by researchers from the United Kingdom (UK) and Bangladesh finding that As concentrations in about 27% of samples exceeded 50 µg/L, the Bangladeshi Drinking Water Standard (Smedley, 2008). In local areas, the percentage of affected wells was higher; in central Bangladesh, As in water from about 75% of 6000 wells exceeded 50 µg/L (van Geen et al., 2003). The sediments of Holocene age are micaceous sands, silts, and clays. Reduced, gray sediments of the upper aquifer, where As concentrations in groundwater are high, are underlain by brown, oxidized sediments where As concentrations in groundwater are low. In some places the two layers are separated by a thick clay layer (Harvey et al., 2002). The mineralogy of both layers, where the clay was absent, was found to be similar for part of the Bengal Basin near Dhaka, Bangladesh. Minerals included quartz, plagioclase and potassium feldspar, micas (biotite, muscovite, and phlogopite), chlorite, and amphibole; carbonaceous material was sparse. Trace amounts of siderite were present in the reduced sediments, but not in the deeper, oxidized sediments (Stollenwerk et al., 2007). In the shallow aquifer in West Bengal, organic carbon is present as petroleum-related compounds (Rowland et al., 2006).

In Cambodia, Vietnam, and Laos, As in young deltaic sediments of the Mekong and Red River basins has also contaminated groundwater, again exposing several million people to unhealthy levels of As (> 1,000 µg/L in some cases) in drinking water (Agusa et al., 2006; Berg et al., 2001; Chanpiwat et al., 2011; Hoang et al., 2010; Luu et al., 2009; Sthiannopkao et al., 2008). The aquifers are composed of quartz sands and clays that host Fe oxide and hydroxide phases, also possibly manganese (Mn) oxides, and organic matter is present. Siderite, pyrite, and orpiment are also found in sediments of the Mekong delta (Quicksall et al., 2008) and siderite, ilmenite, vivianite, gibbsite and boehmite are reported for the Red River delta sediments (Eiche et al., 2008).

In the Pannonian Basin of Hungary and Romania, Quaternary sediments of fluvial and eolian origin have contributed As to groundwater; the sediments are composed of sands and loess. Quartz, feldspar, carbonates (calcite and dolomite, muscovite, chlorite, clays, and humic substances are reported and fine particles of Fe hydroxides are indicated (Varsányi and Kovács, 2006). Other Fe-bearing minerals, from which As may be released, are reported to include goethite, limonite, pyrite, and siderite (Rowland et al., 2011).

Alluvial and lacustrine sediments in the Huhhot Basin of Inner Mongolia form two aquifers, separated by a clay confining layer; some boreholes completed in the deeper aquifer are artesian. The sediments are more fine-grained in the low-lying parts of the basin and it is in these sediments that reducing conditions are present and As concentration in groundwater are highest (1,500 µg/L). Organic matter is found in the aquifers and dissolved organic carbon concentrations in groundwater are high (Smedley et al., 2003).

The Atlantic Coastal Plain is located along the east coast of the USA. In addition to quartz-rich deposits of near-shore origin, the Coastal Plain is composed partly of sediments of marine origin that contain the mineral glauconite, the As contents of which are high (up to 130 mg/kg) in some formations (Dooley, 2001). In the state of Maryland, USA, in an aquifer composed of these marine sediments, As concentrations in groundwater exceed 10 µg/L and have been found as high as 80 µg/L (Pearcy et al, 2011). Farther north, in the state of New Jersey, As concentrations in water from an observation well in a glauconite-bearing aquifer were 110 µg/L (dePaul and Szabo, 2007), and water from several domestic wells in similar aquifers in the same region has exceeded the state MCL of 5 µg/L. In shallow groundwater discharging to New Jersey Coastal Plain streams underlain by the glauconitic sediments, As concentrations have exceeded the MCL, ranging as high as 89.2 µg/L. The sediments below the streambeds also contain other phyllosilicates (illite, smectites, muscovite, biotite, chlorite) and quartz. In addition to the glauconite sands, associated phosphorite deposits were found

Arsenic in Groundwater: A Summary of Sources and the Biogeochemical and
Hydrogeologic Factors Affecting Arsenic Occurrence and Mobility

127

to contain As up to 56.6 mg/kg and siderite that precipitated in sediments beneath a streambed contained 184 mg/kg of As (Barringer et al., 2011; Mumford et al., 2012).

Glauconite sands and clays of Pliocene and Miocene age, overlain by younger non-glauconitic sands, clays, thin coal beds and peats, are also present in the lowlands of xouth Sumatra, Indonesia. Arsenic concentrations exceeded the WHO guidelines in water from several wells completed in both the glauconitic formations and the younger sediments, with the higher concentrations found in water from the youngest (Holocene) sediments (Winkel et al., 2008). Glauconitic sediments are found on several continents (Barringer et al., 2010) and the recent findings in the Atlantic Coastal Plain and Sumatra indicate that these marine sediments, in addition to all the aquifers in other geologic settings, can now also be considered a potential source of As-contaminated groundwater.

# 3. Biogeochemical factors

### 3.1. Oxidation

Oxidation of As-bearing sulfides has been proposed as a mechanism for releasing As from geologic materials (Smedley, 2008). Although originally proposed as a mechanism for As release from the alluvial sediments of West Bengal and Bangladesh, the presence of sulfide minerals in those aquifers is rare and appears limited to biogenic framboidal pyrite, and pyrite in woody peat, and on magnetite (Acharyya, 2002). Oxidation of sulfides in mined areas throughout the world is a well-known phenomenon that has led to high concentrations of As in soils, surface water and groundwater; examples of such occurrences include western Canada, the western USA, and the Bolivian Altiplano (Lopez et al., 2012; Moncur et al., 2005; Welch et al., 2000). Sulfide oxidation has also resulted in high As concentrations (up to 215 µg/L) in groundwater in the eastern USA, in parts of the Piedmont rocks in Pennsylvania and New Jersey (Peters and Burkert, 2007; Serfes, 2005), where an As-oxidizing bacterium was involved in the mobilization of As (Rhine et al., 2008). A broad diversity of microorganisms oxidizes dissolved arsenic for different reasons including dissimilatory respiration, detoxification, and energy needs (Santini and Ward, 2012).

### 3.2. Reduction

Because of the magnitude of the contamination in West Bengal and Bangladesh, these two regions have received substantial attention from the research community. In general, reductive dissolution of Fe hydroxides and release of sorbed As explains much of the observed mobilization of As from sediments to groundwater (e.g., Nickson et al., 2000; Zheng et al., 2004). Organic matter in the alluvial aquifers is likely an important component of the reduction process. Field and experimental studies have shown that metal-reducing microbes can enhance mobilization of As, and that oxidation of the organic matter drives the redox reactions whereby Fe hydroxides are reductively dissolved and sorbed As released (Islam et al., 2004; McArthur et al., 2004). Arsenic also is released from Mn oxides as they reductively dissolve, but may not remain in the groundwater, instead resorbing to Fe hydroxides (McArthur et al., 2004). Other studies indicate that As could also be released from biotite into Bangladeshi groundwater (Hopf et al., 2001; Seddique et al., 2008).

Reductive dissolution of Fe hydroxides has been proposed as a viable mechanism for As release in many other affected aquifers as well—for example, in Croatia, Inner Mongolia, northern China, and the eastern and southeastern USA (Barringer et al., 2010, Barringer et al., 2011; Guo et al., 2010; Haque et al., 2008; Mumford et al, 2012; Pearcy et al., 2011; Sharif et al., 2008; Ujevic et al., 2010; Xie et al, 2008). Although less intensely studied than Bengal Basin sediments, release of As from Fe oxides appears to be an important mechanism in the Inner Mongolian sediments (Smedley, 2008).

### 3.3. Microbially mediated reactions

Given the similarities to the shallow aquifers in Cambodia, Vietnam, Laos, and Myanmar, much of what has been found in West Bengal and Bangladesh may apply to the aquifers in those countries as well. A study of indigenous bacteria in Cambodian sediments indicated arsenic-respiring bacteria that reduce As(V) (arsenate) to As(III) (arsenite) were fueled by inputs of organic carbon (Lear et al., 2007), similar to findings for bacteria in West Bengal sediments. Additionally, results of experiments showed microbially mediated reduction of As(V), and Fe, and release from minerals (clay, glauconite sands, oxides and hydroxides) (Campbell et al., 2006; Dong et al., 2003; Hopf et al., 2009; Kostka et al., 1999; McLean et al., 2006; Pearcy et al., 2011), demonstrating the involvement of microbes in the reduction-oxidation (redox) reactions.

The bacteria involved in the reactions involve several groups. Iron-reducing bacteria of the genus *Geobacter* can reduce Fe in minerals such as hydroxides (Lloyd and Oremland, 2006), thus leading to dissolution of the hydroxides and sorbed As release. Geobacter bacteria have been investigated as As reducers; although *G. uraniumreducens* contains genes for As respiration, it was not conclusively shown to respire As, and *G. sulfurreducens* did not reduce As enzymatically (Islam et al., 2005; Lear et al., 2007). Bacteria known as dissimilatory arsenate respiring prokaryotes (DARPs) are identified as arsenic reducers, which means that As(V) serves as the terminal electron acceptor in dissimilatory reduction of arsenate. In the glauconitic sediments of the New Jersey Coastal Plain, USA, amplification of the arsenic respiratory reductase gene (*arrA*) followed by alignment and gene sequencing revealed clones with close (99%) similarity to *Alkaliphilus oremlandii* (CP000453) (formerly *Clostridium* species strain OhILAs)—a known arsenate-respiring bacterium (Mumford et al., 2012). Also using molecular techniques, an arsenate-respiring proteobacterium *Sulfurospirillum* sp. strain NP4 was identified by Lear et al. (2007) in Cambodian sediments. The bacterium *Desulfotomaculum auripigmentum* reduces As(V) to As(III) as well as sulfate to sulfide, and precipitates orpiment (Newman et al. 1997; Ehrlich and Newman, 2009)

It should be noted, however, that not all microbial reduction of arsenate is the result of bacterial respiration. Many bacteria detoxify arsenate by reducing it to arsenite and expelling it (Oremland and Stolz, 2005) and, in some cases, by methylating and expelling it (Bentley and Chasteen, 2002). Presence of dimethylarsinate and monomethylarsonate in groundwater, as was found in shallow groundwater in glauconitic sediments of the New Jersey Coastal Plain, probably was indicative of such microbial activity (Mumford, et al., 2012). In experiments, As was mobilized from apatite by the bacterium *Burkholderia fungorum*, which utilizes phosphorus from the apatite (Mailloux et al., 2009). There is apatite present in the sediments released from the Himalayas to the alluvial aquifers of South East Asia, and there are apatite-rich phosphorite beds in the glauconitic New Jersey Coastal Plain. In addition to reductive dissolution of iron hydroxides releasing sorbed As, arsenate reduction by As-respirers and other As-reducing bacteria, As may also be released by the mechanism suggested by Mailloux et al. (2009) in some aquifers.

Conditions that support microbial sulfate reduction are reported for the aquifers of the West Bengal, Mekong, and Red River deltas. Where such conditions exist, there is the potential for precipitation of sulfide minerals that could remove As as well as Fe from solution. Buschmann and Berg (2009) found As concentrations to be lower in groundwater from zones where sulfate ($SO_4^{2-}$) and Fe reduction were occurring in the aquifers of the Bengal, Mekong, and Red River deltas. They suggest such conditions, which could result in precipitation of insoluble sulfides, are a control on As levels in groundwater. Some of the chemical reactions that likely affect As mobility are shown in Table 2.

| Reaction description | Equation | Reference |
|---|---|---|
| Oxidation of pyrite | $FeS_2 + 15/4 O_2 + 7/2 H_2O \rightarrow Fe(OH)_3 + 2H_2SO_4$ <br> $10FeS_2 + 30NO_3^- + 20 H_2O \rightarrow 10 Fe(OH)_3 + 15 N_2 + 15 SO_4^{2-}$ <br> $+ 5 H_2SO_4$ | Welch et al., <br> (2000) |
| Oxidation of arsenopyrite | $FeAsS(S) + 11/4 O_2(aq)\, 3/2 H_2O\,(aq) \rightarrow Fe^{2+}(Aq) + SO_4^{2-}(aq) +$ <br> $H_3AsO_3(aq)$ | Morin & Calas, <br> (2006) |
| Oxidation of arsenite | $H_3AsO_3 + \frac{1}{2} O_2\,(aq) \rightarrow H_2AsO_4^- + H^+$ | Morin & Calas, <br> (2006) |
| Reductive dissolution of Fe hydroxides (release of sorbed arsenate not shown) | $4Fe^{III}OOH + CH_2O + 7H_2CO_3 \rightarrow 4Fe^{II} + 8HCO_3^- + 6H_2O$ | Nriagu et al. <br> (2007) |
| Reduction of sulfate, formation of sulfide | $2CH_2O + SO_4^{2-} \rightarrow 2HCO_3^- + H_2S$ | Nriagu et al., <br> (2007) |
| Oxidation of organic carbon (lactate) and reduction of As(V) | $CH_3\text{-}CHOH\text{-}COO^- + 2HAsO_4^{2-} + 3H^+ \rightarrow CH_3COO^- + 2HAsO_2 +$ <br> $2H_2O + HCO_3^-$ | Saltikov et al., <br> (2003) |
| Microbially mediated precipitation of orpiment | $2HAsO_2 + 3HS^- + 3H^+ \rightarrow As_2S_3 + 4H_2O$ | Ehrlich and <br> Newman(2009) |
| Incongruent dissolution of glauconite (with release of arsenic not shown) | $K_2(Fe_{1-x}Mg_x)_2Al_6(Si_4O_{10})_3(OH)_{12(s)} + 3/2 O_{2\,(g)} + 6H^+ \rightarrow K^+$ <br> $+2xMg^{+2}_{\,(aq)} + 6SiO_{2(aq)} + 2(1-X).Fe(OH)_{3(s)} + 3/2 Al_4(Si_4O_{10})$ <br> $(OH)_{8(s)}$ | Chapelle and <br> Knobel <br> (1983) |

**Table 2.** Reactions involved in, or affecting, reduction, oxidation, and (or) precipitation of arsenic in water and sediments.

### 3.4. Sources, sinks, electron donors, and competitive ions

Aqueous sulfide was measured in As-rich groundwater discharging to a New Jersey Coastal Plain, USA, stream (Barringer et al., 2010), but sulfide minerals (pyrite) were rare or not present in cores of streambed sediments. Siderite ($FeCO_3$) was found, however, more than a meter deep in the sediments and, judging from the As content (184 mg/kg), the siderite is an effective sink for As released to the shallow groundwater (Mumford et al, 2012). Siderite apparently forms when bacteria respire organic matter, creating bicarbonate ($HCO_3^-$), in an Fe-rich, reducing environment with circumneutral to alkaline pH (Fredrickson et al., 1998). Siderite is reported for parts of the Southeast Asian alluvial aquifers, and presumably acts as an As sink there as well, because Islam et al., (2005) found both arsenate (As(V)) and arsenite (As(III)) sorbed effectively to the siderite in their experiments. Jönsson and Sherman (2008), however, found that the binding of As(III) with siderite is weak. Clearly, however, siderite plays a role in removing one or both of the prevalent As species from solution. Whether siderite remains a permanent sink for As is not known; it would seem possible that biogeochemical conditions in an aquifer could change such that sorbed As would be released back into solution.

In the microbial release of As from geologic materials to groundwater, the presence of organic matter is seen as a critical factor, as it is an electron donor that fuels microbial activity. Organic acids also may compete with As species, along with oxyanions such as phosphate, molybdate, sulfate, and silicate, for binding sites on solids (Wang and Mulligan, 2006b). Nevertheless, the main role of organic matter in As release appears to be that it provides the necessary substrate to bacterial communities for growth and activity—as part of the process in which it is oxidized by bacteria, the organic matter also may produce quinone-like moities that act as electron shuttles in the resulting redox reactions (Mladenov, et al., 2010).

In the studies of the biogeochemistry of the Southeast Asian aquifers, the source of the organic matter was a matter of controversy. Buried peat was suggested as the source (McArthur et al., 2004), whereas Harvey et al. (2002) indicated that young carbon from the land surface moved to depth by irrigation pumping accounted for the organic matter in the redox reactions. (It may be that both sources are operative in different places.) Rowland et al., (2006, 2007) found naturally occurring hydrocarbons in West Bengal and Cambodian aquifers that could promote the microbial activity involved in arsenic release. Héry et al., (2010) point out that very low organic carbon contents (i.e. ≤1%) in sediments is sufficient to stimulate the microbially mediated reactions that result in metal and arsenic reduction in the aquifers.

Organic matter can come from anthropogenic sources as well as natural sources. Barringer et al. (2010) indicated that the likely source of organic matter in the glauconitic system they studied came from wastewater discharged for many years from farming and other subsequent activities. The issue of whether anthropogenic inputs from agricultural practices have contributed organic matter to shallow groundwater in West Bengal and Bangladesh has received much debate (Farooq et al., 2010; Neumann et al., 2009; Sengupta et al., 2008) and it is not entirely resolved. It has been noted, however, at various contamination sites, that petroleum leaks, organic-rich leachates from landfills, as well as inputs of organic carbon for remediation purposes has led to mobilization of As (Hering et al., 2009). Thus it is clear that inputs of organic carbon from both natural and anthropogenic activities can supply electrons and stimulate the

Arsenic in Groundwater: A Summary of Sources and the Biogeochemical and
Hydrogeologic Factors Affecting Arsenic Occurrence and Mobility

131

microbially mediated processes that lead to As release from geologic materials into groundwater.

# 4. Hydrogeologic factors

## 4.1. Residence time

Several biogeochemical processes that release As from geologic materials have been identified,
as presented in the previous section. Smedley and Kinniburgh (2002) point out that whether
released As remains at problematic levels in groundwater depends not only on whether there
are biogeochemical reactions that retard the transport of As, but also upon the hydrologic and
hydrogeologic properties of the aquifer, such as flow velocity and dispersion. If the kinetics of
As release are slow, and groundwater residence time is short, then As concentrations may not
increase to the point where groundwater would be considered contaminated. Conversely, if
reactions that mobilize As are rapid and residence time is long, then As can accumulate in
groundwater such that concentrations become hazardous—as seen in Bangladesh, for
example. Eventually, if the biogeochemical conditions that lead to release and mobilization of
As continue to be present (within a geologic timeframe), then the source could become
exhausted.

## 4.2. Seasonal changes in recharge

Natural fluctuations can affect the fate and transport of As within groundwater systems.
Seasonal fluctuations in recharge could, during periods of high precipitation, bring dilution
to shallow groundwater, but also transport surficially derived materials to the aquifer. As
mentioned above, the transport of dissolved organic matter from agricultural land in the
Bengal Delta has been suggested to fuel bacterial activity that releases As from the aquifer
materials, and, in that region, seasonal (monsoon) rainfall has an important effect on recharge
rates and the transmission of land-derived substances to depth in the aquifers. On a much
smaller scale, As concentrations in shallow groundwater that discharges to Coastal Plain
streams in New Jersey, USA, varies with season and hydrologic conditions. Increased recharge
during springtime results in more diluted shallow groundwater and low As concentrations,
whereas hot, dry weather results in decreased recharge and higher As concentrations. Where
clay lenses underlie the stream channel shallow groundwater levels above the clay decline
during warm, dry periods, and some stream segments may ultimately lose water to groundwater. Thus, in some stream segments seasonal hydrologic conditions control As-rich groundwater discharges to the stream. In other segments, As-rich groundwater may discharge on a
relatively constant basis with higher concentrations of As being present during warm dry
weather (Barringer et al., 2010).

## 4.3. Effects of pumping

Pumping-induced changes to hydraulic gradients can alter flow paths at regional and local
scales and can lead to introduction of contaminants to otherwise potable water. For As-

contaminated well water in Bangladesh, the contamination is found in the shallower of two aquifers. Although water of the deeper aquifer is generally free of As contamination, the replacement of shallow tube wells with wells in the deep aquifer was thought to have the potential to transport contamination downward to low-As ground water (Ravenscroft et al, 2001; Harvey et al., 2002; van Geen et al., 2003).

Pumping also can result in changes in redox conditions along a flow path. As suggested by Peters and Blum (2003), anoxic water can be drawn upward to oxic zones near the wellhead, resulting in disequilibrium between As species. The converse is also possible—anoxic waters could be introduced by pumping to oxic zones such that Fe(III) in ferric hydroxides is reduced, hydroxides are dissolved, and sorbed As is released. Or, in cases where As is released under anoxic conditions, introduction of oxic water through pumping could slow or terminate the reaction. An example of pumping-induced changes in redox conditions is found at the individual borehole scale in eastern Wisconsin, USA. In a field experiment, longer periods between pumping episodes allowed longer periods of anoxia, which resulted in higher concentrations of As in groundwater (Ayotte et al., 2011). Pumping also could move constituents such as organic carbon from the surface to depth, where the carbon could stimulate microbially mediated redox reactions such as Fe reduction that leads to As release from aquifer materials. Pumping could also move higher pH water into a zone of lower pH water, creating an environment in which As sorbed to aquifer materials can desorb.

Flow rates at the individual borehole scale can also be sufficiently rapid that contact time between groundwater and aquifer material is minimized, and reactions releasing As may be relatively slow. Limited contact time is thought to be the explanation for lower dissolved constituents, including As, in water from wells with high yield compared with those with low yield and higher dissolved constituents in the Hungarian Pannonian Basin ( Varsányi and Kovács, 2006). Similar results are reported for a well in a sandstone aquifer in Wisconsin, USA, where reducing conditions developed with no pumping and As concentrations increased (indicating dissolution of Fe hydroxides and As release), whereas rapid well purging introduced oxic conditions to water near the well bore (Gotkowitz et al., 2004).

## 5. Conclusions

Arsenic contamination of groundwater resources has been identified in many parts of the world. In some cases, as in geothermal fields, the impact of As on drinking-water supplies may not be great, but there are parts of the world where groundwater is a major drinking-water source for millions of people, many of which are in poverty and have limited ability to solve the problem of a contaminated water source. Since the discoveries of widespread groundwater contamination with As, considerable effort has been expended to find suitable, inexpensive methods for removing As. A discussion of those efforts, and the results, is beyond the scope of this chapter; the reader is directed to Feenstra et al. (2007) for an overview of As removal methods.

Geogenic sources of As are numerous—within the various geologic materials, the most common occurrences of As appear to be in sulfides (mainly pyrite) and as sorbed species on Fe hydroxides, although As also appears within some silicate and carbonate rocks. The main processes involved in As release to groundwater are reduction of Fe hydroxides, reduction of As within minerals and as a sorbed species, competitive sorption with other oxyanions, and sulfide oxidation. Increasingly, studies show that these processes can be mediated by microbes.

Although major geologic sources of arsenic include alluvial materials, mineralized sedimentary and metasedimentary rocks and volcanic rocks and related deposits, recent findings indicating that As also is released from glauconitic sediments suggests that not all geologic sources and conditions for the release of As to groundwater have yet been identified. Further, it is apparent from some studies that human activities can increase the rates and amounts of As mobilized and dissolved in groundwater through inputs of organic carbon, and from water withdrawals and other changes to natural hydrologic systems. Thus, while there are human efforts to mitigate the As contamination of drinking-water supplies, there are also human activities that exacerbate the problem. The more fully we understand how, when, and where As is mobilized from geologic materials, or from anthropogenic releases to the environment, the more effectively we can find solutions to this major contamination problem.

## Author details

Julia L. Barringer and Pamela A. Reilly

U.S. Geological Survey, USA

## References

[1] Acharyya, S.K. (2002). Arsenic contamination in groundwater affecting major parts of southern West Bengal and parts of western Chhattisgarh: Source and mobilization process. *Current Science*, 82, 740-744.

[2] Agusa, T., Kunito, T., Fujihara, J., Kubota, R., Minh, T.B., Trang, P.T.K., Iwata, H., Subramanian, A., Viet, P.H. & Tanabe, S. (2006). Contamination by arsenic and other trace elements in tube-well water and its risk assessment to humans in Hanoi, Vietnam. *Environmental Pollution*, 139, 95-106.

[3] Aiuppa, A., Avino, R., Caliro, S., Chiodini, G., D'Alessandro, W., Favara, R., Federico. C., Ginevra, C., Inguaggiato, S., Longo, M., Pegoraino, G. & Valenza, M. (2006). Mineral control of arsenic content in thermal waters from volcanic-hosted hydrothermal systems: Insights from the island of Ischia and Phlegrean Fields (Campanian Volcanic Province, Italy). *Chemical Geology*, 229, 313-330.

[4]  Appleyard, S.J., Angeloni, J., & Watkins, R. (2006). Arsenic-rich groundwater in an urban area experiencing drought and increasing population density, Perth, Australia. *Applied Geochemistry*, 21, 83-97.

[5]  Arnórsson, S. (2003). Arsenic in surface-and in 90°C ground waters in a basaltic area, N-Iceland: processes controlling its mobility. *Applied Geochemistry*, 18, 1297-1312.

[6]  ATSDR (2000). Toxicological profile for arsenic. U.S. Department of Health & Human Services, Public Health Service Agency for Toxic Substances and Disease Registry, 428.

[7]  Ayotte, J.D., Montgomery, D.L., Flangan, S.M., & Robinson, K.W. (2003). Arsenic in groundwater in Eastern New England: occurrence, controls, and human health implications. *Environmental Science and Technology*, 37, 2075-2083.

[8]  Ayotte, J.D., Neilson, M.G., Robinson, G.R., & Moore, R.B. (1999). Relation of arsenic, iron, and manganese in ground water to aquifer type, bedrock lithogeochemistry, and land use in the New England Coastal Basins. *U.S. Geological Survey Water-Resources Investigations Report 99-4162*.

[9]  Ayotte, J.D., Szabo, Z., Focazio, M.J. & Eberts, S.M. (2011). Effects of human-induced alteration of groundwater flow on concentrations of naturally-occurring trace elements at water-supply wells. *Applied Geochemistry*, 26, 747-762.

[10]  Ball, A.L., Rom, W.N. & Glenne, B. (1983). Arsenic distribution in soils surrounding the Utah copper smelter. *American Industrial Hygiene Association Journal*, 44, 341-348.

[11]  Ball, J.W., McCleskey, R.B., Nordstrom, D.K., Holloway, J.M., Verplank, P.L. & Sturtevant, S.A. (2002). Water chemistry data for selected springs, geysers, and streams in Yellowstone National Park, Wyoming, 1999-2000. *U.S. Geological Survey Open-File Report 2002-382*.

[12]  Ball, J.W. Nordstrom, D.K., Jenne, E.A. & Vivit, D.V. (1998). Chemical analyses of hot springs, pools, geysers, and surface waters of Yellowstone National Park, Wyoming and vicinity, 1972-1975. *U.S. Geological Survey Open-File Report 98-182*.

[13]  Barringer, J.L., Barringer, T.H., Lacombe, P.J. & Holmes, C.W. (2001). Arsenic in soils and sediment adjacent to Birch Swamp Brook in the vicinity of Texas road (downstream from the Imperial Oil Company Superfund Site)), Monmouth County, New Jersey. *U.S. Geological Survey Water-Resources Investigations Report 00-4185*.

[14]  Barringer, J.L., Bonin, J.L., Deluca, M.J., Romaga, T., Cenno, K., Alebus, M., Kratzer, T. & Hirst B. (2007). Sources and temporal dynamics of arsenic in a New Jersey watershed, USA. *Science of the Environment*, 379, 56-74.

[15]  Barringer, J.L., Mumford, A, Young, L.Y., Reilly, P.A., Bonin, J.L. & Rosman, R. (2010). Pathways for arsenic from sediments to groundwater to streams: Biogeochemical processes in the Inner Coastal Plain, New Jersey, USA. *Water Research*, 44, 5532-5544.

[16] Barringer, J.L., Reilly, P.A., Eberl, D.D., Blum, A.E., Bonin, J.L., Rosman, R., Hirst, B., Alebus, M., Cenno, K., & Gorska, M. (2011). Arsenic in sediments, groundwater, and streamwater of a glauconitic Coastal Plain terrain, New Jersey, USA—Chemical "fingerprints" for geogenic and anthropogenic sources. *Applied Geochemistry*, 26, 763-776.

[17] Barringer, J.L. Szabo, Z. & Barringer, T.H. (1998). Arsenic and metals in soils in the vicinity of the Imperial Oil Company Superfund site, Marlboro Township, Monmouth County, New Jersey. *U.S. Geological Survey Water-Resources Investigations Report 98-4016.*

[18] Baur, W.H. & Onishi, B-M.H. (1969). Arsenic. In *Handbook of geochemistry*, ed. K.H. Wedepohl, 33-A-1-33-0-5. Berlin, Springer-Verlag.

[19] Beaulieu, B.T. & Savage, K.S. (2005). Arsenate adsorption structures on aluminum oxide and phyllosilicate mineral surfaces in smelter-impacted soils. *Environmental Science and Technology, 39*, 3571-3579.

[20] Bentley, R., & Chasteen,.G. (2002). Microbial methylation of metalloids, arsenic, antimony, and bismuth. *Microbiology and Molecular Biology Reviews, 66*, 250-271,

[21] Berg, M., Tran, H.C., Nguyen, T.C., Pham, H.V., Schertenleib, R. & Giger, W. (2001). Arsenic contamination of groundwater and drinking water in Vietnam: A human health threat. *Environmental Science and Technology, 35*, 2621-2626.

[22] Bhattacharya, P., Mukherjee, A.B., Bundschuh, J., Zevenhoven, R. & Loeppert, R.H. (eds.) (2007). *Trace Metals and other Contaminants in the Environment*, Volume 9. Elsevier BV.

[23] Bottomley, D.J. (1984). Origins of some arseniferous groundwaters in Nova Scotia and New Brunswick, Canada. *Journal of Hydrology*, 69, 223-257.

[24] Bowell, R.J., Morley, N.H. & Din, V.K. (1994). Arsenic speciation in soil porewaters from the Ashanti Mine, Ghana. *Applied Geochemistry*, 9, 15-22.

[25] Boyle, R.W. & Jonasson, I.R. (1973). The geochemistry of arsenic and its use as an indicator element in geochemical prospecting. *Journal of Geochemical Exploration*, 2, 251-296.

[26] Brooks, W.E. (2008). Arsenic. *2007 Minerals Yearbook, U.S. Geological Survey.*

[27] Brown, C.J. & Chute, S.K. (2002). Arsenic in bedrock wells in Connecticut. (Abstract). *Arsenic in New England: A Multidisciplinary Scientific Conference*, National Institute of Environmental Health Sciences, Superfund Basic Research Program, Manchester, New Hampshire, May 29-31, 2002.

[28] Brown, K.L., & Simmons, S.F. (2003). Precious metals in high-temperature geothermal systems in New Zealand. *Geothermics, 32*, 619-625.

[29] Brunt, R., Vasak, L.,& Griffioen, J. (2004). Arsenic in groundwater: Probability of oc-
     currence of excessive concentration on a global scale. *International Groundwater Re-
     sources Assessment Centre: Report nr. SP 2004-1.*

[30] Bundschuh, J., Litter, M.I., Parvez, F., Román-Ross, G., Nicolli, H.B., Jean, J-S., Liu, C-
     W., López, D., Armienta, M.A., Guilherme, L.R.G., Cuevas, A.G., Cornejo, L., Cum-
     bal, L., Toujaguez, R. (2012). One century of arsenic exposure in Latin America: a
     review of history and occurrence from 14 countries. *Science of the Total Environment,*
     429, 2-35.

[31] Burkel, R.S., & Stoll, R.C. (1999). Naturally occurring arsenic in sandstone aquifer
     water supply wells of Northeastern Wisconsin. *Ground Water Monitoring and Remedia-
     tion,* 19, 114-121.

[32] Buschmann, J., & Berg, M. (2009). Impact of sulfate reduction on the scale of arsenic
     contaminations in groundwater of the Mekong, Bengal, and Red River deltas. *Applied
     Geochemistry,* 24, 1278-1286.

[33] Camm, G.. Glass, H.J., Bryce, D.W. & Butcher, A.R. (2004). Characterization of a min-
     ing-related arsenic-contaminated site, Cornwall, UK. *Journal of Geochemical Explora-
     tion,* 82, 1-15.

[34] Campbell, K.M., Malasarn, D. Saltikov, C., Newman, D.K. & Hering, J.G. (2006). Si-
     multaneous microbial reduction of iron (II) and arsenic (V) in suspensions of hy-
     drous ferric oxide. *Environmental Science and Technology,* 40, 5950-5955.

[35] Cancès. B., Juillot, F., Morin, G., Laperche, V., Polya, D., Vaughn, D.J., Hazeman, J.-L,
     Proux, O., Brown, G.E., Jr. & Calas, G. (2008). Changes in arsenic speciation through
     a contaminated soil profile: an XAS based study. *Science of the Total Environment,* 397,
     178-189.

[36] Carrillo-Chávez, A., Drever, J.I. & Martinez, M. (2000). Arsenic content and ground-
     water geochemistry of the San Antonio-El Triunfo, Carrizal and Los Planes aquifers
     in southernmost Baja California, Mexico. *Environmental Geology,* 39, 1295-1303.

[37] Carpenter, R., Peterson, M.L. & Jahnke, R.A. (1978). Sources, sinks, and cycling of ar-
     senic in the Puget Sound Region. In *Estuarine Interactions* ed. M.L. Wiley. New York,
     Academic Press.

[38] Chanpiwat, P., Sthiannopkao, S., Cho, K.H., Kim, K-W, San, V., Suvathong, B. &
     Vongthavady, C. (2011). Contamination by arsenic and other trace elements of tube-
     well waters along the Mekong River in Lao PDR. *Environmental Pollution,* 159,
     567-576.

[39] Chapelle, F.H. & Knobel, L.L. (1983). Aqueous geochemistry and the exchangeable
     cation composition of glauconite in the Aquia aquifer, Maryland. *Ground Water,* 21,
     343-352.

[40] Chen, S-L., Dzeng, S.R., Yang, M-H., Chiu, K-H., Shieh, G-M., Wai, C.M. (1994). Arsenic species in groundwaters of the Blackfoot Disease Area, Taiwan. *Environmental Science and Technol.ogy*, 28, 877-881.

[41] Cravotta, C.A., III. (2008). Dissolved metals and associated constituents in abandoned coal-mine drainages, Pennsylvania, USA: Part I. Constituent quantities and correlations. *Applied Geochemistry*, 23, 166-202.

[42] Cullen, W.R. (2008). Is Arsenic an Aphrodisiac? The Sociochemistry of an Element, Cambridge, UK, RSC Publishing.

[43] Daniele, L. (2004). Distribution of arsenic and other minor trace elements in the groundwater of Ischia Island (southern Italy). *Environmental Geology*, 46, 96-103.

[44] Dangic, A. (2007). Arsenic in surface- and groundwater in central parts of the Balkan Peninsula (SE Europe), Chapter 5,127-156. In *Trace Metals and other Contaminants in the Environment, Volume 9*. Ed. P. Bhattacharya,, A.B. Mukherjee, J. Bundschuh, R. Zevenhoven, & R.H. Loeppert. Elsevier BV.

[45] Davenport, J.R. & Peryea, F.J. (1991). Phosphate fertilizers influence leaching of lead and arsenic in a soil contaminated with lead arsenate. *Water, Air, and Soil Pollution*, 57-58, 101-110.

[46] DePaul, V.T. & Szabo, Z. (2007). Occurrence of radium-224, radium-226, and radium-228 in water from the Vincentown and Wenonah-Mt. Laurel aquifers, the Englishtown aquifer system, and the Hornerstown and Red Bank Sands. *U.S. Geological Survey Scientific Investigations Report 2007-5064*.

[47] Dong, H. Kukkadapu, R.K., Fredrickson, J.K., Zachara, J.M., Kennedy, D.W. & Kostanadrithes, H. (2003). Microbial reduction of structural Fe(III) in illite and goethite. *Environmental Science and Technology*, 37, 1268-1276.

[48] Dooley, J.H. (2001). Baseline concentrations of arsenic, beryllium, and associated elements in glauconite and glauconitic soils in the New Jersey Coastal Plain. *N.J. Geological Survey Investigation Report*. Trenton, NJ, N.J. Department of Environmental Protection..

[49] Drummond, D.D., & Bolton, D.W. (2010). Arsenic in ground water in the Coastal Plain aquifers of Maryland. *Report of Investigations No. 78. Department of Natural Resources, DNR 12-4282010-450*. Resource Assessment Service, Maryland Geological Survey, Baltimore MD.

[50] Durant, J.L., Ivushkina, T., MacLaughlin, K., Lukacs, H., Gawel, J, Senn, D. & Hemond, H.F. (2004). Elevated levels of arsenic in the sediments of an urban pond: sources, distribution and water quality impacts. *Water Research*, 38, 2989-3000.

[51] Ehrlich, H.L. & Newman, D.K. (2009). Geomicrobiology, Fifth ed., Boca Raton, FL, CRC Press

[52] Eiche, E., Neumann, T., Berg, M., Weinman, B., van Geen, A., Norra, S., Berner, Z., Trang, P.T.K., Viet, P.H. & Stüben, D. (2008). Geochemical processes underlying a sharp contrast in groundwater arsenic concentrations in a village on the Red River delta, Vietnam. *Applied Geochemistry*, 23, 3143-3154.

[53] Farooq, S.H., Chandrasekharam, D., Berner, Z., Norra, S. & Stüben, D. (2010). Influence of traditional agricultural practices on mobilization of arsenic from sediments to groundwater in Bengal delta. *Water Research*, 44, 5575-5588.

[54] Feenstra, L., van Erkel, J. & Vasak, L. (2007). Arsenic in groundwater: Overview and evaluation of removal methods. *IGRAC Report nr.SP 2007-2*. Utrecht, International Groundwater Resources Assessment Centre.

[55] Foley, N.K., Ayuso, R.A., Ayotte, J.D., Marvinney, R.G., Reeve, A.S. & Robinson, G.R., Jr. (2002). Mineralogical pathways for arsenic in weathering meta-shales: An analysis of regional and site studies in the northern Appalachians. (Abstract). *Arsenic in New England: A Multidisciplinary Scientific Conference*, National Institute of Environmental Health Sciences, Superfund Basic Research Program, Manchester, New Hampshire, May 29-31, 2002.

[56] Fredrickson, J.K., Zachara, J.M., Kennedy, D.W., Dong. H., Onstott, T.C., Hinman, N.W. & Li, S.M. (1998). Biogenic iron mineralization accompanying the dissimilatory reduction of hydrous ferric oxide by a groundwater bacterium, *Geochimica et Cosmochimica Acta*, 62, 3239-3257.

[57] Goldberg, S., & Glaubig, R.A. (1988). Anion sorption on a calcareous, montmorillonitic soil—Arsenic. *Soil Science Society of America Journal*, 52, 1297-1300.

[58] Gómez, J.J., Lillo, J. & Sahún, B. (2006). Naturally occurring arsenic in groundwater and identification of the geochemical sources in the Duero Cenozoic Basin, Spain. *Environmental Geology*, 50, 1151-1170.

[59] Gotkowitz, M.B., Schreiber, M.E., & Simo, J.A. (2004). Effects of water use on arsenic release to well water in a confined aquifer. *Ground Water*, 42, 568-575.

[60] Gunduz, O. & Simsek, C. (2008). Mechanisms of arsenic contamination of a surficial aquifer in Turkey. *GQ07: Securing Groundwater Quality in Urban and Industrial Environments (Proceedings of the 6th International Groundwater Quality Conference, Freemantle, Australia, 2-7 December 2007, IAHS publ. no. XXX, 2008*.

[61] Gunduz, O., Simsek, C. & Hasozbek, A. (2010). Arsenic pollution in the groundwater of Simav Plain, Turkey: Its impact on water quality and human health. *Water, Air, and Soil Pollution*, 205, 43-62.

[62] Guo, H., Zhang, B., Wang, G. & Shen, Z. (2010). Geochemical controls on arsenic and rare earth elements approximately along a groundwater flow path in the shallow aquifer of the Hetao Basin, Inner Mongolia. *Chemical Geology*, 270, 117-125.

[63] Gurung, J., Ishiga, H. & Khadka, M.S. (2005). Geological and geochemical examination of arsenic contamination of groundwater in the Holocene Terai basin, Nepal. *Environmental Geology*, 49, 98-113.

[64] Gurzau, A.E., and Pop, C. (2012). A new public health issue: Contamination with arsenic of private water sources. *Proceedings of the AERAPA Conference*. Available from aerapa.conference.ubbcluj.ro/2012/Gurzau.htm, accessed 7/6/12.

[65] Haack, S.K., & Rachol, C.M. (2000). Arsenic in groundwater in Washtenaw County, Michigan. *U.S. Geological Survey Fact Sheet FS 134-00*.

[66] Haeri, A., Strelbitskaya, S., Porkhial, S., & Ashayeri, A. (2011). Distribution of arsenic in geothermal waters from Sabalan geothermal field, N-W Iran. *Proceedings 36th Workshop on Geothermal Reservoir Engineering, Standford University, Stanford, California, January 31-February 2, 2011*.

[67] Haque, S., Ji, J. & Johannesson, K.H. (2008). Evaluating mobilization and transport of arsenic in sediments and groundwaters of the Aquia aquifer, Maryland, USA. *Journal of Contaminant Hydrology*, 99, 68-84.

[68] Harvey, C.F., Swartz, C.H., Badruzzaman, A.B.M., Keon-Blute, N., Yu, W., Ali, M.A.l, Jay, J., Beckie, R., Niedan, V., Brabander, D., Oates, R.M., Ashfaque, K.N., Islam, S., Hemond, H.F. & Ahmed, M.F. (2002). Arsenic mobility and groundwater extraction in Bangladesh. *Science*, 298, 1602-1606.

[69] Heinrichs, G. & Udluft, P. (1999). Natural arsenic in Triassic rocks: a source of drinking-water contamination in Bavaria, Germany. *Hydrogeology Journal*, 7, 468-476.

[70] Hemond, H.F. (1995). Movement and distribution of arsenic in the Aberjona watershed. *Environmental Health Perspectives, Supplement 1*, 103, 35-40.

[71] Hering, J.G., O'Day, P., Ford, R.G., He, Y.T., Bilgin, A., Reisinger, H.J. & Burns, D.R. (2009). MNA as a remedy for arsenic mobilized by anthropogenic inputs of organic carbon. *Ground Water Monitoring and Remediation*, 29, 84-92.

[72] Héry, M., van Dongen, B.E., Gill, F., Mondal, D., Baughan, D.J., Pancoast, R.D., Polya, D.A. & Lloyd, J.R. (2010). Arsenic release and attenuation in low organic carbon aquifer sediments from West Bengal. *Geobiology*, 8, 155-168..

[73] Hileman, B. (2007). Arsenic in chicken production. *Chemical Engineering News*, 85, 34-35.

[74] Hoang, T.H., Bang, S., Kim, K-W, Nguyen, M.H. & Dang, D.M. (2010). Arsenic in groundwater and sediment in the Mekong River delta, Vietnam. *Environmental Pollution*, 158, 2648-2658.

[75] Hopf, J., Langenhorst, F., Pollok, K., Merten, D. & Kothe E. (2009). Influence of organisms on biotite dissolution: an experimental approach. *Chemie der Erde*, 69, 45-56.

[76]  Islam, F.S., Gault, A.G., Boothman, C., Polya, D.A., Charnock, J.M., Chatterjee, D. & Lloyd, J.R. (2004). Role of metal-reducing bacteria in arsenic release from Bengal delta sediments. *Nature*, 430, 68-71.

[77]  Islam, F.S., Pederick, R.L., Gault, A.G., Adams, L.K., Polya, D.A., Charnock, J.M. & Lloyd, J.R. (2005). Interactions between the Fe(III)-reducing bacterium *Geobacter sulfurreducens* and arsenate, and capture of the metalloid by biogenic Fe(II). *Applied and Environmental Microbiology*, 71, 8642-8648.

[78]  Jönsson, J. & Sherman, D.M. (2008). Sorption of As(III) and As(V) to siderite, green rust (fougerite) and magnetite: implications for arsenic release in anoxic groundwaters. *Chemical Geology*, 255, 173-181.

[79]  Kahn, B.I., Solo-Gabriele, H.M., Townsend, T.G. & Cai, Y. (2006). Release of arsenic to the environment from CCA-treated wood: 1. Leaching and speciation during service. *Environmental Science and Technology*, 40, 988-993.

[80]  Karczewska, A., Bogda, A. & Kryasiak, A. (2007). Arsenic in soils in areas of former mining and mineral processing in Lower Silesia, southwestern Poland, Chapter 16, 411-440. In *Trace Metals and other Contaminants in the Environment, Volume 9*. Ed. P. Bhattacharya, A.B. Mukherjee, J. Bundschuh, R. Zevenhoven & R.H. Loeppert. Elsevier BV.

[81]  Karydakis, G., Arvanitis, A., Andritsos, N. & Fytikas, M. (2005). Low enthalpy geothermal fields in the Strymon Basin (Northern Greece). *Proceedings World Geothermal Congress 2005*, Antalya, Turkey, 24-29 April 2005.

[82]  Kinniburgh, D.G., Newell, A.J., Davies, J., Smedley, P.L., Milodowski, A.E., Ingram, J.A. & Merrin, P.D. (2006). The arsenic concentrations on groundwater from the Abbey Arms Wood observation borehole, Delamere, Cheshire, UK. 265-284. In *Fluid flow and solute movement in sandstone: the onshore UK Permo-Triassic redbed sequence. Ed. R.D. Barker & J.H. Tellam*. London, Geological Society of London, Special Publication 263.

[83]  Kobayashi, D.S. & Lee, G.F. (1978). Accumulation of arsenic in sediments of lakes treated with sodium arsenite. *Environmental Science and Technology*, 12, 1195-2000.

[84]  Kolker, A., Palmer, C.A., Bragg, L.J. & Bunnell, J.E. (2006). Arsenic in Coal. *U.S. Geological Survey Fact Sheet FS 2005-3152*.

[85]  Kostka, J.E., Haefele, E., Viehweger, R. & Stucki, J.W. (1999). Respiration and dissolution of iron (III)-containing clay minerals by bacteria. *Environmental Science and Technology*, 33, 3127-3133.

[86]  Krüger, T., Holländer, H.M., Boochs, P-W., Billib, M., Stummeyer, J., Harazim, B. (2007). *In situ* remediation of arsenic at a highly contaminated site in Northern Germany. GQ07 Securing groundwater quality in urban and industrial environments,

*Proc. 6th International Groundwater Quality Conference,* Freemantle, Western Australia, 2-7 December 2007.

[87] Lear, G., Song B., Gault, A.G., Polya, D.A. & Lloyd, J.R. (2007). Molecular analysis of arsenate reducing bacteria within Cambodian sediments following amendment with acetate. *Applied and Environmental Microbiology,* 73, 1041-1048.

[88] Lloyd, J.R. & Oremland, R.S. (2006). Microbial transformations of arsenic in the environment: From soda lakes to aquifers. *Elements,* 2, 85-90.

[89] Lopez, D.L., Bundschuh, J., Birkle, P., Armienta, M.A., Cumbal, L., Sracek, O., Cornejo, L. & Ormachea, M. (2012). Arsenic in volcanic geothermal fluids of Latin America. *Science of the Total Environment,* 429, 57-75.

[90] Luu, T.T.G., Sthiannopkao, S. & Kim, K-W. (2009). Arsenic and other trace elements contamination in groundwater and a risk assessment study for the residents of the Kandal Province of Cambodia. *Environment International,* 35, 455-460.

[91] Mailloux, B.J., Aleandrova, E., Keimowitz, A.R., Wovkulich, K.,Freyer, G.A., Herron, M., Stolz, J.F., Kenna, T.C., Pichler, T., Polizzotto, M.L., Dong, H., Bishop, M. & Knappett, P.S.K (2009). Microbial mineral weathering for nutrient acquisition releases arsenic. *Applied and Environmental Microbiology,* 75, 2558-2565.

[92] Mandal, B.K., & Suzuki, K.T., 2002. Arsenic round the world: a Review. *Talanta,* 58, 201-235.

[93] Manning, B.A. & Goldberg, S. (1996). Modeling competitive adsorption of arsenate with phosphate and molybdate on oxide minerals. *Soil Science Society of America Journal,* 60, 121-131.

[94] Martin, A.J. & Pedersen, T.F. (2002). The seasonal and interannual mobility of arsenic in a mine-impacted lake. *Environmental Science and Technology,* 36, 1516-1523.

[95] Matschullat, J. (2000). Arsenic in the geosphere—a review. *Science of the Total Environment,* 249, 297-312.

[96] McArthur, J.M., Banergee, D.M., Hudson-Edwards, K.A., Mishra, R., Purohit, R., Ravenscroft, P., Cronin, A., Howarth, R.J., Chatterjee, A., Talukder, T., Lowry, D., Houghton, S.M & Chadha, D.K. (2004). Natural organic matter in sedimentary basins and its relation to arsenic in anoxic ground water: the example of West Bengal and its worldwide implications. *Applied Geochemistry,* 19, 1255-1293.

[97] McCarthy, K.T., Pichler, T. & Price, R.E. (2005). Geochemistry of Champagne Hot Springs shallow hydrothermal vent field and associated sediments, Dominica, Less Antilles. *Chemical Geology,* 224, 55-68.

[98] McClintock, T.R., Chen, Y., Bundschuh, J., Oliver, J.T., Navoni, J., Olmos, V., Lepori, E.V., Ahsan, H., Parvez, F. (2012). Arsenic exposure in Latin America: Biomarkers, risk assessments and related health effects. *Science of the Total Environment,* 429, 76-91.

[99]   McLean, J.E., DuPont, R.R. & Sorensen, D.L. (2006). Iron and arsenic release from aquifer solids in response to biostimulation. *Journal of Environmental Quality*. 35, 1193-1203.

[100]  Méndez, M., & Armienta, M.A. (2003). Arsenic phase distribution in Zimapan mine tailings, Mexico. *Geofísica Internacional*, 42, 131-140.

[101]  Mladenov, N., Zheng, Y., Miller, M.P., Nemergut, D.R., Legg, T., Simone, B., Hageman, C., Rahman, M.M., Ahmed, K.M. & McKnight, D.M. (2010). Dissolved organic matter sources and consequences for iron and arsenic mobilization in Bangladesh aquifers. *Environmental Science and Technology*, 44, 123-128.

[102]  Moncur, M.C., Ptacek, C.J., Blowes, D.W. & Jambor, J.L. (2005). Release, transport and attenuation of metals from an old tailings impoundment. *Applied Geochemistry*, 20, 639-659.

[103]  Morin, G. & Calas, G. (2006). Arsenic in soils, mine tailings, and former industrial sites. *Elements*, 2, 97-101.

[104]  Mosier, D.L., Beger, V.I. & Singer, D.A. (2009). Volcanogenic massive sulfide deposits of the world: database and grade and tonnage models. *U.S. Geological Survey Open-File Report 2009-1034*. Available from http://pubs.usgs.gov/of/2009/1034.

[105]  Mueller, S., Verplanck, P. & Goldfarb, R. (2001). Ground-water studies in Fairbanks, Alaska—A better understanding of some of the United States' highest natural arsenic concentrations. *U.S. Geological Survey Fact Sheet FS-111-01*.

[106]  Mukherjee, A., Sengupta, M.K., Hossain, M.A., Ahamed, S., Das, B., Nayak, B., Lodh, D., Rahman, M.M. & Chakraborti, D. (2006). Arsenic contamination in groundwater: A global perspective with emphasis on the Asian scenario. *Journal of Health and Population Nutrition*, 24, 142-163.

[107]  Mumford, A.C., Barringer, J.L., Benzel, W.M., Reilly, P.A. & Young, L.Y. (2012). Microbial transformations of arsenic: Mobilization from glauconitic sediments to water. *Water Research*, 46, 2859-2868.

[108]  Nakayama, S.M.M., Ikenaka, Y., Hamada, K., Muzandu, K., Choongo, K., Teraoka, H., Mizuno, N. & Ishizuka, M. (2011). Metal and metalloid contamination in roadside soils and wild rats around a Pb-Zn mine in Kabwe, Zambia. *Environmental Pollution*, 159, 175-181.

[109]  National Research Council (1999). Arsenic in Drinking Water. Washington, D.C. National Academy Press.

[110]  Navarro, A., Font, X., Viladevall, M. (2011). Geochemistry and groundwater contamination in the La Selva geothermal system (Girona, Northeast Spain). *Geothermics*, 40, 275-285.

[111] Neumann, R.B., Ashfaque, K.N., Badruzzaman, A.B.M, Ali, M.A., Shoemaker, J.K & Harvey, C.F. (2009). Anthropogenic influences on groundwater arsenic concentrations on Bangladesh. *Nature Geoscience* 1-7. DOI:10.1038/NGEO685.

[112] Newman, D.K., Beveridge, T.J. & Morel, F.M. (1997). Precipitation of arsenic trisulfide by *Desulfotomaculum auripigmentum*. *Applied and Environmental Microbiology*, 63, 2022-2028.

[113] Ng, J.C., Wang J., Shraim, A. (2003). A global health problem caused by arsenic from natural sources. *Chemosphere*, 52, 1353-1359.

[114] NHMRC (1996). National Health and Medical Research Council. Australian drinking water guidelines.

[115] Nickson, R.T, McArthur, J.M., Ravenscroft, P., Burgess, W.G. & Ahmed, K.M. (2000). Mechanism of arsenic release to groundwater, Bangladesh and West Bengal. *Applied Geochemistry*, 15, 403-413.

[116] Nickson, R.T, McArthur, J.M., Shrestha, B., Kyaw-Myint, T.O., Lowry, D. (2005). Arsenic and other drinking water quality issues, Muzaffargarh District, Pakistan. *Applied Geochemistry*, 20, 55-68.

[117] Nicolli, H.B., Bundschuh, J., Blanco, M, del, C., Tujchneider, O.C., Panarello, H.O., Dapeña, C., Rusnasky, J.E. (2012). Arsenic and associated trace-elements in groundwater from the Chaco-Pampean plain, Argentina: Results from 100 years of research. *Science of the Total Environment*, 429, 36-56.

[118] Nimick, D.A., Moore, J.N., Dalby, C.E. & Savka, M.W. (1998). The fate of arsenic in the Madison and Missouri Rivers, Montana and Wyoming. *Water Resources Research*, 34, 3051-3067.

[119] NJDEP (2009). NJ Department of Environmental Protection Standards for Drinking Water. Ground Water, Soil, and Surface Water—Arsenic (Total). Available from www.state.nj.us/dep/standards/pdf/7440-38-2.pdf.

[120] Noguchi, K. & Nakagawa, R. (1969). Arsenic and arsenic-lead sulfides in sediments from Tamagawa hot springs, Akita Prefecture. *Proceedings of Japan Academy*, 45, 45-50.

[121] Nordstrom, D.K. (2002). Worldwide occurrences of arsenic in ground water. *Science*, 296, 2143-2145.

[122] Nriagu, J.O., Bhattacharya, P., Mukherjee, A.b., Bundschuh, J., Zevenhoven, R., & Loeppert, R.H. (2007). Chapter 1. Arsenic in soil and groundwater: an overview. 3-60. In *Trace Metals and other Contaminants in the Environment, Volume 9*. Ed. P. Bhattacharya, A.B.Mukherjee, J. Bundschuh, R. Zevenhoven & R.H.Loeppert 3-60. Elsevier BV

[123] O'Day, P.A. (2006). Chemistry and mineralogy of arsenic. *Elements*, 2, 77-83.

[124] Onishi, H & Sandell, E.B. (1955). Geochemistry of arsenic. *Geochimica et Cosmochimica Acta*, 7, 1-33.

[125]  Oremland R.,S., & Stolz, J.F. (2005). Arsenic, microbes, and contaminated aquifers. *Trends in Microbiology*, 13, 45-49.

[126]  Palumbo-Roe, B., Klinck, B. & Cave, M. (2007). Arsenic speciation and mobility in mine wastes from a copper-arsenic mine in Devon, UK: a SEM, XAS, sequential chemical extraction study, Chapter 17. 441-471. In *Trace Metals and other Contaminants in the Environment, Volume 9*. Ed. P. Bhattacharya, A.B.Mukherjee, J.Bundschuh,, R. Zevenhoven & R.H. Loeppert. Elsevier BV.

[127]  Pearcy, C.A., Chevis, D.A., Haug, T.J., Jeffries, H.A., Yang, N., Tang, J., Grimm. D.A., & Johannesson, K.H. (2011). Evidence of microbially mediated mobilization from sediments of the Aquia aquifer, Maryland, USA. *Applied Geochemistry*, 26, 575-586.

[128]  Peters, S.C. (2008). Arsenic in groundwaters in the Northern Appalachian Mountain belt: A review of patterns and processes. *Journal of Contaminant Hydrology*, 99, 8-21.

[129]  Peters, S.C. & Burkert, L. (2007). The occurrence and geochemistry of arsenic in groundwaters of the Newark Basin of Pennsylvania. *Applied Geochemistry*, 23, 85-98.

[130]  Peters, S.C. & Blum, J.D. (2003). The source and transport of arsenic in a bedrock aquifer, New Hampshire, USA. *Applied Geochemistry*, 18, 1773-1787.

[131]  Peters, S.C., Blum, J.D., Klaue, B. & Karagas, M.R. (1999). Arsenic occurrence in New Hampshire drinking water. *Environmental Science and Technology*, 33. 1328-1333.

[132]  Piqué, A., Grandia, F., & Canals, A. (2010). Processes releasing arsenic to groundwater in the Caldes de Malavella geothermal area, NE Spain. *Water Research*, 44, 5618-5630.

[133]  Planer-Friedrich, B., London, J., McCleskey, R.B., Nordstrom, D.K., & Wallschläger. (2007). Thioarsenates in geothermal waters of Yellowstone National Park: Determination, preservation, and geochemical importance. *Environmental Science and Technology*, 41, 5245-5251.

[134]  Price, R.E. & Pichler, T. (2006). Abundance and mineralogical association of arsenic in the Suwannee Limestone (Florida): Implications for arsenic release during water-rock interaction. *Chemical Geology*, 228, 44-56.

[135]  Quicksall, A.N., Bostick, B.C. & Sampson, M.L. (2008). Linking organic matter deposition and iron mineral transformations to groundwater arsenic levels in the Mekong delta, Cambodia. *Applied Geochemistry*, 23, 3088-3098.

[136]  Rahman, M.M., Sengupta, M.K., Ahamed, S., Lodh, D., Das, B., Hossain, M.A., Nayak, B., Mukherjee, A., Mukherjee, S.C., Pati, S., Saha, K.C., Palit, S.K., Kaies, I., Barua, K., Asad, K.A. (2005). Murshidabad—One of the nine groundwater arsenic-affected districts of West Bengal, India, Part I: Magnitude of contamination and population at risk. *Clinical Toxicology*, 43, 823-834.

[137] Ravenscroft, P., Brammer, H., & Richards, K. (2009) Arsenic Pollution: A Global Synthesis. Wiley-Blackwell, 588 pp.

[138] Ravenscroft, P., McArthur, J.M. & Hoque, B.A. (2001). Geochemical and palaeohydrological controls on pollution of groundwater by arsenic. 53-78. In. *Arsenic exposure and health effects IV*. Ed. W.R. Chappell, C.O. Abernathy & R. Calderon.. Oxford, Elsevier Science, Ltd.

[139] Rhine, E.D., Onesios, K.M., Serfes, M.E., Reinfelder, J.R. & Young, L.Y. (2008). Arsenic transformation and mobilization from minerals by the arsenite oxidizing strain WAO. *Environmental Science and Technology*, 42, 1423-1429.

[140] Rice, K.C., Conko, K.M. & Hornberger, G.M. (2002). Anthropogenic sources of arsenic and copper to sediments in a suburban lake, northern Virginia. *Environmental Science and Technology*, 36, 4962-4967.

[141] Robinson, B., Clothier, B., Bolan, N.S., Mahimairaja, S., Grevn, M., Moni, C., Marchetti, M., van den Dijssel, C. & Milne, G. (2004). Arsenic in the New Zealand environment. *SuperSoil 2004. 3rd Australian New Zealand Soils Conference*, 509, December 2004, University of Sydney, Australia. Available from www.regional.org.au/au/asssi.

[142] Rosas, I., Belmont, R., Armienta, A., Boaz, A. (1999). Arsenic concentrations in water, soil, milk, and forage in Comarca Lagunera, Mexico. *Water, Air, and Soil Pollution*, 112, 133-149.

[143] Rowland, H., Omoregie, E., Millot, R., Jimenez C., Mertens, J., Baciu, C., Hug, S.J. & Berg, M. (2011). Geochemistry and arsenic mobilization to groundwaters of the Pannonian Basin (Hungary and Romania). *Applied Geochemistry*, 26, 1-17.

[144] Rowland, H.A.L., Pederick, R.L., Polya, D.A., Pancost, R.D., van Dongen, B.E., Gault, A.G., Vaugh, D.J., Bryant, C., Anderson, B., Lloyd, J.R. (2007). The control of organic matter on microbially mediated iron reduction and arsenic release in shallow alluvial aquifers, Cambodia. *Geobiology*, 5, 281-292.

[145] Rowland, H.A.L., Polya, D.A,, Lloyd, J.R. & Pancost, R.D. (2006). Characterization of organic matter in a shallow, reducing, arsenic-rich aquifer, West Bengal. *Organic Chemistry*, 37, 1101-1114.

[146] Ryan, P.C., Kim, J., Wall, A.J., Moen, J.C., Corenthal, L.G., Chow, D.R., Sullivan, C.M., & Bright, K.S. (2011). Ultramafic-derived arsenic in a fractured rock aquifer. *Applied Geochemistry*, 26, 444-457.

[147] Saltikov, C.W., Cifuentes, A., Venkateswaran, K., & Newman, D.K. (2003). The *ars* detoxification system is advantageous but not required for As(V) respiration by the genetically tractable *Shewanella* strain ANA-3. *Applied and Environmental Microbiology*, 69, 2800-2809.

[148] Santini, J.M. and Ward, S.A. (eds.) (2012) Metabolism of Arsenite. CRC Press.

[149] Scanlon, B.R., Nicot, J.P., Reedy, R.C., Kurtzman, D., Mukherjee, A., & Nordstrom, D.K. (2009). Elevated naturally occurring arsenic in a semiarid oxidizing system, Southern High Plains aquifer, Texas, USA. *Applied Geochemistry*, 24, 2061-2071.

[150] Seddique, A.A., Masuda, H., Miamura, M., Shinoda, K., Yamanaka, T., Itai, T., Maruoka, T., Uesugi, K., Ahmed, K.M. & Biswas, D.K. (2008). Arsenic release from biotite into a Holocene groundwater aquifer in Bangladesh. *Applied Geochemistry*, 23, 85-98.

[151] Sengupta, S., McArthur, J.M., Sarkar, A., Leng, M.j., Ravenscroft, P., Howarth, R.J., & Banerjee, D.M. (2008). Do ponds cause arsenic pollution of groundwater in the Bengal Basin? An answer from West Bengal. *Environmental Science and Technology*, 42, 5156-5164.

[152] Senior, L.A. & Sloto, R.A. (2006). Arsenic, boron, and fluoride concentrations in ground water in and near diabase intrusions, Newark Basin, Southeastern Pennsylvania. *U.S. Geological Survey Scientific Investigations Report 2006-5261*.

[153] Serfes, M. (2005). Arsenic occurrence, sources, mobilization, transport and prediction in the major bedrock aquifers of the Newark Basin, Ph.D. Dissertation, Rutgers University, 122 p.

[154] Sharif, M.U., Davis, R.K., Steele, K.F., Kim, B., Hays, P.D., Kresse, T.M. & Fazio, J.A. (2008). Distribution and variability of redox zones controlling spatial variability of arsenic in the Mississippi River Valley alluvial aquifer, southeastern Arkansas. *Journal of Contaminant Hydrology*, 99, 49-67.

[155] Smedley, P.L. (2008). Sources and distribution of arsenic in groundwater and aquifers. Chapter 4, In *Arsenic in Groundwater: a World Problem*, Ed. C.A.J. Appelo. *Proceedings of the IAH Seminar*, Utrecht, Netherlands, Nov. 2006. NERC Open Research Archive. Available from http://nova.nerc.ac.uk, accessed 7/6/12.

[156] Smedley P.L., & Kinniburgh, D.G. (2002). A review of the source, behavior and distribution of arsenic in natural waters. *Applied Geochemistry*, 17, 517-568.

[157] Smedley, P.L., Knudsen, J. & Maiga, D.(2007). Arsenic in groundwater from mineralized Proterozoic basement rocks of Burkina Faso. *Applied Geochemistry*, 22, 1074-1092.

[158] Smedley, P.L., Zhang, M-Y, Zhangm G-Y & Luo, Z-D. (2003). Mobilization of arsenic and other trace elements in fluviolacustrine aquifers of the Huhhot Basin, Inner Mongolia. *Applied Geochemistry*, 18, 1453-1477.

[159] Smith, A.H., Lingas, E.O., & Rahman, M. (2000). Contamination of drinking-water by arsenic in Bangladesh: a public health emergency. *Bulletin of the World Health Organization*, 78, 1093-1103.

[160] Smith, A.H., Lopipero, P.A., Bates, M.N., & Steinmaus, C.M. (2002). Arsenic epidemiology and drinking water standards. *Science*, 296, 2145-2146.

[161] Stauffer, R.E. & Thompson, J.M. (1984). Arsenic and antimony in geothermal waters of Yellowstone National Park, Wyoming, USA. *Geochimica et Cosmochimica Acta*, 48, 2547-2561.

[162] Sthiannopkao, S., Kim, K.W., Sotham, S. & Choup, S. (2008). Arsenic and manganese in tube well waters of Prey Veng and Kandal Provinces, Cambodia. *Applied Geochemistry*, 23, 1086-1093.

[163] Stollenwerk, K.G., Breit, G.N., Welch, A.H., Yount, J.C., Whitney, J.W., Foster, A.L., Uddin, M.N., Majumder, R.K. & Ahmed, N. (2007). Arsenic attenuation by oxidized aquifer sediments in Bangladesh. *Science of the Total Environment*, 379, 133-150.

[164] Tanner, C.C. & Clayton, J.S. (1990). Persistence of arsenic 24 years after sodium arsenite herbicide application to Lake Rotoroa, Hamilton, New Zealand. New Zealand Journal of Marine & Freshwater Research, 24, 173-179.

[165] Thornburg, K. & Sahai, N. (2004). Arsenic occurrence, mobilization, and retardation in sandstone and dolomite formations of the Fox River Valley in Eastern Wisconsin. *Environmental Science and Technology*, 38, 5087-5094.

[166] Tun, T.N. (2003). Arsenic contamination of water sources in rural Myanmar. Proceedings, 29th WEDC International Conference, Abuja Nigeria, 219-221.

[167] Ujevic, M., Duic, Z, Casiot, C., Sipos, L., Santo, V., Dadic, Z. & Halamic, J. (2010). Occurrence and geochemistry of arsenic in the groundwater of eastern Croatia. *Applied Geochemistry*, 25, 1917-1029.

[168] Ure, A. & Berrow, M., (1982). Chapter 3. The elemental constituents of soils. In *Environmental Chemistry* Ed. H.J.M Bowen. 94-203. London, Royal Society of Chemistry.

[169] USEPA (2001). Drinking Water Standard for Arsenic, USEPA Fact Sheet 815-F-00-105.

[170] USEPA (2011). Five-Year Review Report, Vineland Chemical Company Superfund Site, Vineland Township, Cumberland County, New Jersey. U.S. Environmental Protection Agency, Region 2, New York.

[171] van Geen, A., Zheng, Y., Versteeg, R., Stute, M., Horneman, A., Dhar, R., Steckler, M., Gelman, A., Small, C., Ahsan, H., Graziano, J.H., Hussain, I. & Ahmed, K.M. (2003). Spatial variability of arsenic in 6000 tube wells in a 25 km² area of Bangladesh. *Water Resources Research*, 39, (HWC 3) 1-16.

[172] van Siclen, A.P. & Gerry, C.N. (1936). Arsenic. *U.S. Minerals Yearbook*. 495-500.

[173] Varsányi, I. & Kovács, L.Ó. (2006). Arsenic, iron, and organic matter in sediments and groundwater in the Pannonian Basin, Hungary. *Applied Geochemistry*, 21, 949-963.

[174] Walsh, L.M., & Keeney, D.R. (1975). Behavior and phytotoxicity of inorganic arsenicals in soils. Chapter 3, In *Arsenical Pesticides,* Ed. E.A. Woolson. 35-52. Washington, D.C., ACS Symposium series 7, American Chemical Society.

[175]  Wang, S., & Mulligan, C.N. (2006a). Occurrence of arsenic contamination in Canada: Sources, behavior and distribution. *Science of the Total Environment*, 366, 701-721.

[176]  Wang, S., & Mulligan, C.N. (2006b). Effect of natural organic matter on arsenic release from soils and sediments to groundwater. *Environmental Geochemistry and Health*, 28, 197-214.

[177]  Wang, M., Zheng, B., Wang, B, Li, S., Wu, D. & Hu, J. (2006). Arsenic concentrations in Chinese coals. *Science of the Total Environment*, 357, 96-102.

[178]  Webster, J.G., & Nordstrom, D.K. (2003). Geothermal Arsenic; Chapter 4, 101-125. In *Arsenic in Groundwater*, Ed. A.H. Welch & K.G. Stollenwerk. Boston, Kluwer Academic Publishers.

[179]  Welch, A.H., Lico, M.S., Hughes, J.L. (1988). Arsenic in groundwater of the western United States. *Ground Water*, 26, 333-347.

[180]  Welch, A.H. & Stollenwerk, K. G. (eds.) (2003). Arsenic in Groundwater. Boston, Kluwer Academic Publishers.

[181]  Welch, A.H., Westjohn, D.B., Helsel, D.R., Wanty, R.B. (2000). Arsenic in groundwater of the United States: Occurrence and geochemistry. *Ground Water*, 38, 589-604.

[182]  WHO (1993). Guidelines for Drinking Water Quality. Recommendations, 2nd Ed., Geneva, World Health Organization.

[183]  WHO (2001). Arsenic compounds: Environmental health criteria, 224, 2nd ed. Geneva, World Health Organization.

[184]  Winkel, L., Berg, M., Stengel, C. & Rosenberg, T. (2008). Hydrogeological survey assessing arsenic and other groundwater contaminants in the lowlands of Sumatra, Indonesia. *Applied Geochemistry*, 23, 3019-3028.

[185]  Woolson, E.A. (1977). Fate of arsenicals in different environmental substrates. *Environmental Health Perspectives*. 19, 73-81

[186]  Xie, X., Wang, Y., Duan, M. & Liu, H. (2008). Sediment geochemistry and arsenic mobilization in shallow aquifers of the Datong basin, northern China. *Environmental Geochemistry and Health*, DOI 10.1007/s10653-008-9204-7.

[187]  Yudovich, Ya. E. & Ketris, M.P. (2005). Arsenic in coal: a review. *International Journal of Coal Geology*, 61, 141-196.

[188]  Zheng, Y., Stute, M., van Geen, A., Gavrieli, I., Dhar, R., Simpson, H.J., Schlosser, P. & Ahmed, K.M. (2004). Redox control of arsenic mobilization in Bangladesh groundwater. *Applied Geochemistry*, 19, 201-214.

# Chapter 6

# Modeling the Long-Term Fate of Agricultural Nitrate in Groundwater in the San Joaquin Valley, California

Francis H. Chapelle, Bruce G. Campbell,
Mark A. Widdowson and Mathew K. Landon

Additional information is available at the end of the chapter

## 1. Introduction

Nitrate contamination of groundwater systems used for human water supplies is a major environmental problem in many parts of the world. Fertilizers containing a variety of reduced nitrogen compounds are commonly added to soils to increase agricultural yields. But the amount of nitrogen added during fertilization typically exceeds the amount of nitrogen taken up by crops. Oxidation of reduced nitrogen compounds present in residual fertilizers can produce substantial amounts of nitrate which can be transported to the underlying water table. Because nitrate concentrations exceeding 10 mg/L in drinking water can have a variety of deleterious effects for humans, agriculturally derived nitrate contamination of groundwater can be a serious public health issue.

The Central Valley aquifer of California accounts for 13 percent of all the groundwater withdrawals in the United States [1]. The Central Valley, which includes the San Joaquin Valley, is one of the most productive agricultural areas in the world and much of this groundwater is used for crop irrigation. However, rapid urbanization has led to increasing groundwater withdrawals for municipal public water supplies. That, in turn, has led to concern about how contaminants associated with agricultural practices will affect the chemical quality of groundwater in the San Joaquin Valley [2]. Crop fertilization with various forms of nitrogen-containing compounds can greatly increase agricultural yields. However, leaching of nitrate from soils due to irrigation has led to substantial nitrate contamination of shallow groundwater [3]. That shallow nitrate-contaminated groundwater has been moving deeper into the

Central Valley aquifer since the 1960s [3]. Denitrification can be an important process limiting the mobility of nitrate in groundwater systems [4]. However, substantial denitrification requires adequate sources of electron donors in order to drive the process. In many cases, dissolved organic carbon (DOC) and particulate organic carbon (POC) are the primary electron donors driving active denitrification in groundwater. The purpose of this chapter is to use a numerical mass balance modeling approach to quantitatively compare sources of electron donors (DOC, POC) and electron acceptors (dissolved oxygen, nitrate, and ferric iron) in order to assess the potential for denitrification to attenuate nitrate migration in the Central Valley aquifer.

## 1.1. Interactions of dissolved organic carbon, oxygen, and nitrate

There are at least three distinct compartments present in groundwater systems that store natural organic carbon capable of reacting with dissolved oxygen, nitrate, and other electron acceptors. Dissolved organic carbon (DOC) is present at varying concentrations in all groundwater systems [5,6,7], and this dissolved compartment can store substantial amounts of organic carbon. In addition to DOC, many groundwater systems contain particulate organic carbon (POC) in various stages of diagenesis [5,6,8,9,10]. Microbial degradation of POC can be an additional source of DOC to soil interstitial water [10] and groundwater [11]. Finally, silicate, iron oxyhydroxide, and other minerals present in aquifer solids have a capacity to adsorb DOC [12], removing it from the dissolved phase [13,14,9,15]. These adsorption processes are partially reversible, so that desorption of organic carbon from aquifer materials is also a potential source of DOC [16]. A mass balance model of organic carbon dynamics in groundwater systems, therefore, will need to account for each of these carbon-storing compartments and their interactions.

In contrast to the complexity inherent in the multiple sources, sinks, and composition of DOC, atmospheric oxygen carried through the unsaturated zone by infiltrating precipitation is the sole source of dissolved oxygen (DO) to groundwater systems which lack active photosynthesis. In addition, DO's relatively limited solubility in fresh water (10.1 mg/L at 15 °C) provides a convenient upper limit to concentrations of DO that can be delivered to the water table. These characteristics will be useful in constraining a quantitative mass balance between DOC and DO.

The interaction of DO with the three compartments of organic carbon present in groundwater systems determines the transformation or lack of transformation of nitrate. The usual ecological succession of electron acceptor utilization in groundwater systems (oxygen>nitrate>Fe(III)> sulfate> carbon dioxide [17] implies that once concentrations of DO drop below approximately 0.5 mg/L, reduction of nitrate will occur and may coincide with any of the succeeding predominant terminal electron-accepting processes. Constructing a mass balance between the sources and bioavailability of DOC, DO, and nitrate, therefore, is central to assessing the fate and transport of nitrate.

# 2. Methods

## 2.1. Study area

The San Joaquin Valley occupies the southern two-thirds of the Central Valley of California (Figure 1), a large northwest-trending, asymmetric structural trough filled with marine and continental sediments up to 10 km thick [18]. East of the valley the Sierra Nevada rise to an elevation of more than 4,200 m. West of the valley are the Coast Ranges, a series of parallel ridges of moderate elevation. Streams in the northern part of the San Joaquin Valley drain through the San Joaquin River northward to the San Francisco Bay. During predevelopment, groundwater generally moved toward the center of the valley where it discharged to the San Joaquin River. However, extensive development of groundwater for agriculture and public water supply has substantially altered the flow system.

The hydrologic system in the Modesto area (Figure 1) is complex, in part because of the heterogeneous nature of the hydrogeological setting. The primary aquifers in the study area are a complex sequence of overlapping alluvial fan deposits that have been eroded from the Sierra Nevada (Figure 2). These alluvial fan deposits consist of coarse-grained sands and gravels with discontinuous clayey silts and clays [19]. A relevant characteristic of these sediments for the present study is that they contain very low amounts of organic carbon, typically in the range of 0.01 to 0.1 weight percent organic carbon (Figure 2). The low sediment organic carbon content reflects the generally arid climate in the recent geologic past, and the overland transport of alluvial fan deposits prior to final deposition.

The relatively low amounts of available organic carbon have resulted in groundwater, that is predominantly oxic [20] (Figure 1). Most of the individual analyses (Figure 1) of groundwater are either oxic or mixed oxic and anoxic. The specific criteria used to distinguish redox conditions are described in [20]. In general, concentrations of dissolved oxygen tend to decrease and concentrations of dissolved iron tend to increase as groundwater approaches the discharge area near the San Joaquin River.

## 2.2. MODFLOW model of the San Joaquin Valley aquifer

The numerical model in this study is based on a regional model of groundwater conditions in the Central-Eastern San Joaquin Valley [19, 21]. The U.S. Geological Survey (USGS) three-dimensional finite-difference code MODFLOW-2000 [22] was used to simulate groundwater flow and water-level distributions across the study area (Fig. 1). This regional model was constructed using a three-dimensional grid consisting of 153 rows and 137 columns and 16 layers. The 400 m by 400 m cells were uniform in dimension, and model layers varied from 0.5 m to 16 m above layer 8 and from 20 to 74 m below layer 8. The model area extends from the Coast Ranges to the Sierra Nevada foothills, although the area west of the San Joaquin River was not simulated. The external boundaries of the regional model were a no-flow boundary on the northeastern boundary and general head boundaries on the other three sides. Hydraulic conductivities were estimated based on sediment texture documented in drilling logs. Complete details of the model, including input parameters, calibration, flow budget and travel times are described in [21].

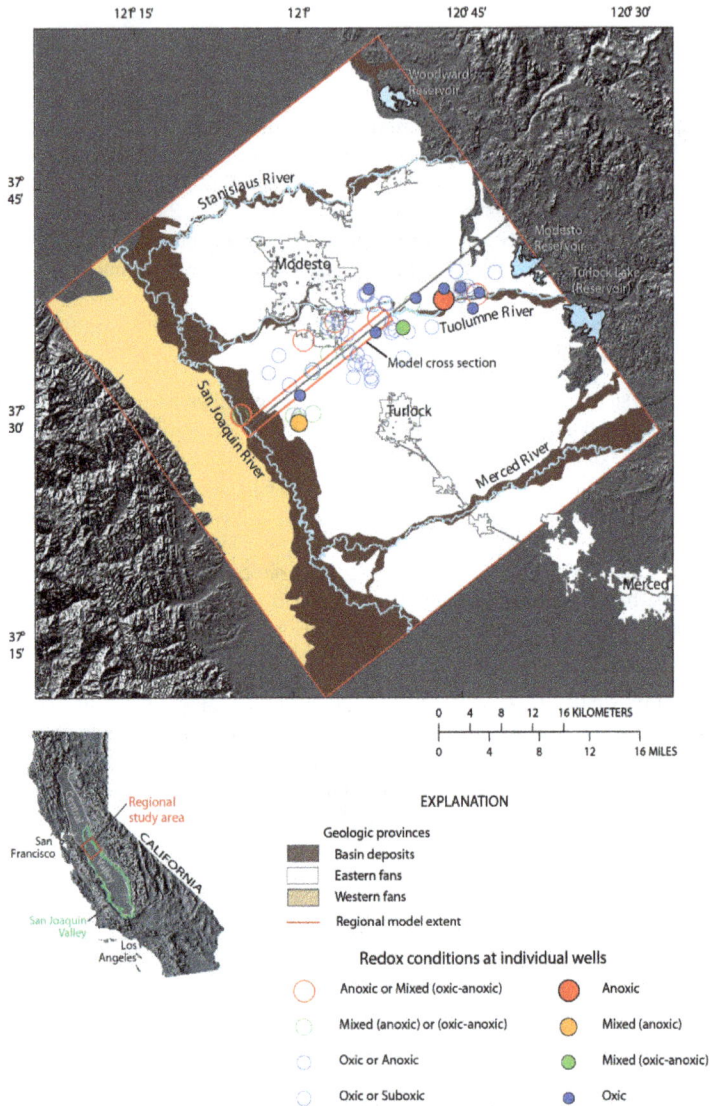

**Figure 1.** Location of study area, orientation of the SEAM 3D model, and redox conditions observed in the Central Valley aquifer from wells located within 5 km of the line of section through the regional model.

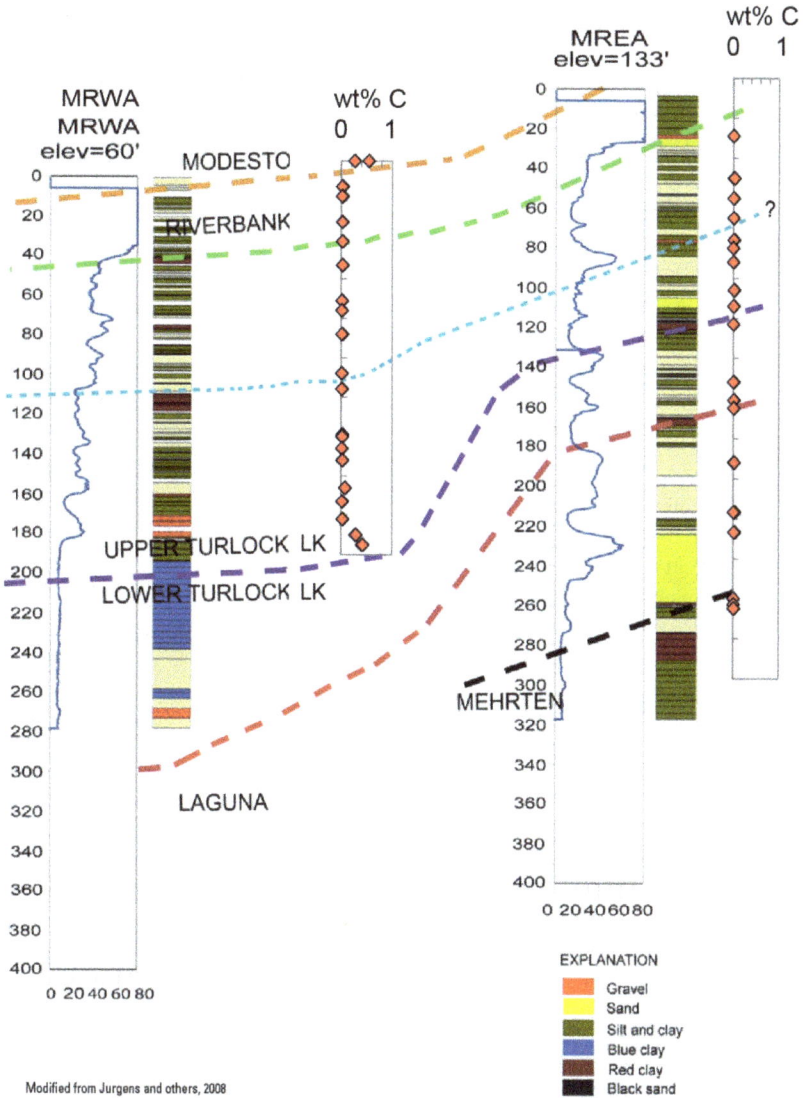

**Figure 2.** Cross section showing the lithology, borehole resistivity logs, and Formations (modified from [19]), and organic carbon content of sediments in the San Joaquin aquifer.

## 2.3. SEAM-3D model of the San Joaquin Valley aquifer

The Sequential Electron Acceptor Model in three dimensions (SEAM-3D) was used to construct a quantitative mass balance between electron donors and acceptors in this study [24]. Because of the computational complexity of this mass balance, it was not feasible to model the entire area of the aquifer (Fig. 1). Rather, a cross-sectional model approximately 25 km in length (Fig. 1) was constructed from the regional model of [21] from the Tuolumne River to the San Joaquin River (Fig. 1). Only the top 11 of the 16 layers of [21] were included in the SEAM cross-sectional simulations; the bottom 5 layers were below a regional confining unit (Corcoran clay) and were likely to contain water that is too old for evaluating changes in redox conditions as a result of anthropogenic processes. Specified heads cells were used at the eastern and western boundaries of the model to approximately reproduce the configuration of the flow system prior to development. The USGS program MODPATH was used to simulate groundwater travel times. MODPATH is a particle tracking post-processing model that computes three dimensional flow paths using output from groundwater flow simulations based on MODFLOW [23]. The simulated head distribution and approximate times-of-travel for the cross-sectional model are shown in Figure 3. In general, the simulated flow system reflects recharge near the Sierra Nevada in the vicinity of the Tuolumne River and discharge at the San Joaquin River. The undulations of the flowpaths shown in Figure 3 reflect lithological heterogeneities that cause variations in the distribution of hydraulic conductivity. The triangles shown on each of the flowpaths delineated in Figure 3 show the lateral distance traveled by groundwater in 10 years. The shallowest flowpaths have travel times on the order of 50 years and the deepest on the order of hundreds of years.

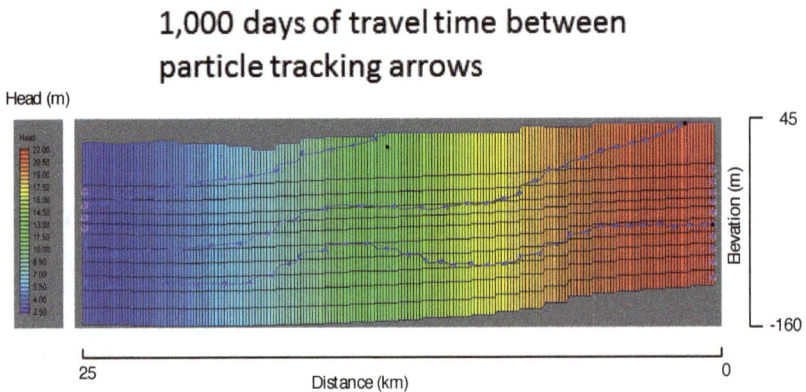

**Figure 3.** Model-derived flowpaths and times of travel in the Central Valley aquifer. See Fig. 1 for the location of the model cross section.

Equations for the mass balance of bioavailable organic carbon and electron acceptors (EAs) assume that DOC serves as the primary electron donor (carbon/energy source) for a hetero-trophic microbial population in the aquifer system. Physical and biogeochemical processes incorporated in equations of transport include advection, dispersion, microbially-mediated biotransformation, rate-limited sorption and desorption. For example, the mass balance equation for bioavailable DOC is given as

$$-\frac{\partial}{\partial x_i}\left(v_i C_{DOC}\right)+\frac{\partial}{\partial x_i}\left(D_{ij}\frac{\partial C_{DOC}}{\partial x_j}\right)+\frac{q_s}{\theta}C_{DOC}^* - R_{\sin k,DOC}^{bio} - \rho_b\frac{\partial \bar{C}_{DOC}}{\partial t} = \frac{\partial C_{DOC}}{\partial t} \tag{1}$$

where $C_{DOC}$ is the concentration of bioavailable DOC in the aqueous phase [M L$^{-3}$]; $\bar{C}_{DOC}$ is the concentration of bioavailable DOC in the solid phase [M M$^{-3}$]; $\rho_b$ is the bulk density of the subsurface material [M L$^{-3}$]; $v_i$ is the average pore water velocity [L T$^{-1}$]; $D_{ij}$ is the tensor for the hydrodynamic dispersion coefficient [L2 T$^{-1}$]; $R_{\sin k,DOC}^{bio}$ is a biodegradation sink term de-pendent on the mode of respiration [M L-3 T$^{-1}$]; $C_{DOC}^*$ is the DOC concentration of the source or sink flux [M L$^{-3}$]; $\theta$ is aquifer porosity [-]; $x_{ij}$ is distance along the respective Cartesian coor-dinate axis [L]; $t$ is time [T]; and $q_s$ is the volume flow rate per unit volume of aquifer repre-senting fluid sources (positive) and sinks (negative) [T$^{-1}$].

Mass balance equations of the aqueous phase EAs (DO and nitrate, respectively) are

$$-\frac{\partial}{\partial x_i}\left(v_i E_{O_2}\right)+\frac{\partial}{\partial x_i}\left(D_{ij}\frac{\partial E_{O_2}}{\partial x_j}\right)+\frac{q_s}{\theta}E_{O_2}^* - R_{\sin k,O_2}^{bio} = \frac{\partial E_{O_2}}{\partial t} \qquad \text{a}$$

$$-\frac{\partial}{\partial x_i}\left(v_i E_{NO_3}\right)+\frac{\partial}{\partial x_i}\left(D_{ij}\frac{\partial E_{NO_3}}{\partial x_j}\right)+\frac{q_s}{\theta}E_{NO_3}^* - R_{\sin k,NO_3}^{bio} = \frac{\partial E_{NO_3}}{\partial t} \qquad \text{b} \tag{2}$$

where $E_{O_2}$ and $E_{NO_3}$ are the aqueous phase concentrations [M L$^{-3}$] of DO and nitrate, respec-tively; $E_{O_2}^*$ and $E_{NO_3}^*$ are the DO and nitrate concentrations [M L$^{-3}$] of source or sink fluxes, respectively; and $R_{\sin k,O_2}^{bio}$ and $R_{\sin k,NO3}^{bio}$ are the EA biodegradation sink terms [M L$^{-3}$ T$^{-1}$], re-spectively. This treatment assumes the effects of sorption on nitrate transport are small. The consumption of the bioavailable Fe(III) concentration [M M$^{-3}$] in the solid phase, $\bar{E}_{Fe}$, is ex-pressed as

$$-R_{\sin k,Fe}^{Bio} = \frac{d\bar{E}_{Fe}}{dt} \tag{3}$$

Biodegradation of DOC is a function of EA availability and is described using modified Monod kinetics [24]. In summary, the overall approach is to write an equation of mass balance for each individual solute and solid-phase constituent considered and then to solve these equations simultaneously in order to compute a true mass balance as a function of time and space.

### 2.4. Parameter estimation

The SEAM-3D model was initially parameterized to reproduce the approximate steady-state distribution of dissolved oxygen prior to large-scale agricultural development. A relatively high (6 mg/L) concentration of dissolved oxygen was assumed to enter the aquifer at the eastern boundary, simulating recharge from the Sierra Nevada foothills (Figure 4). In addition, concentrations of sediment organic carbon were fixed at 0.02 weight percent (wt%) throughout the model domain, concentrations of DOC entering the model with recharge were assumed to be 1 mg/L, the half-saturation constant ($K_s$) and maximum oxygen utilization rate ($V_{max}$) were initially set at 5.0 g m$^{-3}$ and 0.002 d$^{-1}$, respectively. The steady-state distribution of dissolved oxygen given these assumptions is shown in Figure 4. Simulated dissolved oxygen concentrations ranged from 6.0 to about 1.0 mg/L and generally decreased with depth.

The next step in constraining the model was to use parameter estimation (PEST) to further refine the monod kinetic parameters [25]. First, water-quality data estimated from predevelopment observations of dissolved oxygen and nitrate were used to refine the $V_{max}$ values (Table 1). In a second step, water-quality observations measured along the flowpath of the cross sectional model [20] (Figure 1) were used to constrain the $V_{max}$ values (Table 1). In general, the PEST approach tended to lower $V_{max}$ values for the $O_2$-DOC, $NO_3$-DOC, and Fe-DOC reactions. The final $V_{max}$ values as constrained by the flowpath water-quality data are listed in Table 1.

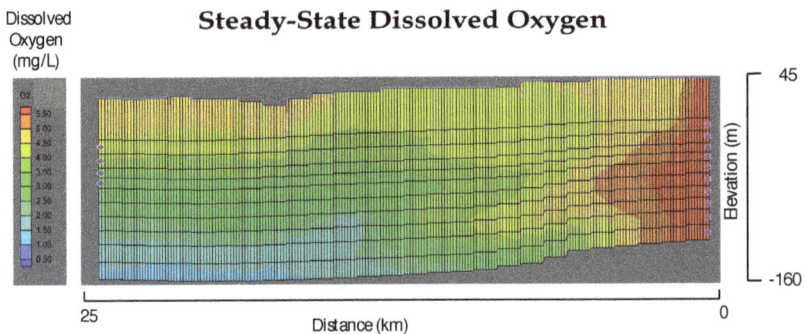

**Figure 4.** Model-derived concentrations of dissolved oxygen in the Central Valley aquifer indicated by the initial parameter estimates. See Figure 1 for the location of the model cross section; Sediment organic carbon = 0.02 wt%

| Parameter | Initial Estimate | Predevelopment observations (PEST) | Flowpath observations (PEST) |
|---|---|---|---|
| $O_2$ half saturation constant (g m$^{-3}$) | 1.0 | 1.85 | 3.0 |
| $NO_3$ half saturation constant (g m$^{-3}$) | 1 | 1 | 1 |
| $O_2$-DOC $V_{max}$ (d$^{-1}$) | 0.0002 | 0.00068 | 0.0032 |
| $NO_3$-DOC $V_{max}$ (d$^{-1}$) | 0.0001 | 0.000029 | 0.00001 |
| Fe-DOC $V_{max}$ (d$^{-1}$) | 0.00002 | 0.00002 | 0.000013 |
| Sediment organic carbon (wt%) | 0.02 | 0.02 | 0.02 |
| Organic carbon dissolution rate (d$^{-1}$) | 0.01 | 0.01 | 0.0041 |
| DOC solubility (mg/L) | 1 | 1 | 1 |

DOC in recharge = 1 mg/L

$O_2$/DOC $V_{max}$ = 0.0002 d$^{-1}$

**Table 1.** Kinetic model parameters used in the SEAM-3D model.

## 2.5. Sediment organic carbon measurements

The sediments used in this study to quantify concentrations of particulate and adsorbed organic carbon were collected by the USGS using a wire-line core barrel driven into the bottom of a borehole drilled with mud rotary coring methods. Cores were collected from two bore holes named MRWA and MREA located west and east of the city of Modesto respectively (Figure 2). The cores were collected on site, logged, and stored in core boxes prior to transporting them to a storage facility. The cores were subsampled for analysis of organic carbon approximately 4 years after collection. In the storage facility, the cores were broken in half and the center of the core subsampled for analysis in order to minimize the effects of residual drilling mud. Concentrations of sediment organic carbon (Fig. 2) were analyzed with an organic carbon analyzer (Costech Analytical Technologies, Inc., Valencia, California) with a detection limit 0.001 w% organic carbon.

## 2.6. Modeling organic carbon dynamics in groundwater systems

The soil science literature has extensively considered the dynamics of DOC formation and transport in soils [10], and this literature is a useful starting point for considering DOC dynamics in groundwater systems. The standard conceptual model of DOC dynamics in soils [10] considers that the total amount of carbon present at any given time and place reflects DOC and POC delivery from surface sources, production of DOC from bioavailable POC, biodegradation of DOC and POC, and the adsorption/desorption of DOC on soil particles.

This conceptual model, in turn, can be used to build a quantitative mass-balance model of organic carbon dynamics in groundwater systems.

DOC mobilized from surface soils is one source of bioavailable DOC to groundwater, but DOC can also be generated from POC present in aquifer material as well [11]. Of all the particulate organic carbon ($POC_{tot}$) present in an aquifer, only a fraction is available to support microbial metabolism:

$$POC_{bio} = fPOC_{tot} \tag{4}$$

where $f$ is the fraction of bioavailable POC. Microbial metabolism of $POC_{bio}$ can generate additional DOC, and this process can be conceived of as a mass-transfer from the particulate to the dissolved phase according to the equation:

$$\rho_b \frac{\partial \overline{C}_{DOC}}{\partial t} = -k_{DOC}\left(C_{DOC}^{eq} - C_{DOC}\right) \tag{5}$$

Where $\rho_b$ is the bulk density of the subsurface material [M L$^{-3}$], $k_{DOC}$ is the rate constant for DOC production [T$^{-1}$]; $C_{DOC}^{eq}$ is the aqueous concentration of DOC in equilibrium with POC at any point in the system [M L$^{-3}$]. The sorption-desorption of DOC onto aquifer materials such as silicate minerals, ferric oxyhydroxides, and POC, can be approximated using a simple linear isotherm:

$$\overline{C}_{DOC} = K_d C_{DOC}^{eq} \tag{6}$$

where $\overline{C}_{DOC}$ is the concentration of DOC adsorbed to aquifer material; and $K_d$ is the distribution coefficient [L$^3$ M$^{-1}$]. Once the fraction of bioavailable $POC_{tot}$ is specified, the numerical model used in this paper uses equations 2-7 to iteratively calculate concentrations of bioavailable DOC and POC in the system as a function of time and space. This bioavailable DOC then drives the sequential utilization of electron acceptors (EAs) such as dissolved oxygen (DO), nitrate, and Fe(III). Note that this approach yields a closed mass balance, as envisioned by the model of [10], for the total amount of organic carbon stored in all three compartments as a function of time. This approach builds on previous numerical methods used to simulate of redox processes in groundwater systems [26].

## 3. Results and discussion

The sediments analyzed for particulate organic carbon in this study (Fig. 2) were predominantly coarse to fine-grained poorly sorted sands with inter-bedded lenses of silt and clay.

Many of the silt and clay sediments showed visible remains of roots indicating that they represented fossilized soils, which developed in between sedimentation events. The modern surface soils contained between 0.2 and 0.5 wt% organic carbon. The alluvial fan sediments, however, showed very low concentrations of organic carbon, typically between 0.01 and 0.05 wt%. Modern fluvial sediments typically contain well in excess of 0.1 wt% organic carbon. Thus, the San Joaquin alluvial fan deposits contain unusually low amounts of organic carbon. That characteristic, in turn, will affect the fate and transport of nitrate.

The simulated transport of nitrate through the Central Valley aquifer, using the flowpath PEST-estimated kinetic parameters (Table 1) is shown in Figure 5. This simulation assumes that recharge coming in from the agricultural areas east of the San Joaquin River contains 20 mg/L of nitrate. This assumption is consistent with ambient conditions that existed approximately in the year 2000 [3]. The starting point for the simulations that follow, therefore, may be thought of as beginning in 2000.

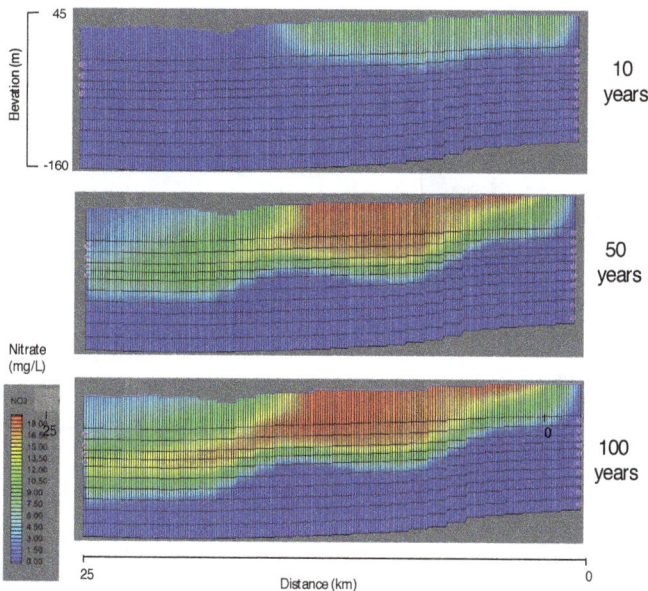

**Figure 5.** Simulated transport of nitrate in the San Joaquin aquifer for 100 years. See Fig. 1 for the location of the model cross section.

After ten years of simulation, nitrate at concentrations of about 10 mg/L have moved into the shallow portions of the aquifer, which is consistent with observed nitrate contamination in this system [3]. After 50 years of simulation, the shallowest portion of the aquifer shows nitrate concentrations ranging from 15 to 20 mg/L, and nitrate has reached the discharge area near the San Joaquin River. After 100 years of simulation, nitrate concentrations in portions of the aquifer near the discharge area have risen above 10 mg/L. The results of these simulations predict that, given the kinetic parameters estimated by PEST, nitrate will penetrate deeper into the aquifer for the foreseeable future. The results also suggest that concentrations of nitrate in the deeper portions of the Central Valley aquifer may increase above the 10 mg/L maximum concentration limit (MCL) established by the U.S. Environmental Protection Agency.

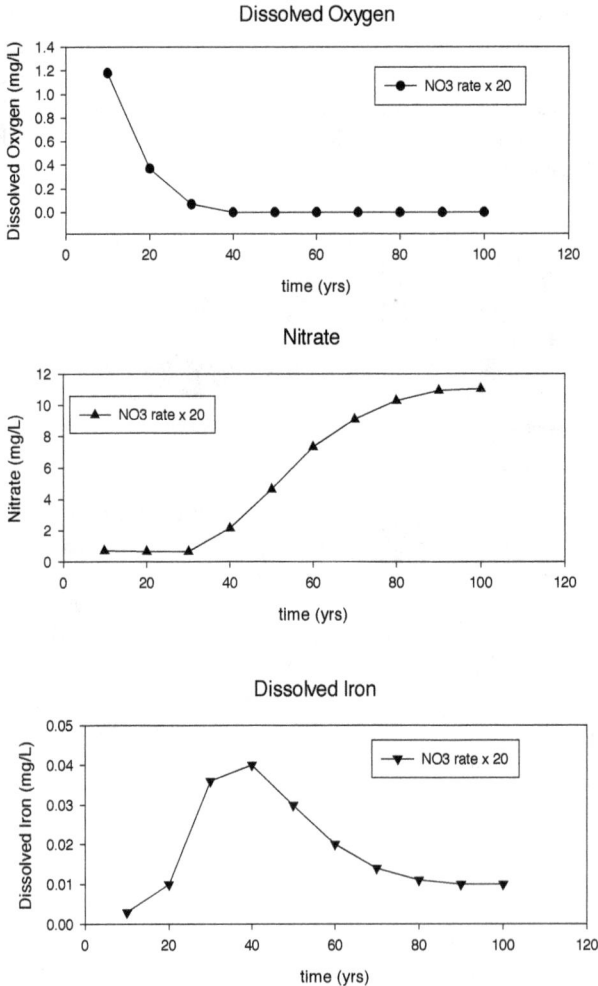

**Figure 6.** Simulated changes in the concentrations of dissolved oxygen, nitrate, and dissolved iron at the San Joaquin River discharge area assuming the nitrate $V_{max}$ is increased by a factor of 20.

## Dissolved Oxygen

## Nitrate

## Dissolved Organic Carbon

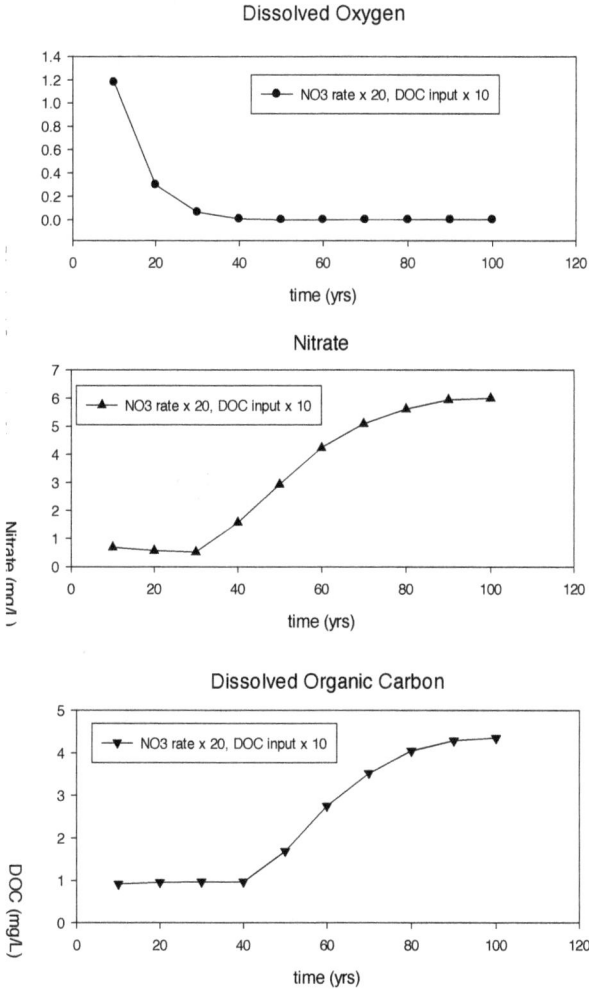

**Figure 7.** Simulated concentration changes of dissolved oxygen, nitrate, and DOC at the San Joaquin River discharge area assuming the nitrate $V_{max}$ is increased by a factor of 20 and DOC concentrations in recharge are increased to 10 mg/L.

Because the kinetic parameters used in the model are subject to uncertainty, the next step in the analysis assessed the sensitivity of the results to changes in those parameters. For this step, the focus was on the cells at the discharge area of the San Joaquin river at the western terminus of the cross-sectional model in layer 6. The first of these simulations assumed that the $V_{max}$ for nitrate reduction coupled to DOC oxidation was increased by a factor of 20. This scenario reflects the possibility that the PEST-derived $V_{max}$ values could be underestimated.

The results are summarized in Figure 6. The decrease in DO concentrations reflects the increase in the the $O_2$-DOC $V_{max}$ from 0.0002 to 0.0032 $d^{-1}$ indicated by the flowpath PEST. As was the case in the simulation shown in Figure 5, nitrate is predicted to begin arriving at the discharge area after about 40 years of simulation. The maximum nitrate concentrations (~10 mg/L) predicted in the simulation of Figure 6, however, were about 20% lower than those shown in the simulation of Figure 5. So, an increase in the $NO_3$-DOC $V_{max}$ by a factor of 20 does increase the attenuation of nitrate. However, nitrate concentrations are still predicted to increase substantially at the discharge area over time.

Note the simulated behavior of nitrate and dissolved iron indicated in Figure 6. As dissolved oxygen concentrations decrease (due to the PEST-indicated increase in $O_2$-DOC $V_{max}$), iron concentrations at the discharge area begin to increase. This is consistent with observed detections of dissolved iron near the discharge area (Figure 1). However, as nitrate encroaches on the discharge area, nitrate metabolism begins to replace iron metabolism and iron concentrations begin to decrease. This reflects the design of SEAM-3D which uses the most efficient available electron acceptor, in this case nitrate, preferentially to Fe(III).

In addition to increasing nitrate concentrations in recharge water, agricultural practices such as fertilization, tilling, and irrigation have the capacity to increase concentrations of DOC as well. The next simulation, therefore, explored the sensitivity of the model to increasing concentrations of DOC in recharge water from 1 to 10 mg/L. The higher $NO_3$-DOC $V_{max}$ used in Figure 6 was also used. The results are shown in Figure 7 and indicate additional attenuation of nitrate, with maximum concentrations decreasing from 10 mg/L (Figure 6) to about 6 mg/L. DOC concentrations at the discharge area also increase from 1 to 4 mg/L. This, in turn, indicates that increasing concentrations of DOC entering the aquifer with recharge may indeed increase nitrate attenuation. However, the model results still indicate substantial nitrate concentrations reaching the discharge area.

# 4. Summary and conclusions

Here, a numerical mass-balance modeling approach was used to simulate the long-term fate and transport of agriculturally-derived nitrate in the aquifer system underlying the San Joaquin Valley in California. The SEAM-3D code (Sequential Electron Acceptor Model in three dimensions) used in this study couples the oxidation of dissolved organic carbon (DOC) to the reduction of dissolved oxygen (DO), nitrate ($NO_3$), and ferric iron (Fe(III)) using monod kinetics and including inhibition functions to force the utilization of electron acceptors in the order DO> $NO_3$>Fe(III). A cross-sectional model 25 kilometers in length was constructed by taking the hydraulic conductivity distribution from a calibrated regional model and providing boundary conditions that approximate the historical steady-state distribution of hydraulic head. The model was initially parameterized by matching model-derived concentrations of DOC and DO to historically measured concentrations. The parameterization was then refined using parameter estimation (PEST) on measured point concentrations of DOC, DO, NO3, and dissolved iron (Fe(II)). The parameterized model was then used as a hypothesis-testing tool to evaluate different possible future scenarios of nitrate transport in the Central Valley aquifer. Model simulations using the PEST-derived model parameters indicate that the amount of dissolved and particulate organic carbon available in the aquifer is inadequate to consume the amount of DO that typically recharges the aquifer, and that $NO_3$ derived from agricultural activities will be drawn deeper into the aquifer in the foreseeable future. Model simulations that increase the assumed rate of nitrate reaction with DOC by a factor of 20 decrease simulated concentrations of nitrate near the discharge area of the aquifer by about 20 percent, but simulated concentrations are still substantial.

Nitrate concentrations have increased substantially in shallow groundwater of the San Joaquin Valley in recent years. This phenomenon has led many public supply well operators to tap progressively deeper groundwater in order to avoid nitrate contamination. That practice, in turn, has led to concern that elevated nitrate concentrations will continue to be drawn deeper into this groundwater system over time. The current study used a mass balance modeling approach to assess the possible transport and attenuation of nitrate in this system. The results indicate that the amount of available organic carbon in this system, either DOC or particulate organic carbon in aquifer solids, is inadequate to fully attenuate the nitrate that is now entering the shallow portion of the aquifer. This finding, in turn, suggests that nitrate contamination in the Central Valley aquifer will continue to move deeper into the system, and may eventually reach the discharge area near the San Joaquin River.

## Acknowledgements

This research was supported by the National Water-Quality Assessment (NAWQA) program of the U.S. Geological Survey. Use of trade names is for identification purposes only and does not constitute endorsement by the U.S. Government.

## Author details

Francis H. Chapelle[1], Bruce G. Campbell[1], Mark A. Widdowson[2] and Mathew K. Landon[1]

*Address all correspondence to: chapelle@usgs.gov

1 U.S. Geological Survey, Columbia, SC, USA

2 Virginia Tech University, Blacksburg, VA, USA

## References

[1]  Maupin, M.A., and N.L. Barber. 2005. Estimated withdrawals prom principal aquifers in the United States, 2000. U.S. Geological Survey Circular 1270. 52 p.

[2]  Belitz, Kenneth, Dubrovsky, N.M., Burow, Karen, Jurgens, Bryant, and Johnson, Tyler, 2003, Framework for a ground-water quality monitoring and assessment program for California: U.S. Geological Survey Water-Resources Investigations Report 03-4166, 78.

[3]  Jurgens, B.C., K.R. Burow, B.A. Dalgish, and J.L. Shelton. 2008. Hydrogeology, water chemistry, and factors affecting the transport of contaminants in the zone of contribution of a public-supply well in Modesto, Eastern San Joaquin Valley, California. U.S. Geological Survey Scientific Investigations Report 2008-5156, 78 pp.

[4]  Korom, S.F., 1992. Natural denitrification in the saturated zone: a review. Water Resources Research, 28(6): 1657-1668.

[5]  Leenheer, J.A. 1974. Occurrence of dissolved organic carbon in selected groundwater samples in the United States. U.S. Geological Survey Journal of Research 2: 361-369.

[6]  Thurman, E.M. 1985. Organic Geochemistry of Natural Waters. Dordrecht/Boston/Lancaster: Martinus Nijhoff/DR W. Junk Publishers, 497 pp.

[7]  Aiken, G.R. 1989. Organic matter in groundwater. U.S. Geological Survey Open File Report 02-89, 7pp.

[8]  McMahon, P.B., D.F. Williams, and J.T. Morris. 1990. Production and carbon isotopic composition of bacterial $CO_2$ in deep Coastal Plain sediments of South Carolina. Ground Water 28(5): 693-702.

[9]  Lilienfein, J., R.G. Qualls, S.M. Uselman, and S.D. Bridgham. 2004. Adsorption of dissolved organic carbon and nitrogen in soils of a weathering chronosequence. Soil Science 68:292-305.

[10]  Kalbiz, K., S. Solinger, J.H. Park, B. Michalzik, B. Matzner. Controls on the dynamics of dissolved organic matter in soils: A Review. Soil Science 165(4):277-304.

[11]  McMahon, P.B. and F.H. Chapelle. 1991. Microbial organic-acid production in aquitard sediments and its role in aquifer geochemistry. Nature 349:233-235.

[12]  Kahle, M., M. Kleber, and R. Jahn. 2003. Retention of dissolved organic matter by phyllosilcate and soil clay fractions in relation to mineral properties. Organic Geochemistry 35:269-276.

[13]  Davis, J.A. 1982. Adsorption of natural dissolved organic matter at the oxide/water interface. Geochimica et Cosmochimica Acta 46:2381-2393.

[14]  Findlay, S. and W.V. Sobczak. 1996. Variability in removal of dissolved organic carbon in hyporheic sediments. Journal of the North American Benthological Society 15 no. 1:35-41.

[15]  Jardine, P.M. M.A. Mayes, P.J. Mulholland, P.J. Hanson, J.R. Tarver, R.J. Luxmoore, J.F. McCarthy, and G.V. Wilson. 2006. Vadose zone flow and transport of dissolved organic carbon at multiple scales in humid regimes. Vadose Zone Journal 5:140-152.

[16] Gu, B., J. Schmitt, Z. Chen, L. Liang, and J.F. McCarthy. 1995. Adsorption and desorption of different organic matter fractions on iron oxide. Geochimica et Cosmochimica Acta 59(2):219-229.

[17] McMahon P.B. and F.H. Chapelle. 2008. Redox processes and the water quality of selected principal aquifer systems. Groundwater 46(2):259-285.

[18] Page, R.W. 1986. Geology of the fresh ground-water basin of the Central Valley, California, with texture maps and sections. U.S. Geological Survey Professional Paper 1401-C, 54 p.

[19] Burow, K.R., Shelton, J.L., Hevesi, J.A., and Weissmann, G.S., 2004, Hydrogeologic Characterization of the Modesto Area, San Joaquin Valley, California: U.S. Geological Survey Scientific Investigations Report 2004-5232, 54 p.

[20] Landon, M.K., Green, C.T., Belitz, K., Singleton, M.J., and Esser, B.K., 2011, Relations of hydrogeologic factors, groundwater reduction-oxidation conditions, and temporal and spatial distributions of nitrate, Central-Eastside San Joaquin Valley, California: Hydrogeology Journal, 19: 1203-1224

[21] Phillips, S.P. C.T. Green, K.R. Burow, J.L. Shelton, and D.L. Rewis. 2007. Simulation of multiscale ground-water flow in part of the northeastern San Joaquin Valley, California.. U.S. Geological Survey Scientific Investigations Report 2007-5009, 43 p.

[22] Harbaugh, A.W. E.R. Banta, M.C. Hill, and M.G. McDonald. 2000. MODFLOW-2000, the U.S. Geological Survey modular ground-water model-User guide to modularization concepts and the ground-water flow process. U.S. Geological Survey Open-File Report 00-92, 121 p.

[23] Pollock, D.W., 1994, User's guide for MODPATH/MODPATH-PLOT, version 3: A particle-tracking post-processing package for MODFLOW, the U.S. Geological Survey finite-difference ground-water flow model: U.S. Geological Survey Open-File Report 94–464, 249 p.

[24] Waddill, D.W. and M.A. Widdowson. 1998. A three-dimensional model for subsurface transport and biodegradation. ASCE Journal of Environmental Engineering, 124(4), 336-344.

[25] Doherty, John, 2005, PEST, model independent parameter estimation users manual, 5th edition: Watermark Numerical Computing, 336 p.

[26] Feinstein, D.T. and M.A. Thomas. 2008. Hypothetical modeling of redox conditions within a complex ground-water flow field in a glacial setting. U.S. Geological Survey Scientific Investigations Report 2008-5066, 28 pp.

# Chapter 7

# Groundwater and Contaminant Hydrology

Zulfiqar Ahmad, Arshad Ashraf, Gulraiz Akhter and
Iftikhar Ahmad

Additional information is available at the end of the chapter

## 1. Introduction

Groundwater and Contaminant Hydrology has a range of research relating to the transport and fate of contaminants in soils and groundwater. The scope of the center includes: 1) the development of new sampling and site characterization techniques; and 2) other improved groundwater remediation techniques.

Contaminant hydrology is the study of processes that affect both ground and surface water pollution. It draws on the principles of hydrology and chemistry. Contaminant hydrology and water quality research seeks to understand the role of soil properties and hydrologic processes on ground and surface water pollution and develop strategies to mitigate their impacts. Research is done at all scales varying from soil pore to basin scale and covers both traditional and emerging contaminants. Groundwater and contaminant hydrology studies include fate and transport of jet fuel leakages from oil depots, producing water injection in shallow wells from the oil and gas exploration field concession areas, veterinary pharmaceuticals from land-applied manure, pathogen losses from manure application, fate and transport of disposal wastes in unlined evaporation ponds from pharmaceutical industries, impacts of tile drainage on sediment and nutrient pollution on Rivers, sediment-turbidity relationships, water quality modeling, and TMDL and paired watershed studies.

Some research institutes address national and international needs for subsurface contaminant characterization and remediation across a spectrum of approaches - laboratory experiments, field tests, and theoretical and numerical groundwater flow and transport investigations. Some of the developing countries most critical subsurface contamination issues, including the chemical evolution of highly alkaline radioactive waste in storage tanks; reduction, re-oxidation, and diffusion of uranium forms in sediments; hydraulic properties of unsaturated gravels; and the natural production of transport-enhancing mobile nanoparticles in the subsurface. Inverse modeling of reactive transport and joint hydrologic and geophysical inversion are investigated to develop new tools and approaches for estimating field-scale reactive transport parameters and characterizing contamination sites.

Contaminants can migrate directly into groundwater from below-ground sources (e.g. storage tanks, pipelines) that lie within the saturated zone. Additionally contaminants can enter the groundwater system from the surface by vertical leakage through the seals around well casings, through wells abandoned without proper procedures, or as a result of contaminant disposal of improperly constructed wells [1].

## 1.1. Governing processes of contaminant transport

Generally three processes can be distinguished which govern the transport of contaminants in groundwater: advection, dispersion and retardation. Dispersion and density/viscosity differences may accelerate contaminant movement, while retardation processes can slow the rate of movement. Some contamination problems involve two or more fluids. Examples include air, water and organic liquids in the unsaturated zone, or organic liquids and water in an aquifer. Tracers are useful for characterizing water flow in the saturated and unsaturated zone.

• Advection

The term advection refers to the movement caused by the flow of groundwater. Groundwater flow or advection is calculated based on Darcy's law. Particle tracking can be used to calculate advective transport paths [2]. Particle tracking is a numerical method by placing a particle into the flow field and numerically integrating the flow path.

• Dispersion

Dispersive spreading within and transverse to the main flow direction causes a gradual dilution of the contaminant plume. The dispersive spreading of a contaminant plume is due to aquifer heterogeneities. Dispersion on the macroscopic scale is caused by variations in hydraulic conductivity and porosity. Solute transport can be influenced by preferential flowpaths, arising from variations of hydraulic conductivity, at a decimetre scale.

• Retardation

Two major mechanisms that retard contaminant movement are sorption and biodegradation. If the sorptive process is rapid compared with the flow velocity, the solute will reach an equilibrium condition with the sorbed phase and the process can be described by an equilibrium sorption isotherm. The linear sorption isotherm can be described by the equation:

$$C^* = K_d C \qquad (1)$$

Where $C^*$= mass of solute sorbed per dry unit weight of solid (mg/kg), $C$= concentration of solute in solution in equilibrium with the mass of solute sorbed onto the solid (mg/l) and $K_d$ = distribution coefficient (L/kg)

- Non aqueous phase liquids (NAPL)

Organic liquids that have densities greater than water are referred to as DNAPL (dense nonaqueous phase liquids). Nonaqueous phase liquids that have densities less than water are called LNAPLs (light nonaqueous phase liquids). Contamination by LNAPL typically involves spills of fuels like gasoline or jet fuel.

### 1.2. Groundwater flow and contaminant transport modeling

The preliminary steps in modeling groundwater flow and contaminant transport include development of a conceptual model, selection of a computer code, and developing model design [3]. Defining a numerical groundwater flow model is based on parameters like (a) sources and sinks of water in the field system; (b) the available data on geohydrologic system; (c) the system geometry i.e. types and extent of model layers; (d) the spatial and temporal structure of the hydraulic properties; and (e) boundary condition. The widely used MOD-FLOW [4] and MT3D solute transport [5] numerical codes use finite differences schemes and are considered very reliable. MODFLOW is a three-dimensional modular finite-difference model of U.S. Geological Survey widely used for the description and prediction of the behavior of groundwater system. The program uses variable grid spacing in x and y directions. Parameter estimation can be approached to find the set of parameter values that provides the best fit of model results to field observations,. At first stage, the model computes drawdown, direction of flow and hydraulic heads on each nodal point using a finite difference grid system. Using the steady-state hydraulic heads calculated by the model as the initial condition, the MT3D model is run to simulate contaminant transport in a groundwater system. Once a model is calibrated, it can be used to make predictions for management or other purposes [6].

Two case studies including i) simulated transport of jet fuel leaking into groundwater, Sindh Pakistan; and ii) deep-seated disposal of hydrocarbon exploration produced water using three-dimensional contaminant transport model, Sindh Pakistan have been discussed to highlight the related issues, implications and concerns.

## 2. Case study — I

### 2.1. Introduction

Groundwater is a major source for domestic and industrial uses in many urban settlements of the world. Effluents from industrial areas as well as accidental spills and leaks from surface and underground storage tanks are the main sources of natural groundwater contamination. When such contamination is detected, it becomes essential to estimate the spatial extent of contamination. Conventionally, determination of the extent of contamination is undertaken by taking many samples within time and budgetary constraints from several points, which in general requires the installation of several observation wells. As a result, the cost of such operations can be very high, especially when measurements with higher resolution are required.

A three-dimensional model of the contaminant transport was developed to predict the fate of jet fuel, which leaked from above surface storage tanks in urban site of Karachi, Pakistan. Since the tanks were situated in a sandy layer, the dissolved product entered the groundwater system and started spreading beyond the site. The modelling process comprised of steady-state simulation of the groundwater system, transient simulation of the groundwater system in the period from January 1986 through December 2015, and calibration of jet fuel that was performed in context of different parameters in groundwater system. The fuel was simulated using a modular three-dimensional finite-difference groundwater model (PMWIN) ModFlow and solute transport model (MT3D) in the 1986-2001 periods under a hypothetical scenario. After a realistic distribution of piezometric heads within the aquifer system, calibration was achieved and matched to known conditions; the solute transport component was therefore coupled to the flow. Jet fuel concentration contour maps show the expanding plume over a given time, which become almost prominent in the preceding years.

Two-dimensional (2-D) solute transport models can be used to predict the effects of transverse dispersion of the contaminant plume (spreading). Additionally, 2-D models are appropriate where the contaminant source may lie within or near the radius of influence of a continuously pumping well. While three-dimensional (3-D) numerical models should only be used if extensive data are available regarding vertical and horizontal heterogeneity, and spatial variability in contaminant concentrations. A localized contaminant transport model for groundwater is developed to gain insight into the dynamics of the leakage of jet fuel from above-ground storage tanks in the metropolitan area of Karachi, Pakistan. Jet fuel consists of refined, kerosene-type hydrocarbons, which are mixtures of benzene, toluene, ethyl benzene, and isomers of xylene [7]. Hypothetical monitoring wells were established to estimate the concentrations of jet fuel over a stipulated time period as a result of continuous seepage from the storage depot. Although, no specific data on the history of seepage were available, in view of the results inferred from an electrical resistivity sounding survey (ERSS) regarding the nature of the subsurface lithologies coupled with the findings of previous investigators from Mott MacDonald Pakistan (MMP) [8], it is envisaged that seepage from the storage tanks occurred for more than a decade. ERSS is used to obtain the subsurface resistivity values that are assigned to different geological material. MMP [8] conducted the study on soil and groundwater assessment of environmental damage due to oil pollution and remedial measures were suggested for depots / installations / airfields of Shell Pakistan, scattered throughout the country. Appendix A provides the composition of jet-fuel (Table 1).

### 2.1.1. Site description

The project site is located between longitudes 67º 07' 20" and 67º 10' 30" and latitudes 24º 52' 20" and 24º 54' 20". The land surface elevation ranges from approximately 15 to 33 meters above mean sea level (masl). In the far south, the Malir River drains into the Arabian Sea (Figure 1). The typical lithology of the site is silty to sandy clay from 0 to 15 ft (4.6 m) bls, gravelly sand from 15 to 43 ft (4.6 to 13 m) bls, and clayey to silty sand from 43 to 80 ft (13 to 24 m) bls. The region is arid with an average annual rainfall of about 200 mm (7.9 in). Out of this, only 10% [9] is considered to recharge the aquifer system (6.34 x 10-9 m/sec).

The vadose zone is contaminated with up to 1300 ppm of total petroleum hydrocarbons (TPH) within the storage site. In the previous study by MMP [8], soil samples were collected from different locations with 3 feet (1 m) below land surface (bls) and analyzed to estimate the concentration of hydrocarbon compound. Soil samples associated with the storage area have indicated higher TPH concentrations. The contaminant plume follows the hydraulic gradient to the southwest.

### 2.1.2. Literature review

Kim and Corapcioglu [7] developed two-dimensional model to describe areal spreading and migration of light nonaqueous-phase liquids (LNAPLs) introduced into the subsurface by spills or leaks from underground storage tanks. The nonaqueous-phase liquids (NAPL) transport model was coupled with two-dimensional contaminant transport models to predict contamination of soil gas and groundwater resulting from a LNAPL migrating on the water table. Simulations were performed using the finite-difference method to study LNAPL migration and groundwater contamination. The model was applied to subsurface contamination by jet-fuel. Results indicated that LNAPL migration was affected mostly by volatilization. Further, the spreading and movement of the dissolved plume was affected by the geology of the area and the free-product plume. Most of the spilled mass remained as a free LNAPL phase 20 years after the spill. The migration of LNAPL for such a long period resulted in the contamination of both groundwater and a large volume of soil.

El-Kadi [10] investigated the US Navy's bulk fuel storage facility at Red Hill located in the island of Oahu. The facility consisted of 20 buried steel tanks with a capacity of about 12.5 million gallons each. The tanks contain jet-fuel and diesel fuel marine. The bottoms of the tanks are situated about 80 feet above the basal water table. The geology of the area is primarily basaltic lava flows. Investigations found evidence of releases from several tanks. Two borings were drilled to identify and monitor potential migration of contamination to the potable water source. A numerical model of the regional hydrogeology at the Red Hill Fuel Storage Facility (RHFSF) was developed to simulate the fate and transport of potential contamination from the jet-fuel tanks and the effect on the saltwater/freshwater transition zone of various pumping scenarios.

Periago et al. [11] investigated infiltration into soil of contaminants present in cattle slurry. Column experiments were performed in order to characterize the release of contaminants at the slurry-soil interface after surface application of slurry with subsequent rainfall or irrigation. The shape of the release curves suggests that the release of substances from slurry can be modeled by a single-parameter release function. They compared prediction of solute transport (a) with input defined by the release function and (b) assuming rectangular-pulse input.

Eric et al. [12] developed a parameter identification (PI) procedure and implemented with the United States Geological Survey's Method of Characteristics (USGS-MOC) model. The test results showed that the proposed algorithm could identify transmissivity and dispersivity accurately under ideal situations. Because of the improved efficiency in model calibration, extended application to field conditions was effective.

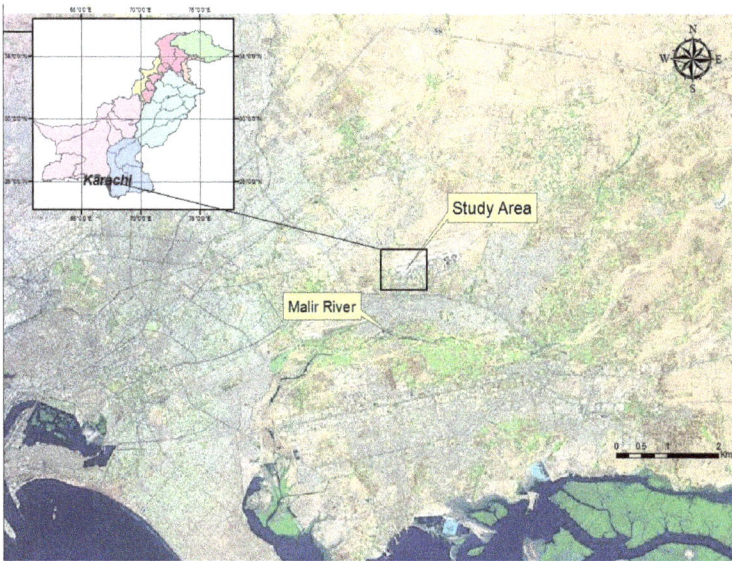

**Figure 1.** Location of the study area in Southern Pakistan

Jin et al. [13] investigated hydrocarbon plumes in groundwater through the installation of extensive monitoring wells. Electromagnetic induction survey was carried out as an alternative technique for mapping petroleum contaminants in the subsurface. The surveys were conducted at a coal mining site near Gillette, Wyoming, using the EM34-XL ground conductivity meter. Data from this survey used to validate with the known concentrations of diesel compounds detected in groundwater. Groundwater data correlated perfectly with the electromagnetic survey data, which was used to generate a site model to identify subsurface diesel plumes. Results from this study indicated that this geophysical technique was an effective tool for assessing subsurface petroleum hydrocarbon sources and plumes at contaminated sites.

## 2.2. Model conceptualization and simulation

The groundwater flow system is treated as a two layer. The upper layer (predominantly silty sand) is bounded above by the water table and is 15 ft (4.6 m) thick, while the lower layer (predominantly clayey sand) is 65 ft (20 m) thick. These are unconfined and recharged from the surface by infiltrating rain, but only over permeable surfaces. A small stream runs along east of model area acts as a drain to groundwater, which flows from the northeast to the southwest. With the exception of the stream, all other boundaries are artificial, that is neither constant head, nor constant flow boundaries. The processes that control the groundwater flow are: (i) recharges from infiltrated rainfall; (ii) flow entering the model across the eastern boundary (also across the northern boundary); (iii) flow reaching the stream; (iv) flow leaving the model across the southern boundary; and (v) pumping from one well near tank no. 9.

Values of hydraulic conductivities (K) for the layers are taken from the literature [14]. The K value for depth range 16-31 ft (4.9-9.4m) is taken as 1.7x10-6 ft/sec (5.2 x10$^{-7}$ m/sec) and for depth range 31-97 ft (9.4-30 m) as 1.5x10$^{-5}$ ft/sec (4.6 x10$^{-6}$ m/sec).

The transport and fate of hydrocarbons depend on multi physical and chemical process-es, including advection, dispersion, volatilization, dissolution, biodegradation, and sorp-tion. When a solute undergoes chemical reactions, its rate of movement may be substantially less than the average rate of groundwater flow. In this study, retardation of the movement of dissolved hydrocarbons is simulated as a sorption process, which in-cludes both adsorption and partitioning into soil organic matter or organic solvents. The MT3D software was used for simulation [5]. It uses a linear isotherm to simulate parti-tioning of a contaminant species between the porous media and the fluid phase due to sorption. This sorption process is approximated by the following equilibrium relationship between the dissolved and adsorbed phases:

$$S = K_d C \tag{2}$$

Where $S$ is the concentration of the adsorbed phase (M/M), $C$ is the concentration of the dissolved phase (M/L3) and $K_d$ is the sorption or distribution coefficient (L$^3$/M). $K_d$ values for organic materials are commonly calculated as the product of the fraction of organic carbon in the soil, *foc*, and the organic carbon partitioning coefficient, $K_{oc}$, or $K_d$ = *foc* Koc. Koc values are contaminant specific and reported in various sources [15-17]. The *foc* in the uncontaminated soil was estimated to range from 0.001 to 0.02 based on guidelines by [18]. Assuming the linear isotherm, the retardation factor ($R$) is expressed as follows:

$$R = 1 + \left( r_b / n_{eff} \right) K_d \tag{3}$$

Where $r_b$ is the bulk density of the porous material (M/L3) and $n_{eff}$ is the effective porosity.

For jet fuel, the distribution coefficient $K_d$ is taken as 0.004415 ft$^3$/kg. With these values [7], $R = 1 + \left[ (48 \text{ kg/ft}^3)/0.25 \right] x\ 0.004415 = 1.848$

### 2.2.1. Numerical ground water flow modeling

Processing ModFlow for Windows (PMWIN5), a modular 3-D finite-difference groundwater model, is used to configure the flow field [4]. The model consists of 41 columns and 39 rows in each layer (Figure 2). The size of cells is 410 ft x410 ft (125 m x 125 m) outside the fuel storage domain and 205 ft x 205 ft (62.5 m x 62.5 m) within the storage domain. Automatic calibration of the water table was made with algorithm - UCODE and a perfect match obtained with the known condition prior to developing the transport model [6]. Using the steady-state hydraulic heads calculated by PMWIN5 as the initial condition, the solute transport model MT3D was run to simulate the dispersion of the dissolved jet-fuel plume [5]. The parameters adjusted were the retardation factor R for each cell within the finite-difference grid, and the dispersion coefficient. Concentration-time curves have been calculated for ten monitoring wells. PMPATH [19] is used to retrieve the groundwater flow model and simulation result from PMWIN5. A semi-analytical particle-tracking scheme is used to calculate the groundwater flow paths, travel times, and time-related capture zones resulting from pumping a neighboring well at the storage facility [20]. As a preprocessor to modeling and creating input data files, the PMWIN5 utility package was used. Prior to initiating the modeling work, a groundwater information system was established with all data in binary and / or ASCII files that could be exported to other softwares.

### 2.2.2. Locations of hypothetical wells

The dissolved phase jet-fuel plume was traced using a combination of ten hypothetical monitoring wells (Figure 2) known as MW-1 through MW-10. The wells served to identify lithology, observe water levels, and monitor concentrations of organic compounds. The wells extend to a depth of 80 ft (24 m). In addition, actual well was completed to a depth of 100 ft (30.5 m) near storage tank no.9 in case of emergency need. In the modeling study, this well was used to track the time-related capture zone. The general layout of the storage tanks over the finite-difference grid is shown in Figure 3. The location of the pumping well is marked as a small red square in Figure 2 and Figure 3.

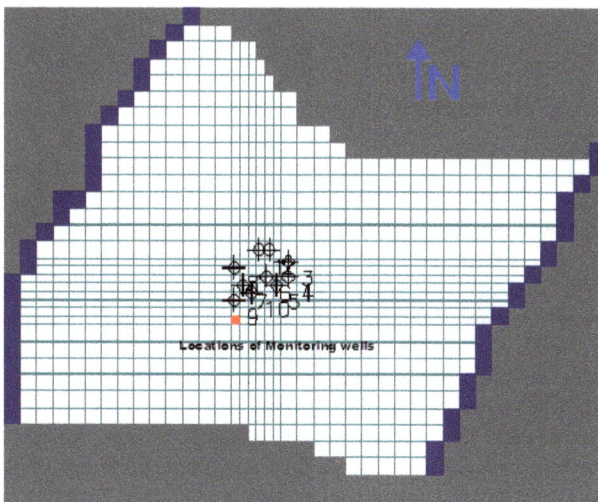

**Figure 2.** Model design indicating finite difference grid and locations of hypothetical monitoring wells

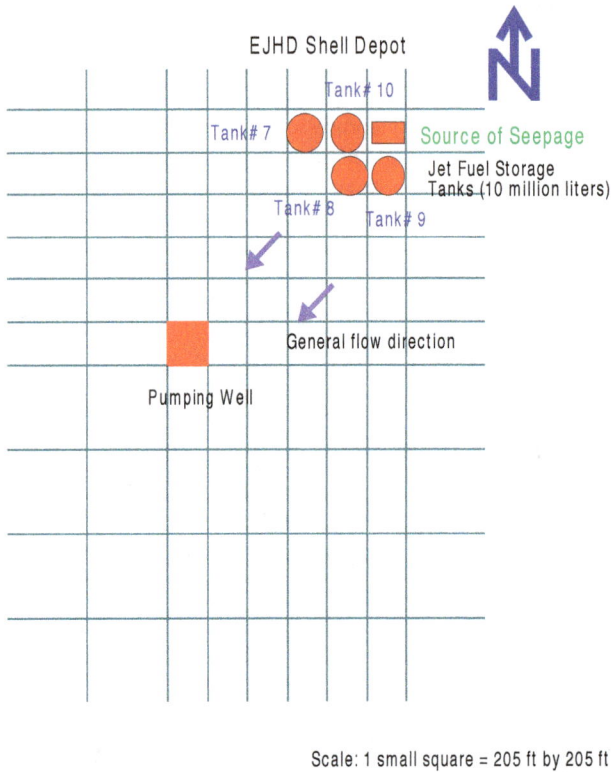

Scale: 1 small square = 205 ft by 205 ft

**Figure 3.** Layout of the storage tanks marked on a finite-difference grid

## 2.2.3. Model calibration

The model was calibrated for steady-state conditions. Since it was speculated that the seepage of the fuel might have started as early as 1986, simulation of the groundwater flow was begun in January 1971, the opening of storage facility. In the steady-state phase the only input comes from constant head boundaries (along the east and west of the model) and from infiltrated rainfall. All output goes into constant head boundaries (the stream and the southwest boundary of the model). The differences in the known and simulated heads were calibrated to less than 0.30 ft (0.091 m) by making slight adjustments in the $K$ values of both the layers. Transient calibration of groundwater flow was accomplished using the time-variant hydraulic head values. Parameters such as recharge rates during each stress period, hydraulic heads in the stream and along the model boundaries, aquifer storage properties, pumping rates, and time-dependent capture zone were adjusted during the calibration. To be objective and consistent, the recharge from infiltration was made equal to 10% of rainfall in each month. Effective porosity of the aquifer was varied between 15% and 25% until the value of 25% was determined to be the best predictor for the model. Constant pumping rates of 1500 US gallons/hr (1.58 x $10^{-3}$ m³/sec) and 1000 US gallons/hr (1.05 x $10^{-3}$ m³/sec) were used in layer 1 and layer 2 respectively. The period from 1986 through 2000 was divided into seven stress periods, each of 2 years in duration. From 2000 to 2001, one stress period was assigned. The length of a stress period was made equal to the number of days in that month.

## 2.3. Results and discussion

The effect of the pumping well is clearly visible as a cone of depression (Figure 4). The drawdown was determined to be 14.0 ft (4.3 m) near the storage facility.

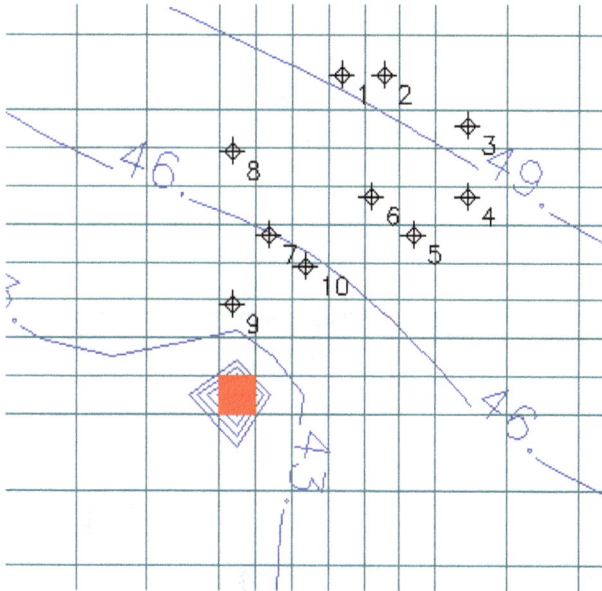

**Figure 4.** Cone of Depression visible around pumping well developed in layer 1

The model assumes uniform recharge from infiltrated rainfall to every "recharging" cell. Although effective porosity, hydraulic conductivity, and recharge may vary in space and time, the model is expected to have produced a reasonable configuration of the groundwater flow pattern throughout the whole period of simulation. The time-related capture zones produced due to constant pumping are shown in Figure 5 and Figure 6. Water balances, which was calculated for each year of the simulated period, showed a perfect match between the input and output components.

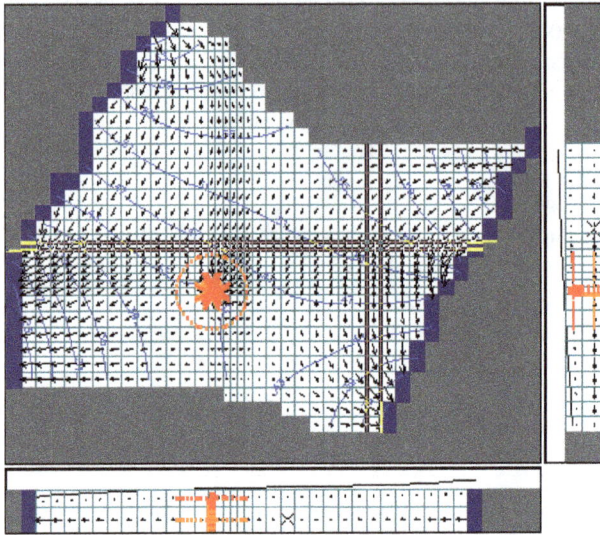

**Figure 5.** Capture zone of the pumping well with arrows indicating flow directions

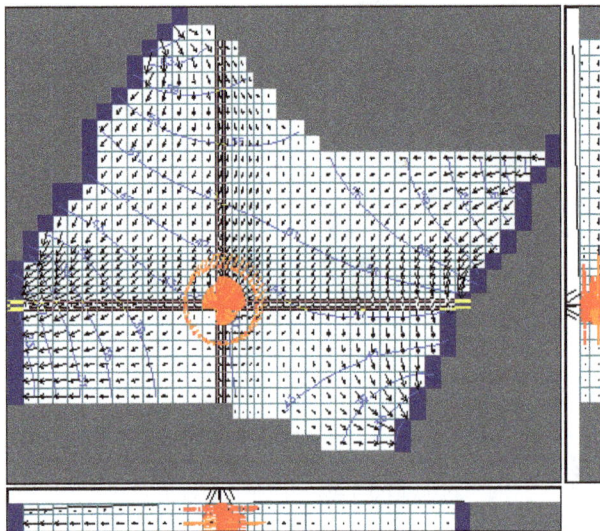

**Figure 6.** Days-capture zone calculated by PMPATH

## 2.3.1. Calibration of plume dispersion

For simulation of the movement of dissolved jet fuel, lateral hydraulic conductivity values equal to 0.864 ft/day (0.263 m/day) for layer 1 and 1.29 ft/day (0.393 m/day) for layer 2 were accepted, while the vertical hydraulic conductivity were taken as 0.0864 ft/day (0.0263 m/day) and 0.129 ft/day (0.0393 m/day) for each layer, respectively [14]. The hydraulic gradient and flow-net were obtained by running the flow component of the model derived from water level information in the previous study.

The United States Environmental Protection Agency (USEPA) and the Georgia Environmental Protection Division (GAEPD) recommend that the value for longitudinal dispersion should be one-tenth of the distance from the place where a contaminant enters the groundwater system to the down-gradient receptor (a well, stream, or other point of compliance). The distance from the storage facility (tank no. 7) to the pumping well is approximately 100 ft (30.48 m). In all calibration runs, as recommended, the value for longitudinal dispersion was set at 10 ft (3 m). USEPA and GAEPD also recommend for a solute transport model that the value for transverse dispersion equal one-third for the longitudinal dispersion. For this model, transverse dispersion would equal 3.3 ft (1.0 m). In the simulation of the fate of jet fuel, the transverse dispersion coefficient was varied within a range of 2.0 ft to 3.3 ft (0.61 m to 1.0 m). In the model a value of 0.001 ft$^2$/day (1 x 10$^{-5}$ cm$^2$/sec) was used for molecular diffusion. With a retardation factor of 1.80, dissolved jet fuel takes 1.33 years to travel a lateral distance of about 70 to 80 ft (21 to 24 m) in groundwater beneath tank no. 8. The best value of the microbial decay coefficient for jet fuel is estimated to be 1-10 / day with a microbial yield coefficient for oxygen of 0.52 [14].

## 2.3.2. Strategy development for release of Jet fuel

The previous integrity test run on the storage tanks containing 10 million liters of jet fuel indicated no loss. The date when the leak initially began is unknown, although inventory records indicated that the leak was not present before tank integrity testing. The product has been detected in several hypothetical-monitoring wells (notably in MW-2 and MW-4) and in many soil samples taken within several tens of feet of the tank. The initial concentration of jet fuel entering the system is not of prime concern for the modeling. The product of the influx (in $L^3/T$) and the concentration (in $M/L^3$) gives the total mass of jet fuel entering the system in a certain time interval. For the purpose of calibrating the jet fuel input, the initial concentration used, based upon field data [8], varied from 0.095 to 0.19 g/ft$^3$ (0.0027 to 0.0054 g/m$^3$). The initial mass of jet fuel, as simulated by the model, was equal to each of four cells "injecting" at a mass rate of 95 to 190 g/ft$^3$ (2.7 to 5.4 g/m$^3$) following the initial period of 15 years during which no groundwater contamination was assumed (Table 1).

| Phase | Stress period | Condition |
|---|---|---|
| Safe Period | 15 years (1971 to 1986) | No leakage |
| Hazardous Period | 10 years (1986 to 1996) | Low to moderate leakage |
| Risk Assessment | 5 years (1996-2001) | Moderate leakage |
| Future Prediction | 14 years (2001 – 2015) | Accretion in leakage |

**Table 1.** Strategy developed for the plume modeling scenarios

Using steady-state hydraulic heads as initial conditions, the evolution of the plume was modeled over nine stress periods as a result of continuous seepage from cells (18,16,1; tank7), (19,16,1; tank 10), (19,17,1; tank 8), and (20,17,1; tank9) as shown in Table 2.

| Stress Period | Time interval (years) | Elapsed Time (sec) | Period |
|---|---|---|---|
| 1 | 2 | 6.30 x 10⁷ | 1986 – 88 |
| 2 | 2 | 12.60 x 10⁷ | 1988 – 90 |
| 3 | 2 | 18.92 x 10⁷ | 1990 – 92 |
| 4 | 2 | 25.23 x 10⁷ | 1990 – 94 |
| 5 | 2 | 3.15 x 10⁸ | 1994 – 96 |
| 6 | 2 | 3.78 x 10⁸ | 1996 - 98 |
| 7 | 2 | 4.41 x 10⁸ | 1998 –00 |
| 8 | 1 | 4.73 x 10⁸ | 2000 – 01 |
| 9 | 14 | 9.14 x 10⁸ | 2001 – 15 |

**Table 2.** Stress period used in time-dependent solute transport modeling of jet fuel

### 2.3.3. Calibration scenario

Parameters describing various processes are used after calibration with different combination of parameters (Table 3).

| Parameters | Value |
|---|---|
| Longitudinal Dispersion | 10 ft |
| Transverse Dispersion | 3.3 ft |
| Molecular Diffusion | 0.001 ft²/day |
| Distribution Coefficient | 0.004415 ft³/kg |
| Retardation Factor ($R$) | 1.80 to 1.848 |
| Decay Coefficient | 1 x10⁻⁹ day⁻¹ |
| Hydraulic Conductivity $K$ (Layer-1) | 1x10⁻⁵ ft/sec (0.864 ft/day) |
| Hydraulic Conductivity $K$ (Layer-2) | 1.49x10⁻⁵ ft/sec (1.29 ft/day) |
| Effective Porosity (Layer-1) | 0.25 |
| Effective Porosity (Layer-2) | 0.30 |

**Table 3.** Preliminary and final values of parameters used in modeling

The release of the jet fuel is simulated in four cells, all along columns 18 to 20 from row 16 to row 17. The area of injection is equal to 42025 ft$^2$ (3,904 m$^2$). The concentration of jet fuel at the source (95 to 190 mg/ft$^3$ [2.7 to 5.4 g/m$^3$]) maintained constant throughout the designated "hazardous period" simulation period (1986-1996). The concentration was increased slightly (about 0.01 %) from 1996 through 2011 and further up to longer time duration of 4 years i.e., up to 2015. The plume simulations are shown in Figure 7.

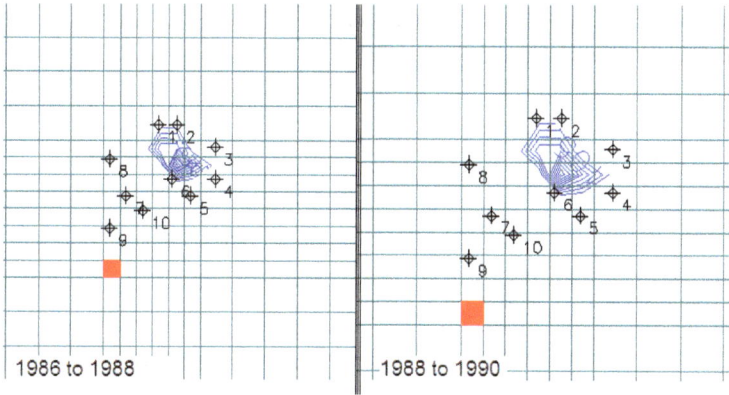

**Figure 7.** Simulated Jet-fuel plumes(1986 to1988 and 1988 to 1990)

The jet-fuel break-through curves for the hypothetical monitoring wells are shown in Figure 8. Conventionally, determination of the extent and level of contamination is undertaken by taking multiple measurements in wells [21-23]. However, higher spatial resolution generally requires installation of monitoring wells, which is costly [24].

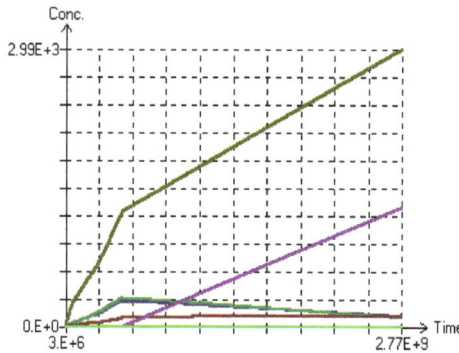

**Figure 8.** Concentration versus time based on data from 10 monitoring wells

The modeled concentration at MW-1, MW-2, and MW-6 was much higher than the concentrations in the remaining wells. The maximum level of concentrations recorded in MW-6 was 2990 μg/ft³. Figure 9 reflects the plume spreading of year 2001.

The shape of the plume is elliptical, with the major axis in the direction of groundwater flow. This shape results from advection and longitudinal dispersion. The lateral spread of the plume results from transverse dispersion and molecular diffusion. Upgradient spread of the plume results from molecular diffusion [25-26]. The plume travels toward the stream, which is still far away in the west. By the end of 2015 the effect of the plume becomes evident and monitoring wells MW-4, MW-5, and MW-6 indicated increased concentration of jet fuel (Figure 10). The resultant plume appears to be spreading more in the elliptical path but in the direction of groundwater flow.

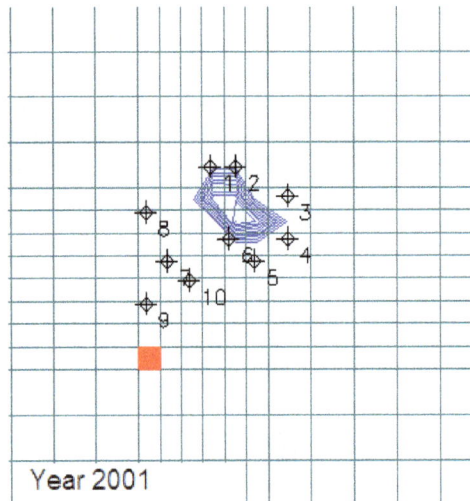

**Figure 9.** Extent of simulated plume in 2001

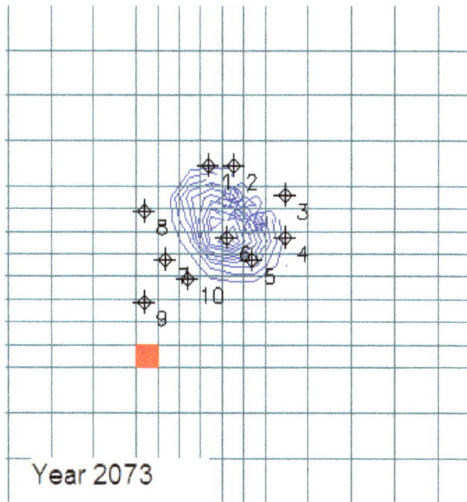

**Figure 10.** Extent of plume spreading in year 2015

## 2.4. Conclusions and recommendation

Based on the modeling, it is concluded that the jet-fuel plume has neither expanded nor moved considerably. It is less than 250 ft (76.2 m) beyond the storage tanks and is oriented northeast to southwest. The level of concentration found in the simulated monitoring wells is significant, but because groundwater is brackish and thus unlikely to be used, no harmful effects are expected. However, with the continuous process of leaking jet-fuel from the storage depot, the level of concentration is expected to increase over the period of time. Regionally, the jet-fuel expansion will have on prominent effect over a longer areal coverage and will be confined to a localized area.

An interdisciplinary investigation of the processes controlling the fate and transport hydro-carbons in the subsurface is needed. Concentrations should be performed in wells within and down-gradient of the plume, as field data would help develop a stronger argument for the fate of jet fuel in groundwater. Periodic observations need to be carried out in wells to have good control on the changes of groundwater chemistry. Defective storage depots need to be mended to stop the release of jet-fuel in future.

## 2.5. Appendix A

Composition of contaminant (Jet fuel)

Knowledge of the geochemistry of a contaminated aquifer is important to understand the chemical and biological processes controlling the migration of hydrocarbon contaminants in the subsurface. Originally, the jet fuel (kerosene oil) is the name assigned to a material with a biological origin, but now it is used to describe materials most of which contain carbon and hydrogen and which may contain oxygen, nitrogen, the halogens, and lesser amounts of other elements. The simplest of these are the hydrocarbons, molecules of hydrogen and carbon, many of which are the components of natural gas, petroleum, and coal. Petroleum, however, has a very large number of components ranging from methane to the high molecular weight materials asphalt and paraffin. Typical fractions into which crude oil is separated in an oil refinery and some principal molecular species are shown in Table 4.

| Fraction from distillation | Boiling range | Product of secondary treatment | Typical molecular components |
|---|---|---|---|
| Gas | Below 20° C | Gas<br>Liquefied Pet. Gas (LPG) | $CH_4$ methane, $C_2H_6$ ethane<br>$C_3H_8$ propane,<br>$C_4H_{10}$ butane |
| Naphtha | 20° – 175° C | Naphtha gasoline | $C_{11}H_{24}$ $C_{18}H_{38}$ |
| Kerosene | 175° – 400° C | Kerosene diesel fuel | $C_{11}H_{24}$ $C_{18}H_{38}$ |
| | | Lubricating oil | $C_{15}H_{32}$ $C_{40}H_{82}$ |
| Residue | above 400° C | Asphalt | Heavier hydrocarbons |

**Table 4.** The Fractions and Representative Components obtained from Crude Oil

# 3. Case study — II

## 3.1. Introduction

It is important to maintain the existing quality of groundwater because once contamination occurs; it is sometime difficult or rather impossible to clean the aquifer. There is high probability associated with certain landuses like agriculture, industrial/urban land and drainage wells for contaminating the groundwater. The early detection of such contamination can be executed through proper monitoring of the groundwater quality. Many oil & gas companies are disposing off their producing water into the deep Ranikot formation in Bhit oil-field area of southern Pakistan. The producing water contains Total dissolved solid (TDS) within range of 18,000 - 22,000 mg/l besides oil condensate. There were concerns that the producing water is affecting the fresh water aquifers belonging to the overlying Nari and Kirthar formations. This phenomenon has been studied by the utilization of groundwater contaminant transport model. Injection has been monitored at 2100 meters depth in the Pab sandstone formation. A three-dimensional contaminant transport model was developed to simulate and monitor the migration of disposal of hydrocarbon exploration produced water in Injection well at 2000 meters depth in the Upper Cretaceous Pab sandstone in the study area. Framework of regional stratigraphic and structural geology, landform characteristics, climate and hydrogeological setup were used to model the subsurface aquifer. The shallow and deep-seated characteristics of geological formations were obtained from electrical resistivity sounding surveys, geophysical well-logging information and available drilling data. The modeling process comprised of steady-state and transient simulations of the prolific groundwater system and, predictive simulation of contaminants transport after 1-, 10- and 30-year of injection. The contaminant transport was evaluated from the bottom of the injection well and its short and, long-term effects were determined on aquifer system lying in varying hydrogeological and geological conditions.

### 3.1.1. Description of study area

The study area of Bhit oil field is located about 43 km south of the Manchhar Lake within longitudes 67° 25' - 67° 48' E and latitudes 26° 01' - 26° 30' N in Dadu district in Sindh province of Pakistan (Figure 11). Manchhar Lake, one of the largest lake in Pakistan and in Asia, is formed in a depression in the western side of the Indus River in Sindh province. The total catchment area of the lake is about 97,125 km² [27]. The surface area of the lake fluctuates with the seasons from as little as 350 km² to as much as 520 km² [28]. The lake is fed by two canals, the Aral and the Danister emerging from the river Indus in their eastern side. Due to less rainfall and contamination of surface water, the Manchhar Lake contains brackish water. The elevation ranges between 45 m at the Manchhar Lake and 163 m towards Bhit study area. On regional scale the area is a part of Gigantic Indus river basin composed of alluvium transported by the river and its tributaries. The main surface water sources are Naig Nai stream originating from Bhit Mountain range in the west, Dhanar Dhoro stream passing close to the oil-field plan, besides other small intermittent streams which remain dry in most parts of the year. The discharge data of these streams were not available. The water supply for local communities is maintained by the springs originating from the nearby Limestone Mountains of the Kirthar Range. According to reconnaissance studies conducted in the area, fresh groundwater source is available at different locations in the alluvium. Presently, there is no significant groundwater development in the area. Only few water wells were constructed to fulfill the requirements of local communities. The major sources of potable water to humans and livestock and for irrigation are unconsolidated aquifers. A water supply scheme consisting of four tube wells at Jhangara village is providing water to the villages between Manchhar Lake and Jhangara. The rainfall is scanty. Average annual rainfall is about 200 mm. It is higher in summer months like July and August due to prevalence of monsoon conditions. The aquifer area is located in the alluvial deposits along the Naig Nai stream.

The water table is generally in phreatic to semi-unconfined conditions. The observation wells drilled in the area indicated water table depth of about 12 m and hydraulic head value of 148.8 meter above sea level (masl). The groundwater flow is generally from southwest towards northeast direction. The flow direction of groundwater is true replica of the flow direction of Naig Nai stream draining the area. The groundwater level is mainly influenced by seasonal floods, stream flows and tubewell (water well) discharge. The fluctuations are small in the deep water table in the piedmont plain.

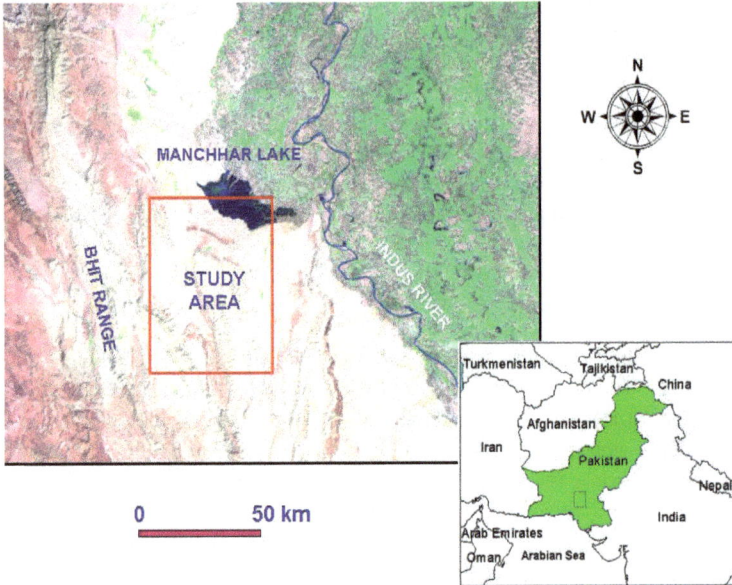

**Figure 11.** Location of study area in southern Pakistan

## 3.2. Material and methods

The data of subsoil properties, aquifer characteristics and existing groundwater conditions were collected through reconnaissance level field investigations including geophysical survey. The electrical resistivity soundings survey (ERSS) was conducted primarily to collect required input data for the modeling. The surface drainage and topographic information were extracted from the topo-sheets prepared by Survey of Pakistan on 1:50,000 scale. The climate data i.e. precipitation and temperature, of 1961-1976 period were acquired of nearby meteorological stations like Karachi, Nawabshah, Moenjodaro and Khuzdar from Meteorological Department and Water and Power Development Authority (WAPDA) Pakistan. The published literatures of the region i.e. see [29-32] had been used to firm up the study results.

### 3.2.1. Geophysical data analysis

The geophysical/drill logs of the injection well field suggest that the subsurface material is composed of layers of sandstone, limestone, dolomite and clay-stone of different formations. One of the interpretative seismic sections of the Bhit concession area (shown in Figure 12) indicates the deeper Chiltan limestone formation beyond the depth of injection well. The hard rock aquifers are mainly composed of partially fractured limestone and sandstone belonging to Nari and Kirthar formations. Limestone, which is the dominant formation, has solution channels due to water action having secondary permeability characteristics. Further, chances of transport contamination could take place much significantly through fracture zone of limestone and dolomitic formations. The hydraulic properties of the underlying overburden and rocks evaluated through geophysical well log data are shown in Table 5.

| S.No. | Formation | Depth (m) | K (m/d) | Transmissivity T (m²/d) |
|---|---|---|---|---|
| 1 | Nari (sandstone) | 113 | 5 | 130 (S=.01) |
| 2 | Kirthar (limestone) | 616 | 24 | 2000 (S=.04) |
| 3 | Ghazij (claystone) | 762 | zero | Regional Seal |
| 4 | Laki (limestone & dolomite) | 1,259 | 0.165 | 4 (S=0.00005) |
| 5 | Dunghan (dolomite and claystone) | - | - | - |
| 6 | Ranikot (Lakhra+Bara+Khadro) | 1,799 | 0.68 | 14.97 (S=0.000007) |
| 7 | Pab (sandstone) | 2,000 to onward | 0.138 | 17.5 (S=0.000005) |

**Table 5.** Summary of the Aquifer characteristics of hard rock formations

The hydraulic conductivity of the unconsolidated deposits is about 19 m/day and the effective porosity is 0.25. Several hydrocarbon wells are producing gas in the Bhit concession area. The Ghazij (claystone) was found to be a cap rock (regional seal) over Pab sandstone - an enriched reservoir of hydrocarbon. The Ranikot formation is a prolific rock unit having good transmissivity and storativity to accumulate disposal waste of the producing gas Bhit concession.

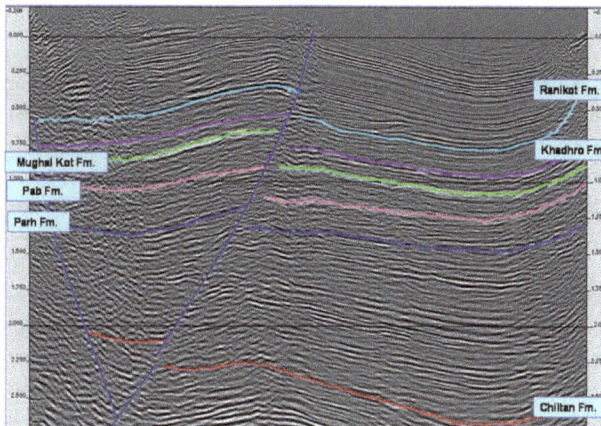

**Figure 12.** Interpretative seismic depth section indicating several geological formations

### 3.2.2. Model conceptualization and simulation

The groundwater flow system was treated as a multi-layered system. The upper layer aquifer is mainly unconfined and at depths where silt and clay horizons are present, the sand could probably cause partial confinement in some areas. The chances of transport contamination could take place much significantly through fracture zone of limestone and dolomitic formations. Seven aquifer layers were defined on the basis of physical characteristics of lithological formations (Table 1). The disposal of produced water in the injection well is set at 2,100 m depth in the Pab sandstone formation. The main source of recharge to groundwater is rainfall which is highly variable. During rains, different nullahs carry flows and infiltrations through the piedmonts and alluvial fans and cause subsequent lateral movement at depth. The recharge is higher in summer especially in the months of July and August. It may occur during rainy month of March in winter, a mean value for the limestone recharge may be taken as 200 mm/year based on the recharge data of rainfall. The main discharge components are groundwater extraction from water wells/dug wells, evapotranspiration, and spring discharge. The abstraction of groundwater becomes higher during months of little recharge to groundwater i.e. November to January, which may affect the storage of groundwater for a limited period.

The MODFLOW code and the MT3DMS code were used to solve the flow and transport equations. The model domain comprised of 40 x 30 grid network with total area of 1200 sq km (Figure 13). First, the model was run for steady state condition. The model took up groundwater extraction from the oil-field water wells (TW1 & TW2) and the community tube wells (TW3 & TW4). The discharge rate of the water wells was 0.0083 $m^3$/sec and discharge rate of deep seated injection well (DW) as supplied was 0.00152 $m^3$/sec. The injection of the produced wastewater from the injection well was considered during simulation. Once the flow model was completed and run was carried out, the contaminant transport model was set and simulated to evaluate the groundwater contamination and movement of plume.

### 3.2.3. Simulation of contaminant transport of injection well

The steady-state hydraulic heads were used as initial condition in MT3DMS option available in PM5 to simulate the dispersion of plume. The MT3DMS model simulates the processes i.e. advection, mechanical dispersion, retardation, decay and molecular diffusion related to the fate of contaminant. Initial concentration was set to zero in all the layers. In Advection package, 3rd order TVD scheme [12, 33] was selected. This method is considered as a good compromise between the standard finite difference and particle tracking approaches. In dispersion package, TRPT (Horizontal transverse dispersivity/Longitudinal dispersivity) was set to 0.3 for all the layers except in layer 3, where it was set to 0.1. In dispersion package, TRPV (Vertical transverse dispersivity/Longitudinal dispersivity) was set to 0.3 for all the layers except in layer 3. For this layer, it was set to 0.1. Longitudinal dispersivity was set to 10 m. There was no sorption selected in chemical reaction package. The injection well was set at layer 7 in the sink/source menu. The concentration of the injection well liquid was considered to be 100 ppm. The model was then simulated for 1-, 10-, and 30-year period for studying the behaviors produced wastewater in and around the injection well.

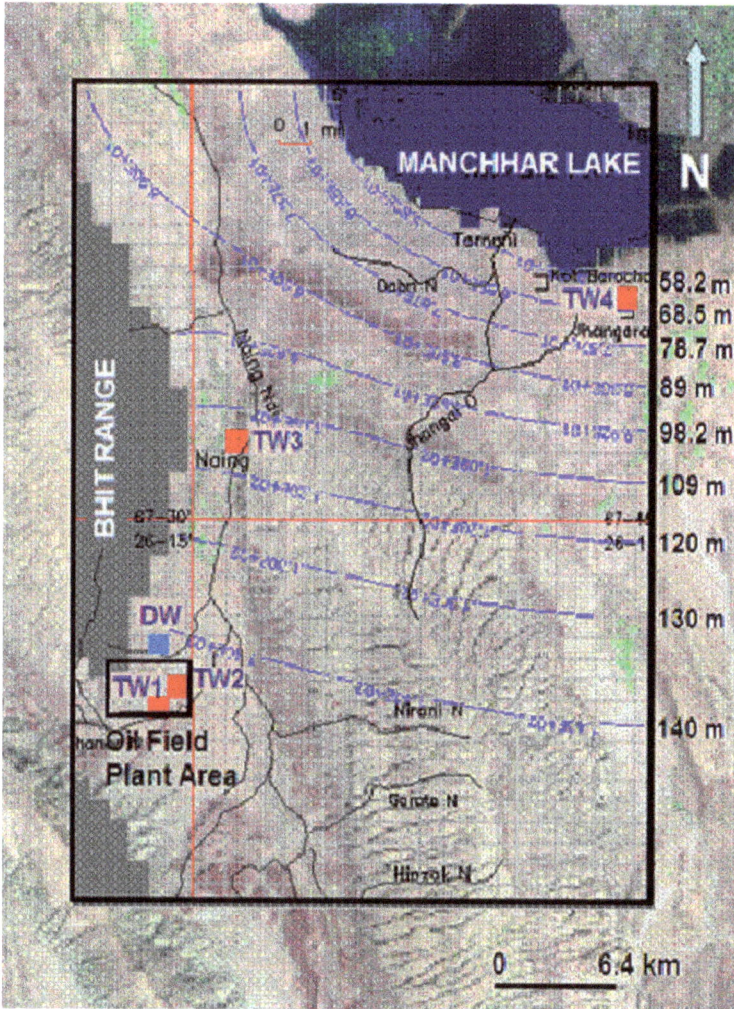

**Figure 13.** Manchhar Lake treated as constant head boundary; Bhit Range as impervious boundary. *DW* deep injection well, *TW* tube well

## 3.3. Results and discussion

The computations were carried out for three cases i.e. in case-I, Injection well was continuously discharged for one-year period, in case-II it well was simulated for 10-year period and in case-III for 30-year period. The hydraulic heads and drawdown were computed in all three cases. The velocity vectors prominent in layer-1 tend to move in the northeast direction towards

Manchhar Lake. The groundwater flow had shown decline in confining layers like 4 and 7. The two tube wells of the oil-fields each discharging at the rate of $8.3 \times 10^{-3}$ m$^3$/sec along with community wells were used for the study. The results obtained from the 3-D transport model are shown in Table 6 and transport of contaminant plumes in three simulation periods in Figures 14-16.

| Layer | 1 Year | 10 Years | 30 Years |
|-------|--------|----------|----------|
| 1 | - | $2.21 \times 10^{-23}$ | $1.96 \times 10^{-20}$ |
| 2 | $3.16 \times 10^{-27}$ | $3.88 \times 10^{-21}$ | $1.32 \times 10^{-18}$ |
| 3 | $3.14 \times 10^{-21}$ | $3.54 \times 10^{-16}$ | $4.01 \times 10^{-14}$ |
| 4 | $3.16 \times 10^{-15}$ | $3.18 \times 10^{-11}$ | $1.06 \times 10^{-09}$ |
| 5 | $9.58 \times 10^{-10}$ | $9.53 \times 10^{-07}$ | $2.45 \times 10^{-05}$ |
| 6 | $8.04 \times 10^{-05}$ | $7.95 \times 10^{-03}$ | $7.21 \times 10^{-02}$ |
| 7 Pab (Sandstone) | $6.28 \times 10^{-02}$ | 0.626 | 1.861 |

**Table 6.** Maximum concentration observed in different simulation periods (ppm)

After 30 years of simulation period, only traces of contamination were found in Ghazij Formation. Moreover, it is found that after 1-year period of simulation the produced wastewater will reach upward in layer-5 (Ranikot Formation) emerging from layer 7 (Pab sandstone) as shown in Figure 14. In this period, no contamination was found in layer 1 and 2. In 10-year simulation a plume of produced water moved from layer 7 to layer 5 (Figure 15). Only traces of contamination were found in layer 3 (Ghazi Formation). In Figure 6, plume of produced water contamination indicates movement from layer 7 to layer 4 after 30 years simulation. The layer 3 was found to be acting as a regional confining seal. In this layer, only traces of contamination were present. The movement of produced wastewater was found within a radius of 3 km at the bottom of injection well in the Pab sandstone. The upper aquifers in the alluvial deposit, Nari sandstone, and Kirthar limestone was remain safe from the effects of produced wastewater disposal from the deep seated injection well. The community wells tapping in the upper few tens of meters, naturally oozing springs and the Manchhar Lake located about 43 km from the injection well were also found to be safe from the effects of produced water injection even after contaminant transport simulation of 30-year period. The development of plume was significant in layer 7 and upward in the three cases (shown in Figures 14-16).

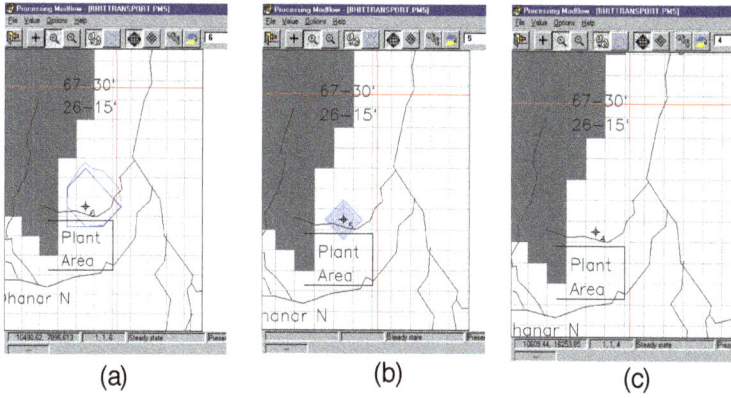

**Figure 14.** Upward movement of the plume from layer 7 to layer 6 (a); from layer 6 to layer 5 (b) and from layer 5 to layer 4 (c) after 1 year simulation period

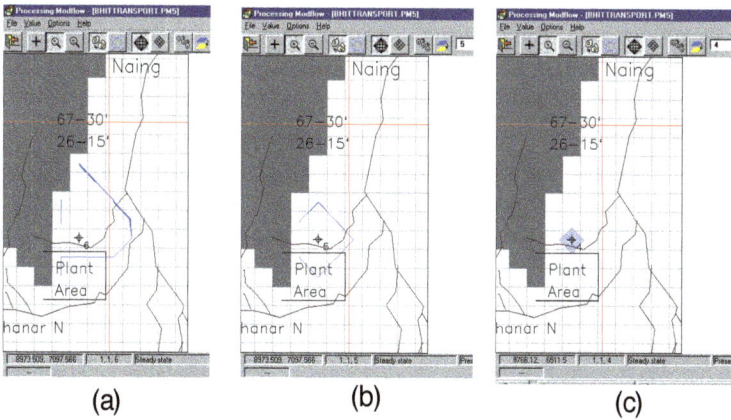

**Figure 15.** Upward movement of the plume from layer 7 to layer 6 (a); from layer 6 to layer 5 (b) and from layer 5 to layer 4 (c) after 10 year simulation period

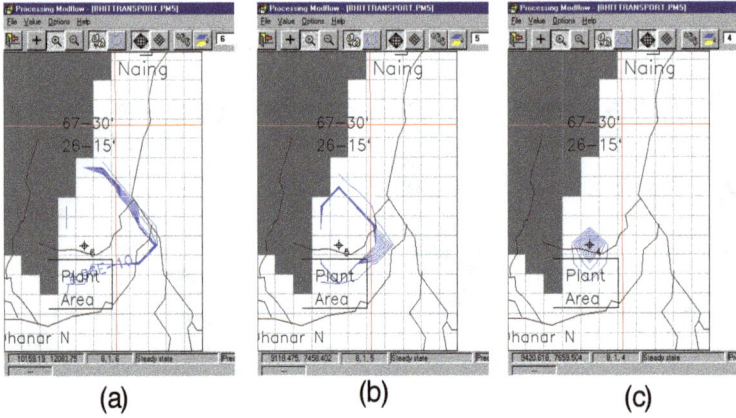

**Figure 16.** Upward movement of the plume from layer 7 to layer 6 (a); from layer 6 to layer 5 (b) and from layer 5 to layer 4 (c) after 30 year simulation period

## 4. Conclusions

The results of the three-dimensional groundwater modeling study highlighted the hydrogeo-logical characteristics and features of the contaminant transport of the deep injected tube well in the Bhit oil-field area. The groundwater contaminant transport modeling technique has proved to be effective in simulating produced wastewater plume from the deep seated injection well. The study would provide base for evaluating risks of contaminants on long term basis in similar conditions in future. Risk of expansion of plume regionally does not exist as the disposal of wastewater is made in the deeper horizon well below the aquifers and also the quantity is quite limited.

Thorough understanding of surface hydrology, hydrogeological conditions and contaminant behavior in the aquifer system coupled with application of reliable modeling techniques could be helpful in dealing with water management issues related to contaminant hydrology.

## Author details

Zulfiqar Ahmad[1], Arshad Ashraf[2], Gulraiz Akhter[1] and Iftikhar Ahmad[3]

1 Department of Earth Sciences, Quaid-i-Azam University, Islamabad, Pakistan

2 National Agricultural Research Center,Islamabad, Pakistan

3 College of Earth and Environmental Sciences, Punjab University, Lahore, Pakistan

# References

[1] Boulding, J. R, & Ginn, J. S. Practical handbook of soil, vadose zone and groundwater contamination: assessment, prevention and remediation, CRC Press; (2004).

[2] Walter, D. A, & Masterson, J. P. Simulation of Advective Flow under Steady-State and Transient Recharge Conditions, Camp Edwards, Massachusetts Military Reservation, Cape Cod, Massachusetts, Water-Resources Investigations Report USGS; (2003). , 03-4053.

[3] Anderson, M. P, & Woessner, W. W. Applied Groundwater Modeling, Simulation of Flow and Advective Transport. Academic Press, Inc San Diego, California; (1992).

[4] Mcdonald, M. G, & Harbaugh, A. W. ModFlow, A modular three-dimensional finite-difference ground-water flow model, U.S. Geological Survey Open-File Report Chapter A1, Washington D.C.; (1988). , 83-875.

[5] Zheng, C, & Wang, P. P. A Modular Three-Dimensional Multispecies Transport Model, US Army Corps of Engineers, Washington, DC: (1999).

[6] Poeter, E. P, & Hill, M. C. UCODE, a computer modelling. *Computers & Geosciences* (1999). , 25, 457-462.

[7] Kim, J, & Corapcioglu, M. Y. Modeling dissolution and volatilization of LNAPL sources migrating on the groundwater table, *Journal of Contaminant Hydrology* (2003). , 65(1-2)

[8] Mott McDonald Pakistan (pvt) LtdA report on soil and groundwater Assessment of depots/ Installations/ airfields, Shell Pakistan Limited; (2000).

[9] Ashraf, A, & Ahmad, Z. Regional groundwater flow modeling of upper Chaj Doab, Indus Basin. *Geophysical Journal International* (2008). , 173, 17-24.

[10] El-Kadi, A. Groundwater Modeling Services for Risk Assessment Red Hill Fuel Storage Facilities, NAVFAC Pacific, Oahu, Hawaii; (2007).

[11] Periago, E. L, Delgado, A. N, & Diaz-fierros, F. F. Groundwater contamination due to cattle slurry: modeling infiltration on the basis of soil column experiments, *Water Research* (2000). , 34(3)

[12] Eric, W, Strecker, E. W, & Chu, W. Parameter Identification of a Ground-Water Contaminant Transport Model, *Groundwater* (1986). , 24(1)

[13] Jin, S, Fallgren, P, Cooper, J, Morris, J, & Urynowicz, M. Assessment of diesel contamination in groundwater using electromagnetic induction geophysical techniques, Journal of Environmental Science and Health (2008). , 43(6)

[14] Fetter, C. W. Applied hydrogeology (4th edition), Prentice Hall; (2000).

[15] KeyGroundwater Fate and Transport Evaluation Report: South Cavalcade Superfund Site, Houston, Texas, prepared by KEY Environmental, Inc. on behalf of Beazer East, Inc. for the U. S. Environmental Protection Agency; (1997).

[16] EPADetermining Soil Response Action Levels Based on Potential Contaminant Migration to Groundwater: A Compendium of Examples, EPA/540/(1989). , 2-89.

[17] ASTMStandard Guide for Risk-Based Corrective Action Applied at Petroleum Release Sites, American Society for Testing and Materials, ASTM E-Philadelphia, PA; (1995). , 1739-95.

[18] Newell, C. J, Mcleod, R. K, & Gonzales, J. R. BIOSCREEN: Natural Attenuation Decision Support System, User's Manual Version 1.3, EPA/600/R-96/087, National Risk Management Research Laboratory, Office of Research and Development, U. S. Environmental Protection Agency, Cincinnati, 45268, Ohio; (1996).

[19] Chiang, W-H, & Kinzelbach, W. Processing MODFLOW. A simulation system for modeling groundwater flow and pollution; (2001).

[20] Pollock, D. W. Documentation of a computer program to compute and display path lines using results from the U.S Geological Survey modular three-dimensional finite-difference groundwater flow model: U.S Geological Survey, open-file report Denver; (1989). , 89-381.

[21] Zhou, Y. Z. Sampling frequency for monitoring the actual state of groundwater systems, *Jour of Hydrology* (1996). , 180, 301-318.

[22] Plus, R. W, & Paul, C. J. Multi-layer sampling in conventional monitoring wells for improved estimation of vertical contaminant distribution and mass. *Jour of Contam. Hydrology* (1997). , 25, 85-111.

[23] Fisher, R. S, & Goodmann, P. T. Characterizing groundwater quality in Kentucky: from site selection to published information. Proceedings of 2002 National Monitoring Conference, national water Quality Monitoring council, May Madison, Wisconsin; (2002). , 19-23.

[24] Zeru, A. Investigations numériques sur l'inversion des courbes de concentration issues d'un pompage pour la quantification de la pollution de l'eau souterraine / Numerical investigations on the inversion of pumped concentrations for groundwater pollution quantification. PhD Thesis, Université Louis Pasteur (France); (2004). , 192.

[25] Bauer, S, Bayer-raich, M, Holder, T, Kolesar, C, Muller, D, & Ptak, T. Quantification of groundwater contamination in an urban area using integral pumping tests *Jour of Contam. Hydrology* (2004). , 75, 183-213.

[26] Bockelmann, A, Zamfirescu, Z, Ptak, T, Grathwohl, P, & Teutsch, G. Quantification of mass fluxes and natural attenuation rates at an industrial site with a limiten monitoring network: a case study site. *Jour of Contam. Hydrology* (2003). , 60, 97-121.

[27] NESPAKMaster Feasibility studies for flood management of hill-torrents of Pakistan, Supporting Vol-IV Sindh Province. National Engg. Services Pak. (Pvt) ltd. Lahore, Pakistan; (1998).

[28] http://enwikipedia.org/wiki/Manchar_Lake (accessed 10 September (2012).

[29]  Ahmad, Z, Ahmad, I, & Akhtar, G. A report on groundwater reserve estimation of the Ahmad Khan well field and its safe utilization for the 4000 TPD Luck Cement plant Pezu, D.I.Khan; (1994).

[30]  Ahmad, N. Groundwater resources of Pakistan, Ripon printing press; (1974).

[31]  WAPDALower Indus Report. Physical Resources, groundwater, supl.6.1.6, Tube wells and Boreholes, Nara Command; (1965).

[32]  WAPDARohri Hydropower Project. Feasibility study, HEPO (1989). (97)

[33]  Harten, A. High resolution schemes for hyperbolic conservation laws. *Jour of Comput. Phys.* (1983). , 49, 357-393.

# Water Resources Sustainability

# Chapter 8

# Geospatial Analysis of Water Resources for Sustainable Agricultural Water Use in Slovenia

Matjaž Glavan, Rozalija Cvejić, Matjaž Tratnik and
Marina Pintar

Additional information is available at the end of the chapter

## 1. Introduction

Global population growth has greatly increased food demand. This, in turn, has intensified agricultural production, already the biggest consumer of water in the world [1]. Development of irrigation techniques has contributed to the global food production [2]. However, climate change simulations predict repeated droughts and deteriorating crop production, illustrating the critical need for sustainable irrigation [3]. Thus, a proactive water management strategy is a priority of any government in the world.

Globally, only 10% of estimated blue water (surface water, groundwater, and surface runoff) and 30% of estimated green water (evapotranspiration, soil water) resources are used for consumption. Nevertheless, water scarcity is a problem due to high variability of water resources availability in time and space [4]. Model results suggest that severe water scarcity occurs at least one month per year in almost one half of the world river basins [5]. One third of the water volume currently supplied to irrigated areas is supplied by locally stored runoff [6]. It is estimated that small reservoirs construction could increase global cereal production in low-yield regions (i.e. Africa, Asia) by approximately 35% [6]. Global water scarcity problems can now be, due to advances in hydrology science in the last decades, easily assessed on fine temporal and spatial scale [4].

Irrigation development and management in Slovenia have completely stagnated in the last decade due to financial shortages. In 1994 the Slovenian government adopted a strategy for agricultural land irrigation (i.e. National Irrigation Programme) [7]. In 1999, the World Bank prepared a feasibility study of this program. However, economic constraints and lack of political will limited the implementation of the program [8].

Slovenia is experiencing periodic droughts of varying intensities in different parts of the country. According to the Court of Audit, the total costs in the agricultural sector due to the droughts in 2000, 2001, 2003 and 2006 were 247 million euro (EUR) [9]. During the same period the government spent 85.9 million EUR for the elimination of the consequences of droughts, and only 3.3 million € on drought prevention measures. This figure is particularly worrisome, because Slovenia is relatively rich in water, with 800-3,000 mm of precipitation per year. With appropriate technical measures, water could be redistributed temporally and spatially, limiting water scarcity and drought effects. Recurring droughts and the results of global and regional climate scenarios [10] predict a tightening of crop production conditions in Slovenia, illustrating the urgent need to address the availability of water resources [11-17].

The Ministry of Agriculture and the Environment has identified the current lack of irrigation infrastructure as a serious obstacle to prevention of agricultural damage and improvement of crop production. Therefore, the Ministry called for two research projects of the Target Research Program, as preparation for the establishment of a new Irrigation Strategy. The first project, Water Perspectives of Slovenia and the Possibility of Water Use in Agriculture (V4-0487) [9, 18], had two objectives, (a) to determine the current water quantity of Slovenian water resources (ground and surface waters, wastewater and sewage treatment plants discharges, existing large reservoirs) potentially available for use, with emphasis on irrigation and (b) to determine the extent to which these water resource meet current irrigation needs.

In 2012 the second project, Projections of Water Quantities for Irrigation in Slovenia (V4-1066) was completed, with the objective to determine to what extent the surface runoff water retained in small water reservoirs along with the rest of the available water, from other water resources, covers irrigation needs. The project also took into account the irrigation norms for different crops, soils, climate zones and climate change scenarios [19]. Analyses of the available water quantities, potential irrigation areas, technical possibilities of construction of small reservoirs, legislation, irrigation norms for crops, climate change impacts were made as part of the agricultural drought risk assessment.

The purpose of this chapter is to present a novel and globally applicable approach for identification of agricultural lands that are at risk for drought. Spatial analysis of available water resources and their quantities for the sustainable irrigation of agricultural land is the key to an efficient integrated water management strategy. Knowing the spatial distribution, accessibility, abundance and availability of water resources is an important element of national security, with regards to the production of sufficient quantities of quality food. Assessing water resources is especially critical in the light of empirical meteorological data and climate model results showing clear changes in the allocation of precipitation and in seasonal patterns.

| Water resource | Data type | Description/properties | Data source |
|---|---|---|---|
| Surface watercourse | River network | Polyline layer | Slovenian Environmental Agency (SEA) |
| | River flow | Geo-referenced tabular data - river flow gauging stations ($m^3 s^{-1}$) | |
| | Water abstraction | Geo-referenced tabular data | |
| | Available water quantities | Geo-referenced tabular data - water available for irrigation and ecologically acceptable flow ($m^3 s^{-1}$) | Institute of Water of the Republic of Slovenia (IWRS) |
| Large water reservoirs | Reservoirs | Polygon layer | IWRS |
| | Reservoir characteristics | Tabular data - reservoir type, volume, purpose of water use, share of water designated for use in agriculture | Slovenian National Committee on Large Dams (SNCLD), SEA |
| Groundwater | Water body | Polygon layer Hydrogeology, water availability | Geological Survey of Slovenia (GSS) |
| | Borehole | Drilling price | |
| | Water rights | Geo-referenced tabular data – water abstraction ($m^3 s^{-1}$) and % of all estimated water in groundwater body | SEA, GSS |
| Accumulated surface runoff | Runoff | Raster layers (mm year$^{-1}$) | SEA |
| | Mean monthly flow | $m^3 s^{-1}$ | |
| | Mean monthly specific runoff | $l s^{-1} km^2$ | IWRS |
| | Soil data | Polygon layer Soil properties (texture, horizons, bedrock, hydraulic conductivity, soil water capacity, hydrological group) | University of Ljubljana - Biotechnical Faculty (UL-BF) |
| | Curve number | Share of precipitation as surface runoff defined by land use and soil hydrological group, slope | |
| | Surface runoff yield and abundance | Quantity of water in millimetres and $m^3 ha^{-1}$ | IWRS UL-BF |
| | DEM | Raster layer - Digital elevation model - 25m | The Surveying and Mapping Authority of the Republic of Slovenia (SMARS) |
| Irrigation | Irrigation area | Polygon layer | UL-BF |
| | Irrigation norm | Gross irrigation norm in millimetres, litres or $m^3$ of water per hectare for defined crop and soils in one year for optimal growing conditions | IWRS, UL-BF |
| | Hydro-module | Qualities of water used in litres per second per hectare of crop in one irrigation cycle | UL-BF |
| | Irrigation systems | Polygon layer Total area and actually irrigated land | Statistical Office of Slovenia (SOS) |
| Land use | Graphical Units of Agricultural Land - GERK | Polygon layer Land cover classification and spatial representation | Ministry of Agriculture and the Environment of the Republic of Slovenia (MAERS) |

**Table 1.** Input data sources for water resources availability assessment

# 2. Materials and methods

## 2.1. Input data

Table 1 provides an overview of the data used for spatial analysis (data type, name, source - location, description). If certain type of map was not available we created maps from tabular data provided from different sources. This type of spatial analysis requires a wide range of data starting with land use classes and soil types and their position in space as these have primary impact on surface runoff, percolation of water to groundwater and on soil water content.

Input data also includes river network, river flow, water abstraction and available water quantities for irrigation and ecologically acceptable flow to represent surface watercourses. Additional inputs include data on reservoir characteristics for spatial representation of large water reservoirs. Groundwater data includes hydrogeology and water availability layers, borehole drilling prices and water rights. The widest range of data was needed to spatially represent accumulated surface runoff. We included in the analysis runoff, mean monthly flow, mean monthly specific runoff, soil data, curve number and irrigation areas and norms which resulted in surface runoff yield and water abundance calculations. Geographic Information System ArcGIS software version 9.3 was used for all spatial analyses. Due to the characteristics of the spatial analysis with the raster layers (raster cells) we used extension build in the ArcGIS program toolbox called Spatial Analyst Tool.

## 2.2. Study area agricultural land

The case study area is the Republic of Slovenia (2,020,318 ha), situated in central Europe between Italy, Austria, Hungary and Croatia. A land use analysis showed that agricultural land potentially suitable for irrigation covers 221,355 ha or 10.3 % (Figure 1, Table 2) of the country.

Based on a land use map, the following agricultural land use classes [20] were identified as suitable for irrigation:

a.   fields and gardens, hops plantations, permanent crops on fields, greenhouses, nurseries, intensive orchards, extensive orchards, other permanent crops,

b.   olive groves,

c.   plantations of forest trees, uncultivated agricultural land.

Fields and gardens are the most suitable areas for irrigation, especially when crop production is being intensified. Irrigation in areas planted with hops, permanent crops on fields (asparagus, artichokes, rhubarb, etc.), intensive orchards (apple trees, pear trees, etc.), nurseries (fruit trees, vines, olive trees, etc.,) and in greenhouses, is particularly critical for sustainable crop production. Extensive orchards are potential areas where new intensive fruit plantations could be planted or old extensive orchards renewed, both could be irrigated to secure more reliable yield. Olive groves are not generally irrigated in Slovenia. An experimental irrigation system was installed within the project: Adapting technology of production to climatic conditions for achieving high quality yield of olives and olive oil (V4-0557). There are several reasons for the absence of olive grove irrigation in Slovenia: relatively high annual precipitation, grower's belief in the relatively low sensitivity of olives trees to drought, lack of reliable water sources, and the terrain, which makes installation of irrigation equipment expensive. Plantations of forest trees with fast growing species like poplar are usually situated on agricultural land. The reasons for growing forest trees on agricultural land are different (paper industry, hydro-meliorations, land reclamations, ameliorations). However, after harvesting these areas could be allocated for agricultural production. Their suitability is even greater because these areas are usually near water resources. Uncultivated agricultural land is usually excluded from production, due to different types of construction sites, only for a certain time period. After completion of works these areas in the majority of cases return back to agricultural production.

| Agricultural land use classes | Area | | |
|---|---|---|---|
| | Hectare (ha) | Percent (%) of agricultural land | Percent (%) of Slovenia |
| Fields and gardens | 182,146.76 | 82.29 | 8.98 |
| Hops | 1,977.91 | 0.89 | 0.10 |
| Permanent crops on fields | 335.95 | 0.15 | 0.02 |
| Greenhouses | 130.01 | 0.06 | 0.01 |
| Nurseries | 47.84 | 0.02 | 0.00 |
| Intensive orchards | 4,385.30 | 1.98 | 0.22 |
| Extensive orchards | 23,929.25 | 10.81 | 1.18 |
| Olive groves | 1,810.83 | 0.82 | 0.09 |
| Other permanent crops | 416.53 | 0.19 | 0.02 |
| Plantations of forest trees | 271.39 | 0.12 | 0.01 |
| Uncultivated agricultural land | 5,903.37 | 2.67 | 0.29 |
| Total | 221,355.15 | 100.00 | 10.92 |

**Table 2.** Agricultural land potentially suitable for irrigation in Slovenia

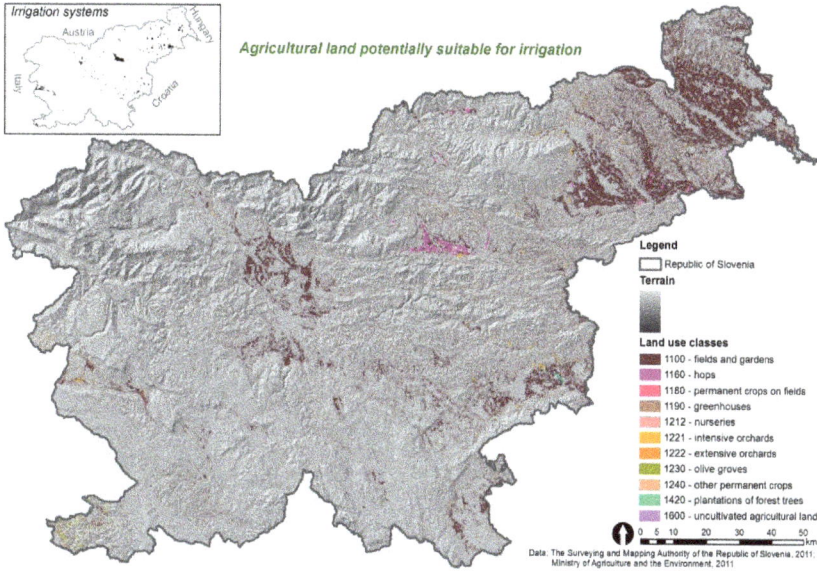

**Figure 1.** Geographic location of Slovenia, agricultural land potentially suitable for irrigation and locations of irrigation systems

In Slovenia in 2010, of 8,299 ha was prepared for irrigation and, only 3,851 ha was actually irrigated [21], accounting for less than 4 % and 2 % of total agricultural land potentially suitable for irrigation (221,355 ha), respectively (Table 3).

| | Year | | | | | | | | |
| --- | --- | --- | --- | --- | --- | --- | --- | --- | --- |
| | **2003** | **2004** | **2005** | **2006** | **2007** | **2008** | **2009** | **2010** | **2011** |
| Land prepared for irrigation (ha) | 6,339 | 5,303 | 4,727 | 5,395 | 7,876 | 7,732 | 7,841 | 7,604 | 8,299 |
| Actually irrigated land (ha) | 2,741 | 2,329 | 1,812 | 2,837 | 3,759 | 3,651 | 3,732 | 3,501 | 3,851 |

**Table 3.** Total area (ha) of agricultural land prepared for irrigation and actually irrigated in Slovenia

## 2.3. Surface watercourses and large water reservoirs

Water accessibility classes for surface watercourses or water reservoirs were spatially defined and created from the percentage (%) of defined agricultural land use areas suitable for irrigation (Figure 2 and 3). The project on water perspectives (V4-0487) [8, 18] defined the percentage of area that can be directly irrigated from existing water reservoirs. Dry water reservoirs were excluded from the analysis. The analysis was supported with field work (questionnaires) checking the status and operational management of reservoirs and with analysis of regulations on operation and maintenance of reservoirs.

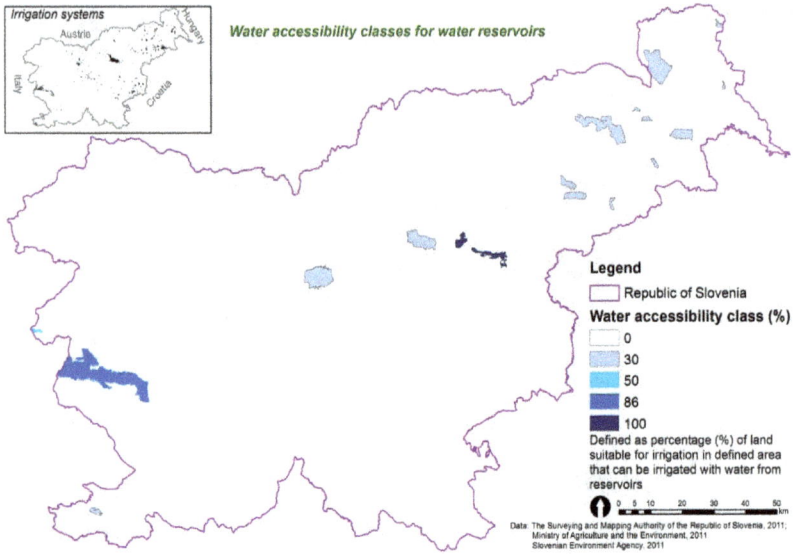

**Figure 2.** Water accessibility classes for water reservoir in Slovenia

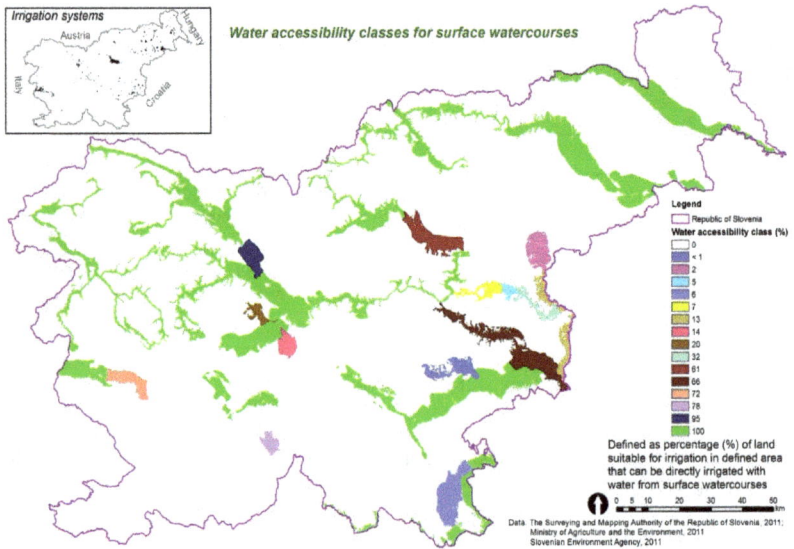

**Figure 3.** Water accessibility classes for surface watercourses in Slovenia

The project identified eighteen (18) water reservoirs, from which at least part of the accumulated water could be allocated for irrigation of agricultural land. In all of the large water reservoirs impact areas were determined, where water quantities are sufficient for direct irrigation of at least 30 % of the agricultural land potentially suitable for irrigation (Figure 2). It follows that the use of water from certain water reservoirs is quantitatively limited to water available for irrigation of agricultural land.

The percentage (%) of the area that can be directly irrigated from surface watercourses was determined on the basis of the available water quantity for irrigation at the last point downstream of individual surface watercourse water body. The project defined seventy (70) areas suitable for irrigation (Figure 3).

Area determination followed the criteria [18] below:

*   maintenance of ecologically acceptable flow (Official Gazette RS, No. 97/2009),

*   water abstraction within each catchment area must not be greater than the available water quantity at the last point downstream of individual catchment area of surface watercourse water body,

*   total water abstraction within a system of catchment areas must not be greater than the total capacity of a set of catchment areas, which is the same size as the availability of water quantity in the final (outflow) node of the concerned system of catchment areas;

*   irrigation area of each watercourse is located in the catchment area of the surface watercourse water body (some exceptions);

*   horizontal distance from the river to the border of agricultural land area potentially suitable for irrigation is not greater than 3 km (some exceptions);

*   difference in height between the watercourse and agricultural land suitable for irrigation does not exceed 100 m.

Water accessibility points for water reservoirs and watercourses were determined by the extent of agricultural land (ha, %), which may be irrigated with the water assigned for the agricultural use from both sources. It is important that the use of water from a reservoir is quantitatively limited to the water available for agricultural land irrigation, and water from watercourses is limited to ecologically acceptable flows.

Large water reservoirs and surface waters are attributed with 100 points of availability if the water resource supplies sufficient quantities of water for irrigation of all potentially suitable agricultural land for irrigation in the defined area of the water body (Table 4). If water quantities are insufficient (0 to 99%) for irrigation of a whole defined area of water body adequate for irrigation, the water resource is attributed with availability points between 0 and 99.

## 2.4. Groundwater

Water accessibility classes for groundwater were determined based on a hydrogeological map [22] which defines three classes of groundwater availability (hard, medium and easy) which were linked with three classes of average cost for borehole (well) drilling. The areas with easily accessible groundwater and the lowest price for borehole drilling were attributed with 100% availability of water (Table 4). The other two accessibility classes have smaller or higher number of percentages (Figure 4), in proportion to the price of borehole drilling and the accessibility of groundwater.

It is important to note that groundwater is priority reserved for drinking water. A relatively small percentage of groundwater is actually abstracted; with the highest rate (35%) in the Savska kotlina with Ljubljansko barje in central Slovenia. However, the analysis of the officially assigned abstraction rates from granted water rights showed that three groundwater bodies are 100% utilized (Savska kotlina with Ljubljansko barje. Kamniško-Savinjske Alpe in central Slovenia and Vzhodne Slovenske gorice in eastern Slovenia).

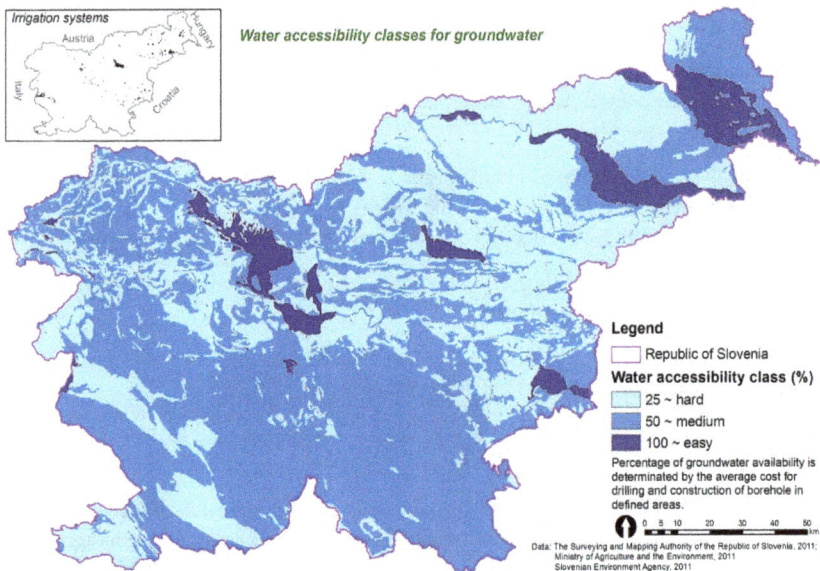

**Figure 4.** Water accessibility classes for groundwater in Slovenia

The average price for borehole drilling in 2010 in an area with easily accessible groundwater (diameter 100 mm, the average rate of flow of 5.5 l s⁻¹, depth 50 m) was estimated to be 11,000 EUR. The average price for borehole drilling in an area with medium accessible groundwater (diameter 100 mm, yield up to 5.5 l s⁻¹, depth of 70 m to 150 m) was estimated to be 15,000 and 30,000 EUR. The average price for borehole drilling in areas with hard accessible groundwater (diameter 100 mm, the average yield of 1 l s⁻¹, at least 200 m depth) was estimated to be 44,300 EUR. Accessibility of groundwater and price of borehole drilling is highly dependent on geology, groundwater levels, aquifer layer thickness and type of aquifer.

| Water resource | Water accessibility and abundance | | | | Water availability points | | | |
|---|---|---|---|---|---|---|---|---|
| **Large water reservoirs** | | | | | | | | |
| | unrestricted (irrigation of 100% of area[1]) | | | | 100 | | | |
| | restricted (irrigation of 0 to 99% of area) | | | | 0-99 | | | |
| **Surface watercourses (rivers, streams)** | | | | | | | | |
| | unrestricted (irrigation of 100% of area) | | | | 100 | | | |
| | restricted (irrigation of 0 to 99% of area) | | | | 0-99 | | | |
| **Groundwater** - customized to geology and borehole drilling costs | | | | | | | | |
| | easy | | | | 100 | | | |
| | medium | | | | 50 | | | |
| | hard | | | | 25 | | | |

**Surface runoff as small water reservoirs** - customized to winter yield, maximal irrigation norm, light soils and drip irrigation

| | abundance | 1 | 2 | 3 | 4 | 5 | 6 |
|---|---|---|---|---|---|---|---|
| | (m³ ha⁻¹) | MED | PAN | SMED | SPAN | CENT | ALPS |
| | > 6000 | -[2] | - | 100[3] | - | 100 | 100 |
| | 4000-6000 | 75 | - | 100 | 100 | 100 | 100 |
| | 2000-4000 | 50 | 75 | 75 | 75 | 100 | 100 |
| | 1000-2000 | 25 | 50 | 50 | 50 | 75 | 100 |
| | 500-1000 | 25 | 25 | 25 | 25 | 50 | 100 |
| | < 500 | 25 | 25 | - | 25 | - | - |
| | Relative slope < 6 % | 0 | 0 | 0 | 0 | 0 | 0 |
| **No accessible water resource** | | | | 0 | | | |

MED – Mediterranean irrigation area; PAN – Pannonian irrigation area; SMED – Sub-Mediterranean irrigation area; SPAN – Sub-Pannonian irrigation area; CENT – Central Slovenian irrigation area; ALPE – Alpine-Dinaric irrigation area

[1] irrigation of x% of area identified as suitable for irrigation from large water reservoir and surface watercourses

[2] class of winter yield abundance does not exist for certain irrigation area

[3] winter yield abundance of surface runoff from 1 ha of land is sufficient for irrigation of 1 ha of permanent crop (orchard)

**Table 4.** Determination of potential availability of water resources for irrigation based on water direct accessibility from (1) water reservoirs, (2) surface watercourses, (3) groundwater and (4) abundance of surface runoff yield as small water reservoirs

Areas with easily accessible groundwater and therefore with the lowest price of borehole drilling are attributed with 100 points of availability (Table 4). Medium and hard accessible groundwater areas are attributed with 50 and 25 availability points, respectively. The price of borehole drilling for those two classes is two or four times higher than for easily accessible groundwater.

## 2.5. Accumulated surface runoff

To create classes of potential abundance of surface runoff for accumulation in small reservoirs, we had to gather information on the maximum irrigation norm for drip irrigation on light soils for several groups of plants per one hectare (vegetables - low norm, vegetables – high norm, strawberries and permanent crops). This was for all irrigation areas and based on the average quantity of water available for irrigation (Table 4 and 5) [6, 23]. The definition was also based on the optimum volume of a small reservoir for the irrigation of one hectare (of accumulated surface runoff) defined by agro-meteorological stations in different irrigation areas for a dry year with a five-year return period (Table 4). Classes of potential winter yield of surface runoff (mm) (1971 - 2000) (Figure 5) were merged with a map of irrigation areas creating classes with assigned attributed points of surface runoff yield abundance [24].

| Irrigation area | Volume of reservoir (m³) | | | Groups of crops | Maximal irrigation norm (m³) | Water availability points |
|---|---|---|---|---|---|---|
| | Optimal | Average loss | Average available for irrigation | | | |
| MED | 1500 | 531,8 | 968,2 | strawberries | 878 | 25 |
| | 2000 | 692,4 | 1307,6 | vegetables – low norm | 1292 | 50 |
| | 4500 | 1477,4 | 3022,6 | vegetables – high norm | 2871 | 75 |
| | 6000 | 1941,2 | 4058,8 | permanent crops | 3720 | 100 |
| PAN | 1500 | 219,5 | 1280,5 | strawberries | 1125 | 25 |
| | 2000 | 284,2 | 1715,8 | vegetables – low norm | 1625 | 50 |
| | 4000 | 536,7 | 3463,3 | vegetables – high norm | 3097 | 75 |
| | 4500 | 598,9 | 3901,1 | permanent crops | 3482 | 100 |
| SMED | 1000 | 131,1 | 868,9 | strawberries | 588 | 25 |
| | 1500 | 187,3 | 1312,7 | vegetables – low norm | 1031 | 50 |
| | 2500 | 296,8 | 2203,2 | vegetables – high norm | 2271 | 75 |
| | 3000 | 350,7 | 2649,3 | permanent crops | 2359 | 100 |
| SPAN | 1000 | 168,4 | 831,6 | strawberries | 951 | 25 |
| | 1500 | 241,0 | 1259,0 | vegetables – low norm | 1299 | 50 |
| | 3000 | 452,1 | 2547,9 | vegetables – high norm | 2568 | 75 |
| | 3500 | 521,3 | 2978,7 | permanent crops | 3024 | 100 |
| CENT | 500 | 45,0 | 455,0 | / | / | 25 |
| | 1000 | 80,8 | 919,2 | strawberries | 552 | 50 |
| | | | | vegetables – low norm | 848 | |
| | 1500 | 115,3 | 1384,7 | vegetables – high norm | 1697 | 75 |
| | 2500 | 182,3 | 2317,7 | permanent crops | 2157 | 100 |

MED – Mediterranean irrigation area; PAN – Pannonian irrigation area; SMED – Sub-Mediterranean irrigation area; SPAN – Sub-Pannonian irrigation area; CENT – Central Slovenian irrigation area; ALPE – Alpine-Dinaric irrigation area

**Table 5.** Determination of availability points for accumulated surface runoff water in small water reservoirs based on average available water for irrigation in reservoir and maximal irrigation norm for drip irrigation and light soils

The magnitude of the abundance points was based on the maximum irrigation norm (drip irrigation) for one hectare of permanent crops (orchards) on light soils and its corresponding optimal reservoir volume for irrigation. If there was enough water for the irrigation of this type of crop (orchard, light soils, drip irrigation, maximum irrigation norm) in an irrigation area it was given 100 availability points (Table 4). Each subsequent class was determined by

25 availability points less, as it does not facilitate sufficient quantities of surface runoff water for irrigation of all groups of agricultural plants.

The determination of abundance points in the case of irrigated land for the Mediterranean and central Slovenian irrigation areas was as follows.

For the drip irrigation of one hectare of permanent crop with maximum irrigation norm on light soils (3,720 m³ ha⁻¹ per year) and water balance for a dry year with a five-year return period we need a small reservoir with optimal volume of 6,000 m³ of water (Table 5). This means that in the Mediterranean area, where potential accumulated surface runoff yield is more than 6,000 m³ ha⁻¹, it is possible to irrigate most of the crops. Therefore this abundance class was attributed with 100 availability points (Table 4). If it is possible to accumulate only up to 1000 m³ ha⁻¹ of surface runoff yield in the Mediterranean irrigation area in 'dry year with five-year return period', then only a small share of crops can be irrigated. This means that the water quantity is insufficient to meet the water needs of the majority of crops in this area. Irrigation of strawberries in the Mediterranean area requires 1,500 m³ of water. Accordingly, this abundance class was attributed with 25 availability points (Table 4 and 5). In central Slovenia, for the drip irrigation of one hectare of permanent crop with maximum irrigation norm on light soil, 2,157 m³ ha⁻¹ per year of water (dry year with five-year return period) is needed. If we include the water balance of the area, a small reservoir with volume of 2,500 m³ would be needed. This means that in central Slovenia where potentially accumulated surface runoff yield exceeds 2,000 m³ ha, the abundance classes were attributed with 100 availability points (Table 4 and 5).

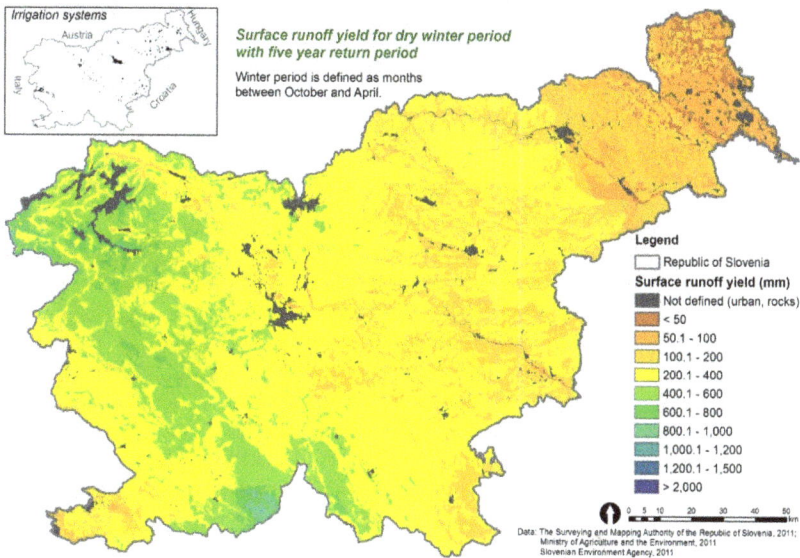

**Figure 5.** Potential surface runoff yield (mm) for dry winter period with five year return period in Slovenia

The final product of assembly and reclassification of individual data resulted in a map of abundance classes of potential surface runoff yield for the dry winter period and irrigation norm by irrigation areas (Figure 6). Also excluded from further analysis was data with a relative slope of less than 6%, and undefined areas (urban, rocky, surface waters). These areas were attributed with 0 availability points.

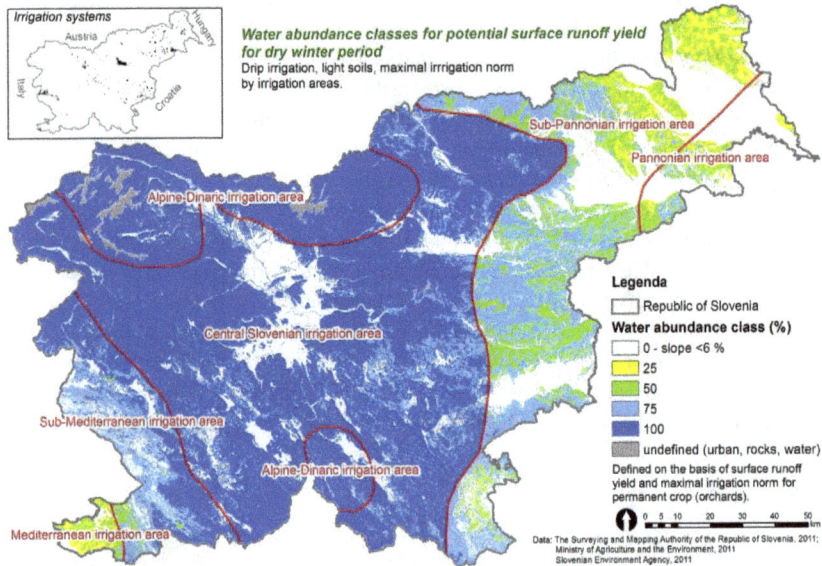

**Figure 6.** Water abundance classes for potential surface runoff yield for dry winter period in Slovenia

## 2.6. Drought risk classes definition

The determination of drought risk classes of agricultural land suitable for irrigation is the sum of the attributed availability points of each individual water resource suitable for irrigation of agricultural land (Table 6). Water resources (large water reservoirs, surface watercourses, groundwater and surface runoff yield) are spatially defined and interrelated (Figures 2 - 6). The analysis was conducted with raster layers whose spatial resolution was 100×100 m (1 ha) for the entire study area.

Drought risk assessment for agricultural land suitable for irrigation is divided into 6 classes (Table 6). Class 1 is attributed with zero points and indicates areas with potential absence of available water resources for irrigation and is defined as an area with 'distinct drought risk'. Class 6 is attributed with 400 availability points, as all water resources (included in the research) are potentiality available for irrigation and is defined as area with virtually no drought risk if proper measures are undertaken. Intermediate classes between 2 and 5 have one or more restricted water resources and/or one or more of the unlimited water resources suitable for irrigation.

| Class | | Sum of points | Definition of water resources availability |
|---|---|---|---|
| Number | Drought risk | | |
| 1 | Distinct | 0 | No available water resources |
| 2 | Very high | 1 - 99 | Only water resources with limited availability |
| 3 | High | 100 - 199 | One water resource with unlimited availability and/or more with limited availability |
| 4 | Medium | 200 - 299 | Two water resources with unlimited availability and/or more with limited availability |
| 5 | Low | 300 - 399 | Three water resources with unlimited availability and/or more with limited availability |
| 6 | None | 400 | All water resources with unlimited availability |

**Table 6.** Determination of risk classes of agricultural land suitable for irrigation in case of drought from the sum of availability points of water resources for irrigation

## 3. Results

Due to the characteristics of the spatial analysis of the raster layers (raster cells) with the ArcGIS program tool (Spatial Analyst Tools), areas of certain land use classes and total area of agricultural land suitable for irrigation were slightly lower in comparison with the real situation. However, in the results section we primarily operate with shares of areas, describing availability points of water resources and drought risk classes.

### 3.1. Water resources availability assessment

Slovenia has unevenly distributed water resources suitable for irrigation as can be seen from the spatial analysis of availability points (Figure 7) in terms of the dry year with five-year return period.

We detected high availability (151-399 points) of water resources for irrigation in river valleys with alluvial soils (rivers Sava, Drava, Mura, Krka and Vipava), where there is, in addition to surface watercourses, also an easily accessible groundwater and in certain areas (river Vipava) large reservoirs (10 % of case study area) (Table 7). In more than 69 % of the case studies water resources for irrigation is rather poorly available (only 100-151 points), which are mostly a combination of groundwater and surface runoff. On more than 17 % of case study areas, available water resources are extremely low (25 - 99 points), with nearly 3 % of area having only low available groundwater (less than 25 points), whose availability for irrigation is in question due to the high costs associated with borehole drilling.

| Availability points classes | Area | |
| --- | --- | --- |
| | ha | % |
| Undefined (urban, rocks, water) | 60.896,5 | 3,01 |
| 0 | 0 | 0,00 |
| 1 - 25 | 54.137,5 | 2,68 |
| 26 - 50 | 173.185,7 | 8,57 |
| 51 - 99 | 122.795,3 | 6,08 |
| 100 - 150 | 1.406.312,6 | 69,61 |
| 151 - 199 | 34.017,6 | 1,68 |
| 200 - 250 | 154.698,3 | 7,66 |
| 251 - 299 | 7.758,8 | 0,38 |
| 300 - 350 | 6.401,5 | 0,32 |
| 351 - 399 | 114,3 | 0,01 |
| 400 | 0 | 0,00 |
| **Total** | **2.020.318,1** | **100,00** |

**Table 7.** Areas (%, ha) classes of availability of water resources for irrigation based on figure 7 for total area of Slovenia

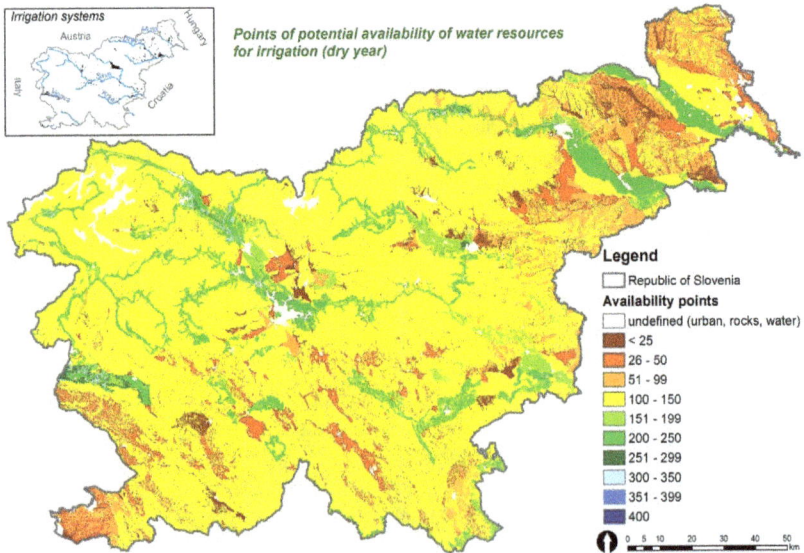

**Figure 7.** Points of potential availability of water resources for irrigation (based on table 3) in Slovenia at 1 ha resolution (100×100 m); dry year with five years return period (80-90 % probability of occurrence)

## 3.2. Drought risk assessment

The map of potential availability of water resources for irrigation was further adjusted and classified in accordance to the potential drought risk (Table 6), thus creating a map of agricultural land suitable for irrigation yet exposed to drought risk at the dry year with five year return period (Figure 8).

We conducted a spatial analysis of agricultural land suitable for irrigation in the case study area to define the availability of water resources for irrigation, and to define potential areas of drought risk. We identified areas of agricultural land at none, low, medium, high, very high and distinct drought risk. Analysis of the potential drought risks of agricultural land suitable for irrigation showed that more than 34 % (75,868 ha) of the case study agricultural land suitable for irrigation is located in areas of very high drought risk (1-99 points). Nearly 50 % of agricultural land (109,231 ha) is located in areas of high drought risk (100-199 points) and almost 15 % (33,010 ha) in areas of medium drought risk (200-299). Low drought risk (300-399) is present in only 0.2 % of agricultural land (442 ha) and is therefore negligible at the macro scale. Based on this analysis we argue that areas of medium and low drought risk should not suffer from water scarcity or drought causing damage in crops production and limiting crop yield, if appropriate infrastructure and systems for water transport and irrigation are installed, maintained and used in these areas. Research analysis did not detect any areas of agricultural land use suitable for irrigation at either absolute extremity of drought risk (0 points and 400 points).

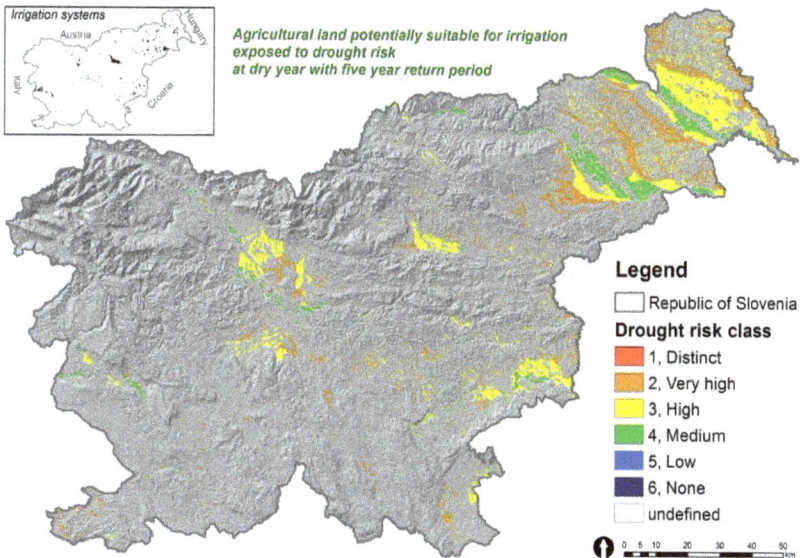

**Figure 8.** Agricultural land potentially suitable for irrigation and exposed to drought risk at dry year with five year return period in Slovenia at 1 ha resolution (100×100 m) (based on table 4)

## 4. Conclusions

This chapter presents a novel methodological approach and findings which substantially contribute to the understanding of spatial water resources availability and drought risk assessment of agricultural land. The methodology is clear, practical and therefore generally applicable in any region or on a global level. Methodology is open to adding other water resources, not presented here (e.g. waste water), in to the water resources availability assessment.

When the spatial analysis of available water quantities for irrigation from water resources is prepared for a certain area (region, state, catchment), it is essential to cooperate with all organizations engaged in regulating water management (e.g. environmental agencies, water and geological institutes and responsible governmental bodies). Water quantities available for irrigation from different water resources are usually regulated by state legislation defining minimal water quantities in the surface watercourses or reservoirs to sustain ecological acceptable flows, for the survival of the organisms in the water bodies. Legislation should also include consideration of the share of total water quantity in the water body which can be abstracted for irrigation of agricultural land, and the share of the water quantity in the water body at ecological acceptable flow that is especially reserved for agriculture and can be abstracted for irrigational purposes. Water reservoirs usually have, in addition, operational regulations defining the share of water reserved for agriculture, recreational activities, or for the conservation of wildlife habitats. Legislation and regulation are key factors to preventing over exploitation of water resources.

Spatial analysis of potentially needed water quantities for irrigation should be based on land use classes, types of crops and crop management. This is especially important in the case of crops with high water demand. Furthermore, spatial analysis should include physical and hydrological properties of soils in the area. This is important if soils in the area are light, with a high share of sand, high hydraulic conductivity and low available water capacity. Finally, it is crucial to define the irrigation norm (maximum, average and minimum) for all types of soils and crops grown in the area. This kind of analysis has to be done in cooperation with soil hydrologists, plant physiologists, agro-meteorologists and specialist technicians in irrigation systems.

To define accessibility or abundance of water resources in this study, we choose to use availability points as a number from 0 to 100. Water accessibility points for water reservoirs and watercourses were determined by the extent of agricultural land (%), which may be irrigated with the water assigned for agricultural use from both sources (0 to 100 points). Water accessibility classes for groundwater were determined on the basis of the hydrogeological map and average cost for borehole drilling, and put into three classes: hard (25 points), medium (50 points) and easy (100 points), defining the availability of groundwater. The determination of abundance points was based on the maximum irrigation norm (drip irrigation) for one hectare of permanent crops (orchards) on light soils and its corresponding optimal reservoir volume for irrigation. If in irrigation area was enough of water for irrigation of orchard on light soils with drip irrigation and maximum irrigation norm, it was given 100 availability points (Table 4). Each subsequent class was determined by 25 availability points less, as it does not facilitate sufficient quantities of surface runoff water for irrigation of all groups of agricultural plants.

Drought risk classes have to be developed in a careful manner with a clear distinction between classes. A maximum of six classes is recommended, to maintain comprehensibility and transparency for the reader. Aggregation of classes is useful, but must include sufficient information for the reader to understand the data. The scale needs to have extreme classes which represent areas without potentially available water resources for irrigation and areas with all potential water resources fully available.

Practical applications of the geospatial analysis of water resources for sustainable agricultural water use are numerous. The results are important for identifying areas on regional and global level which are best suited for irrigation development in terms of water resources availability. Results are important as they help areas suffering from periodic droughts to draw governmental attention. This is important as these areas require financial investment in irrigation equipment and irrigation technologies. It helps small growers in remote hilly or karst areas to identify reliable water resources. The results define areas suitable for building small water reservoirs for accumulated surface runoff water, which can help small farm businesses with vegetable or fruit production to be water independent in the drought periods. This is especially important for the population and agriculture businesses in dry, temperate and continental climates with high seasonal differences in precipitation and evapotranspiration.

## Acknowledgements

Financial support for this study was provided by the Ministry of Agriculture and the Environment of the Republic of Slovenia and Slovenian Research Agency as part of Targeted Research Program - The competitiveness of Slovenia 2006 - 2013 in 2010. Research project number: V4-1066.

## Author details

Matjaž Glavan, Rozalija Cvejić, Matjaž Tratnik and Marina Pintar

University of Ljubljana, Biotechnical Faculty, Agronomy Department, Chair for Agrometeorology, Agricultural Land Management, Economics and Rural Development, Ljubljana, Slovenia

## References

[1] Muralidharan D, Knapp KC. Spatial dynamics of water management in irrigated agriculture. Water Resources Research 2009;45 1-13.

[2] Turral H, Svendsen M, Faures JM. Investing in irrigation: Reviewing the past and looking to the future. Agricultural Water Management 2010;97(4) 551-560.

[3] Wriedt G, Van der Velde M, Aloe A, Bouraoui F. Estimating irrigation water requirements in Europe. Journal of Hydrology 2009;373(3-4) 527-544.

[4] Oki T, Kanae S. Global hydrological cycles and world water resources. Science 2006; 313 (5790) 1068-1072.

[5] Hoekstra AY, Mekonnen MM, Chapagain AK, Mathews RE, Richter BD. Global Monthly Water Scarcity: Blue Water Footprints versus Blue Water Availability. Plos One 2012; 7 (2) 1-9.

[6] Wisser D, Frolking S, Douglas EM, Fekete BM, Schumann AH, Vorosmarty CJ. The significance of local water resources captured in small reservoirs for crop production - A global-scale analysis. Journal of Hydrology 2010;384(3-4) 264-275.

[7] Matičič B, Kravanja N, Jug M, Lobnik F, Prus T, Rupreht J, Šporar M, Vrščaj B, Kočar I, Knapič M, Hočevar A, Kajfež Bogataj L, Avbelj L, Feguš M, Jarc A, Vrevc S, Bitenc D, Četina A, Pajnar N, Vadnal K, Mikluš I, Snučič F, Pavlovčič M, Tajnšek T, Osvald J, Štampar F, Korošec Koruza Z, Čop J, Vončina S, Vuga I, Ozbič. F, Pišot M, Jereb V, Komel L, Skalin B, Udrih R, Škafar L, Maruša T, Mesarec S, Kovačič I, Pirc V, Juvan S, Burja D, Pintar M, Anzeljc D. Nacionalni program namakanja Republike Slovenije - National Irrigation Programme of the Republic of Slovenia. Ljubljana: University of Ljubljana - Biotechnical Faculty, Ministry for Agriculture, Forestry and Food; 1994.

[8] Pintar M, Tratnik M, Cvejić R, Bizjak A, Meljo J, Kregar M, Zakrajšek J, Kolman G, Bremec U, Drev D, Mohorko T, Steinman F, Kozelj K, Prešeren T, Kozelj D, Urbanc J, Mezga K. Ocena vodnih perspektiv na območju Slovenije in možnosti rabe vode v kmetijski pridelavi - Water perspectives of Slovenia and the possibility of water use in agriculture (V4-0487) Ljubljana: University of Ljubljana - Biotechnical Faculty, Ministry for Agriculture and the Environment; 2010. http://www.bf.uni-lj.si/agronomija/o-oddelku/katedre-in-druge-org-enote/za-agrometeorologijo-urejanje-kmetijskega-prostora-ter-ekonomiko-in-razvoj-podezelja/urejanje-kmetijskega-prostora/ (accessed 1 December 2010).

[9] CARS. Audit report on the efficiency management of the Republic of Slovenia in the prevention and elimination of consequences of drought in agriculture - Revizijsko poročilo o smotrnosti ravnanja Republike Slovenije pri preprečevanju in odpravi posledic suše v kmetijstvu - No. 1207-3/2006-22. Ljubljana: Court of Audit of the Republic    of    Slovenia;    2007.    http://www.rs-rs.si/rsrs/rsrs.nsf/I/K99638A13FF506FB3C1257322003D2E6B (accessed 15 January 2010).

[10] EEA. Water resources across Europe — confronting water scarcity and drought European Environment Agency. http://www.eea.europa.eu/publications/water-resources-across-europe (accessed 1 October 2011).

[11] Bergant K, Kajfež-Bogataj L. N-PLS regression as empirical downscaling tool in climate change studies. Theoretical and Applied Climatology 2005;81(1-2) 11-23.

[12] Christensen JH, Christensen OB. A summary of the PRUDENCE model projections of changes in European climate by the end of this century. Climatic Change 2007;817-30.

[13] van der Linden P, Mitchell JFB. ENSEMBLES: Climate Change and its Impacts: Summary of research and results from the ENSEMBLES project. Exeter: Met Office Hadley Centre; 2009. http://www.eea.europa.eu/data-and-maps/indicators/global-and-european-temperature/ensembles-climate-change-and-its (accessed 10 July 2011).

[14] Bergant K. The climate in the future - what do we know about it? - Podnebje v prihodnosti - koliko vemo o njem?. Slovenian Environmental Agency. http://www.arso.gov.si/podnebne%20spremembe/Podnebje%20v%20prihodnosti/ (accessed 13 November 2010).

[15] Medved-Cvikl B, Ceglar A, Kajfez-Bogataj L. Interoperability in Drought Monitoring. Geodetski Vestnik 2011;55(1) 70-86.

[16] Ceglar A, Kajfez-Bogataj L. Simulation of maize yield in current and changed climatic conditions: Addressing modelling uncertainties and the importance of bias correction in climate model simulations. European Journal of Agronomy 2012;37(1) 83-95.

[17] Glavan M, Pintar M. Strengths, Weaknesses, Opportunities and Threats of Catchment Modelling with Soil and Water Assessment Tool (SWAT) Model. In: Nayak P (ed.) Water Resources Management and Modeling. Rijeka, Croatia: InTech; 2012. p310. Available from http://www.intechopen.com/books/water-resources-management-and-modeling/strengths-weaknesses-opportunities-and-threats-of-catchment-modelling-with-soil-and-water-assessment (accessed 20 June 2012)

[18] Cvejić R, Tratnik M, Meljo J, Bizjak A, Prešeren T, Kompare K, Steinman F, Mezga K, Urbanc J, Pintar M. Permanently Protected Agricultural Land and the Location of Water Sources Suitable for Irrigation. Geodetski Vestnik 2012;56(2) 308-324.

[19] Pintar M, Glavan M, Meljo J, Zupan M, Fazarinc R, Podboj M, Tratnik M, Cvejić R, Zupanc V, Kregar M, Krajčič J, Bizjak A. Projekcija vodnih količin za namakanje v Sloveniji - Projections of water quantities for irrigation in Slovenia (V4-1066) Ljubljana: University of Ljubljana - Biotechnical Faculty, Ministry for Agriculture and the Environment 2012. http://www.bf.uni-lj.si/agronomija/o-oddelku/katedre-in-druge-org-enote/za-agrometeorologijo-urejanje-kmetijskega-prostora-ter-ekonomiko-in-razvoj-podezelja/urejanje-kmetijskega-prostora/ (accessed 10 April 2012).

[20] MAE. Graphical spatial data of actual land use. Ministry for Agriculture and the Environment http://rkg.gov.si/GERK/ (accessed 30 January 2010).

[21] SORS. Total area (ha) of agricultural land prepared for irrigation and actually irrigated in Slovenia. Statistical office of the Republic of Slovenia. http://pxweb.stat.si/

pxweb/Dialog/varval.asp?ma=2722202S&ti=&path=../Database/Okolje/27_okolje/
03_27193_voda/04_27222_namakanje/&lang=2 (accessed 15 July 2012).

[22]  Prestor J. Nacionalna baza hidrogeoloških podaktov za opredelitev teles podzemne
      vode Republike Slovenije - National database of hydrogeological information for the
      identification of groundwater bodies of the Republic of Slovenia. Ljubljana: Geologi-
      cal Survey of Slovenia; 2006.

[23]  Pintar M. Določitev izhodiščnih parametrov za rabo vode za namakanje kmetijskih
      površin glede na klimo, tla in tipične kulture (C-769) - Determining basic parameters
      for the use of water for irrigation of agricultural land in relation to climate, soil and
      the typical culture. Ljubljana: Water Management Institute, Ministry of Environment
      and Spatial Planning of the Republic of Slovenia, The Slovenian Environmental Pro-
      tection Administration Office; 1998.

[24]  Bat M, Frantar P, Dolinar M, Fridl J. Vodna bilanca Slovenije 1971-2000 = Water bal-
      ance of Slovenia 1971-2000 Ljubljana: Ministry for Environment and Spatial Plan-
      ning,Slovenian Environmental Agency; 2008. http://www.arso.gov.si/vode/poro
      %C4%8Dila%20in%20publikacije/ (accessed 2 March 2011).

# Chapter 9

# Impact of Drought and Land – Use Changes on Surface – Water Quality and Quantity: The Sahelian Paradox

Luc Descroix, Ibrahim Bouzou Moussa, Pierre Genthon,
Daniel Sighomnou, Gil Mahé, Ibrahim Mamadou,
Jean-Pierre Vandervaere, Emmanuèle Gautier,
Oumarou Faran Maiga, Jean-Louis Rajot,
Moussa Malam Abdou, Nadine Dessay, Aghali Ingatan,
Ibrahim Noma, Kadidiatou Souley Yéro, Harouna Karambiri,
Rasmus Fensholt, Jean Albergel and Jean-Claude Olivry

Additional information is available at the end of the chapter

## 1. Introduction

West Africa has been experiencing drought conditions since the end of the 1960s. This pattern has been particularly evident in the Sahel, but appears to have attenuated in the last decade in the eastern and central parts of this region. On the other hand, annual rainfall remains very low in the western part of the Sahel [1].

A corresponding decrease has also been observed in the mean annual discharge of the Senegal and Niger rivers, which are the largest in the region and primarily fed by water originating from tropical humid areas. However the percentage decrease in mean annual discharge was almost twice as large as the decrease in rainfall [2] for the period 1970-2010. Similar trends have been observed on smaller river systems.

In contrast, even though the Sahel and most of West Africa also have experienced substantial drought over the past 40 years, runoff coefficients and stream flows have increased in most Sahelian areas. This phenomenon has been named "The Sahelian Paradox" after the increase of the groundwater table in Niger since the 1960s was named the Niamey paradox and attributed to substantial changes in land-use. The HAPEX-Sahel (Hydrological and Atmospheric Pilot Experiment) and the AMMA (African Monsoon Multidisciplinary Analysis) programs have provided, among many comprehensive results, valuable measurements dealing with the spatial and temporal variations in Sahelian soil water content as well as with the infiltration of water through deep soil layers of the vadose zone.

The purpose of this chapter is to provide an overview of hydrological behaviour throughout West Africa based on point, local, meso and regional scales observations.


## 2. Background

The paradoxical increase in runoff despite drought conditions in sub-Saharan Africa was first noted in a paper by Albergel [3], analysing decadal series of runoff measurements in experimental sites of Burkina Faso. He noticed that this increase was observed in Sahelian areas, but not in the more humid Sudanian regions.

*"The decrease in rainfall during the 1969-1983 period seems to be largely offset by the evolution of surface features in the functioning of small catchments. These changes favoured the conditions of runoff in the Sahelian basins; there are due to both the human actions and the climatic conditions. The reduction of vegetation cover and the widespread crops areas cause soil surface settling and the appearance of impervious superficial layers, as well as the extension of eroded areas. Some sahelian basins have nowadays [in 1987] the common characteristics of basins located northward, with great areas of bare soils; perennial graminaceae are replaced with annual ones, and combretaceae with prickly bush species" [3].*

Albergel [3] attributed the contrasting behaviour of Sudanian (mean annual rainfall > 750 mm) and Sahelian (mean annual rainfall < 750 mm) areas to increasing bare soils and decreasing vegetation cover in Sahelian basins.

This hypothesis was confirmed in 1999 by Mahé and Olivry [4] and then in 2002 by Olivry [2], who remarked that the discharge of right bank tributaries of Middle Niger River had been increasing since the beginning of the Drought (1968). Similarly, Amani and Nguetora [5] noted that runoff coefficients were increasing significantly in right bank tributaries and showed that the onset of the annual flood was occurring earlier than in previous decades.

Mahé *et al.* [6] analysed the runoff evolution of eight right bank tributaries of the Middle Niger River and noted that the decrease in rainfall did not lead to a decrease in runoff under the Sahelian climate as commonly observed in other basins in the world. Rather, these tributaries exhibited increasing runoff coefficients and in discharges, while "Sudanian" climate tributaries suffered a decrease in discharge and in runoff coefficient [6].


## 3. Material and methods

This study is mainly based on two sources of data:

- field measurements and observations made during the AMMA (African Monsoon Multidisciplinary Analysis) experiment at the Niger experimental site (Niger River middle stretch and Niamey square degree), and:

- rainfall and discharge data collected on the operational network of Niger basin, provided by the Met Offices of Niger, Mali and Burkina Faso and by NBA (Niger Basin Authority).

The methods included the following:

- Analyse of runoff and river discharge data (in order to characterize the trends in the river discharge records) at several scales:

   ○ At the point scale: infiltration tests (using disk infiltrometers at multiple suctions) and soil water content monitoring provides data on soil hydraulic conductivity and other physical properties [7] [8];

   ○ At the local scale: Tondi Kiboro and Wankama catchments, as well as 20 experimental plots of 10 and 100 m², located in the same catchments. These data were collected during the AMMA experiment (2004-2010). On the plots, the measurements were made after each event; on the catchment, stream gauges allowed the monitoring of the discharges[9];

   ○ At the meso scale:

      ▪ Some small direct tributaries of the Niger River; we overall use here the regional balance allowed by the stations located in the Niger River upstream (Kandadji) and downstream (Niamey) the studied stretch; however, some discharge data of small direct tributaries were collected for this study [10];

      ▪ The main tributaries of Niger River's middle stretch.

   ○ At the regional scale: the Niger River, the Senegal River, etc, existing data, allowed these analyses [10] [11].

- Analyses of land cover data (including agricultural data, NDVI etc.) and a map of land cover in the square degree of Niamey were realised during AMMA experiment;

- Analyses of precipitation trends across the Sahel ;

- Analysis of endorheism breaks was carried out in the region of Niamey [11].

# 4. Decrease in rainfall, increase in runoff

## 4.1. Increasing streamflows

The Great Drought of the Sahel is considered one of the most significant climatic events worldwide [12]. For at least 25 years, more than 3 millions km² of semi-arid Sub-Saharan area has suffered a rainfall deficit ranging from 10 to 30%, depending on the location. Figure 1 shows a partial offset of the deficit since the mid-1990s. However, the overall deficit remains and the interannual variability has increased during this period. Not shown in this figure, the intensity of the drought has been largely attenuated in the eastern part of West Africa since the 2000s, but it persists in the western part of the region. In spite of the severe drought, a significant increase in the runoff coefficient and a general increase in stream discharge, have been observed (see background) since the 1960s, in the Sahelian basins (Figure 2).

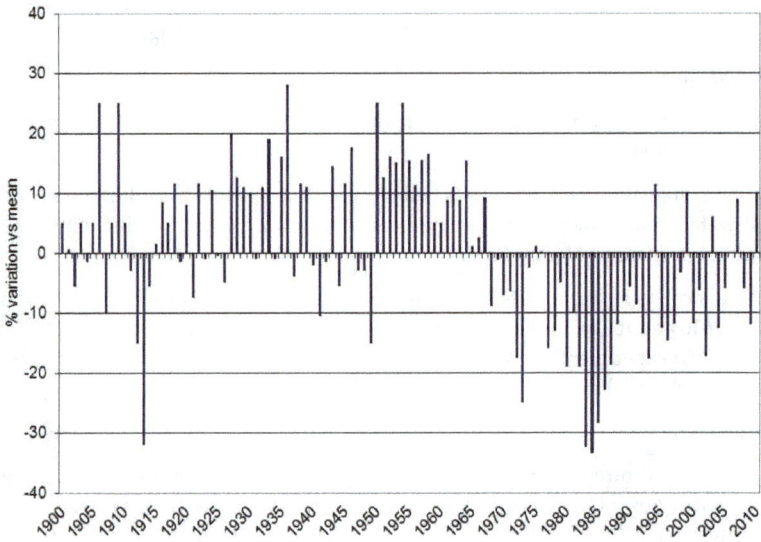

**Figure 1.** Evolution (1900-2010) of the Standard Precipitation Index for the whole Niger River basin

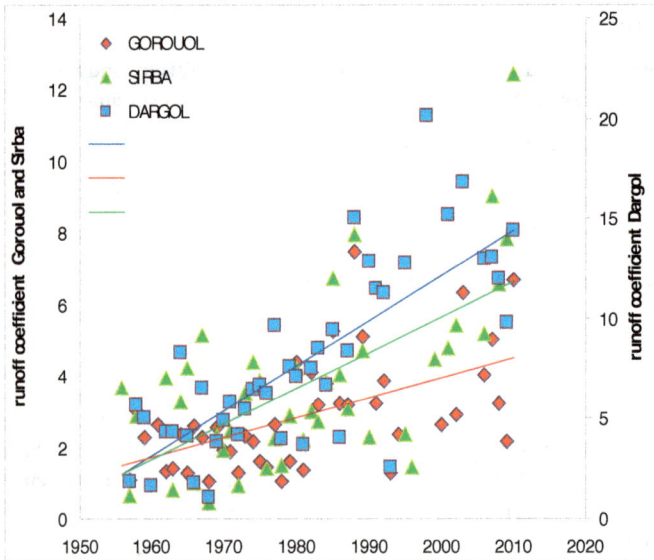

**Figure 2.** Evolution (1955-2010) of the runoff coefficient of three main right bank tributaries of Niger River.

Figure 3 shows the location of the middle Niger River basin in West Africa. Tributary rivers accounted here are located in the eastern blue circle representing the middle basin. Due to the great loop created by the Niger river northward to the margin of the Saharan desert, the annual flood downstream from the Niger Inner Delta -a large humid area located in the northern reach of the river-, in the Middle Niger River, has two flood peaks. The first one, termed the red or local flood, arises from local rainfall draining through a series of tributaries (Fig. 3, enlargement) and occurs between August and September. The second, termed the Guinean flood, originates from precipitation in the Fouta Djallon (Guinea) area during the rainy season (June–September). Delayed by the crossing of the Niger Inner Delta (see Fig. 3) in Malian territory, the Guinean flood takes place around January. The clear separation between these two flood events makes it possible to distinguish the local Sahelian effect from the more remote trend.

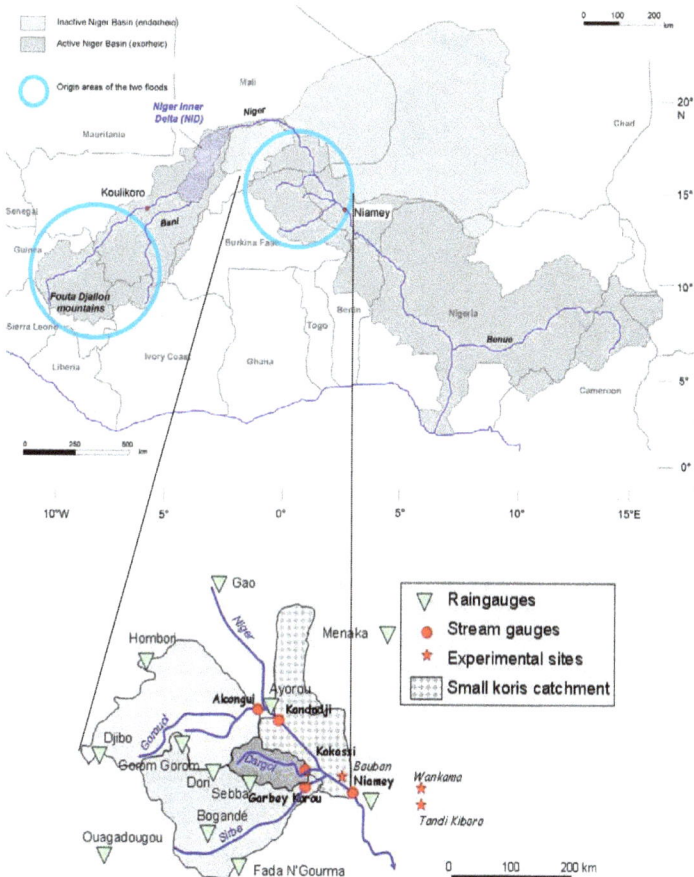

**Figure 3.** The Niger River basin and (enlargement) its middle basin

## 4.2. Earlier flooding

Another change observed during the drought is the earlier onset of yearly flooding, compared with previous periods. Figure 4 shows that the first flood is now arriving approximately forty days earlier than it did forty years ago. This observation is consistent with a decreased soil water holding capacity in the river basin.

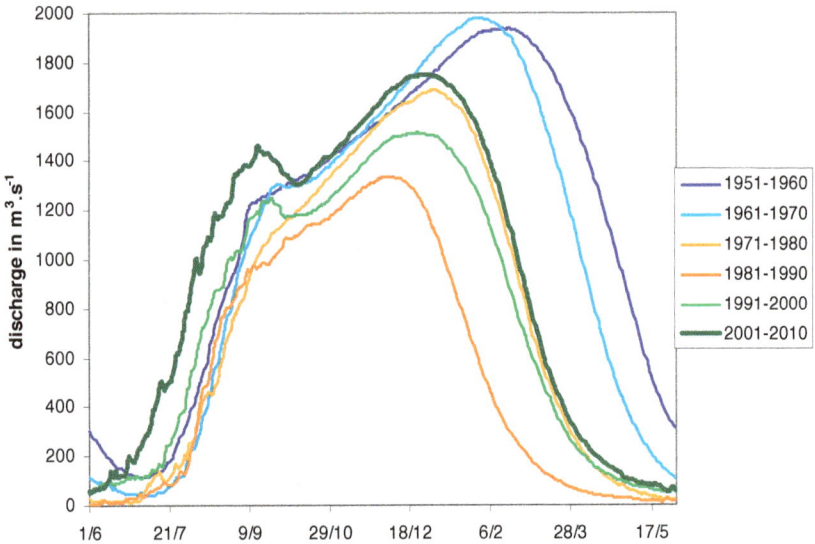

**Figure 4.** The Niger River decadal hydrographs at Niamey station (Republic of Niger)

## 4.3. Sahelian paradox is not due to rainfall

Because rainfall has decreased significantly since 1968, rainfall amount does not explain the increase in runoff and stream flow or the earlier flood occurrence. For example, the three main tributaries affected by the semi-arid Sahelian climate (The Gorouol, the Dargol and the Sirba rivers: see enlargement of figure 3) experienced a significant increase in runoff coefficient during the drought, despite a 20% reduction in rainfall (Table 1).

To identify the drivers of increased stream discharge and early flood onset, we analysed the trends and evolution of rainfall for twelve stations with daily rainfall data from 1950 onward. These stations are indicated in figure 3 [11].

A forward shift in the timing of the monsoon rains does also not appear to explain the Sahelian paradox. Nicholson [13] as well as Ali and Lebel [1] observed in recent decades a reduction of rainfall in August and a relative increase in rainfall amount in June and July. A similar forward shift in monsoon timing was observed in the Middle Niger River basin (figure 5). However the total amount of rainfall in June and July remains lower during the last two decades than during the 1950s and 1960s (figure 6).

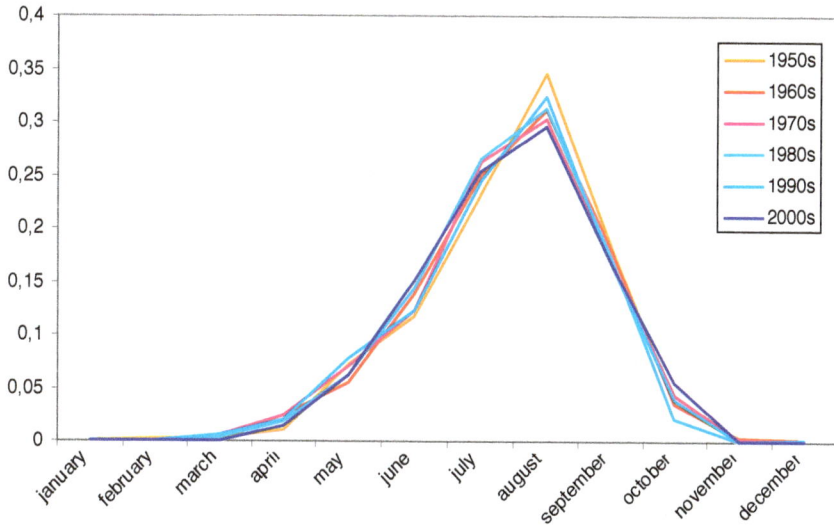

**Figure 5.** Evolution per decades of the rainfall monthly index (middle Niger River basin)

- Because the runoff coefficient increases with increasing rainfall intensity and amount, a rise in the number of extreme events might explain the higher volume of the floods. However a general decrease in the number of extreme rainfall events has been observed since the 1950s, for each class of events (figure 7), ranged by total amount of the event (classes 20-30 mm; 30-40 mm; 40-60 mm and more than 60 mm). However at the whole Sahel scale, a current study shows an increase in the rainfall amount for events upper than 30 mm during the 2000s.

- An increase in extreme events at the beginning of the rainy season also could explain the early flood occurrence. An increase in the total amount of rainfall fallen in events > 40 mm has been observed in the last decade, but only in June. However, the runoff coefficient is nowadays two or three times higher in the Middle Niger River basin than during the 1950s. Thus, the modest increase in rainfall amount (event > 40 mm) observed in June during the 2000s cannot alone explain the timing and magnitude of the recent floods.

| Basin | Gorouol | Sirba | Dargol |
|---|---|---|---|
| Period 1957-1979 | 1.9 | 2.6 | 5.0 |
| Periods 1980-1994 | 3.6 | 4.2 | 8.8 |
| Periods 1995-2010 | 4.3 | 6.0 | 14.2 |

**Table 1.** Evolution of runoff coefficients of the three main Niger River right bank tributaries from 1957.

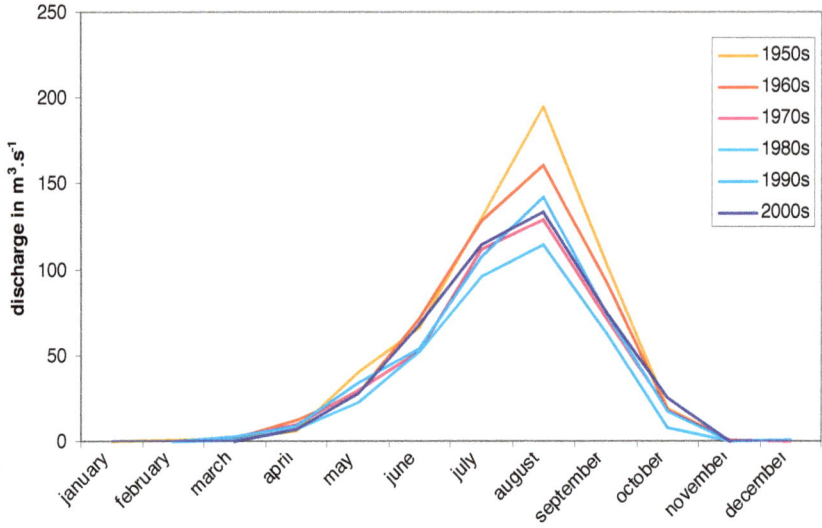

**Figure 6.** Evolution per decades of the monthly mean rainfall amount (middle Niger River basin)

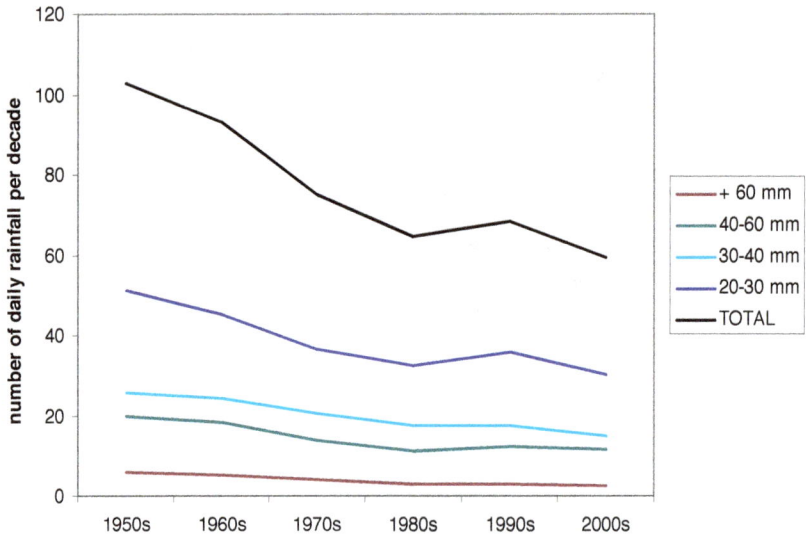

**Figure 7.** Evolution per decades of the number of extreme events (ranged by total rainfall amount of the rainy event), middle Niger River basin

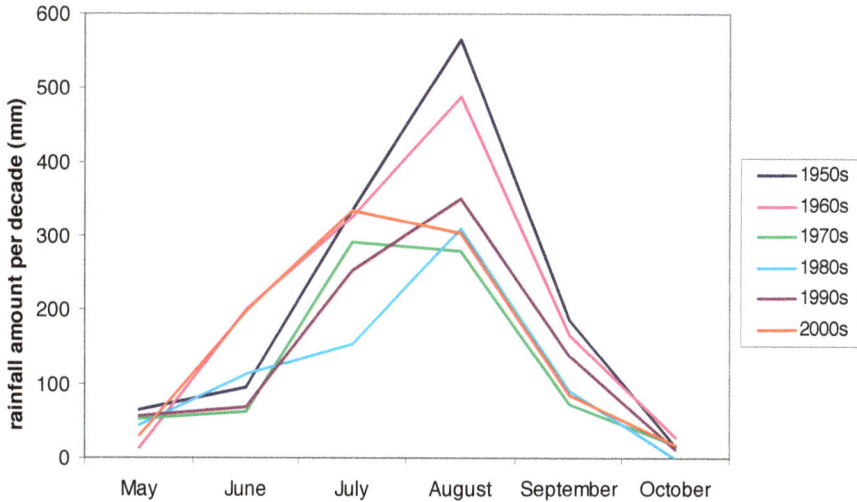

**Figure 8.** Evolution per decades of monthly rainfall amount of events > 40 mm Middle Niger River basin

## 5. Contribution of changing soil characteristics to Sahelian Paradox?

In Sahelian areas at all scales, runoff coefficient generally increased along with river discharges (Table 2). In most cases, these changes correlate with a decrease in vegetation cover, due to land use changes including increasing crop area, overgrazing and wood harvesting. At the regional scale, the change in vegetation cover is not obvious. The runoff coefficient of degraded soils in the Sahel (close to 60% for the ERO type crusted soils –as defined by Casenave and Valentin [14])- is much higher than that observed for millet crops (4%) and for bush and fallows (10%; see Table 1 and figure 9). Small crusted soil areas can alone explain increased stream flows. Therefore increasing runoff coefficients may be consistent with the re-greening documented in remote sensing studies (higher values of NDVI), the remaining soils being covered by higher millet crops density, graminaceae, herbaceae and annual plants than during previous periods.

Runoff coefficients are measured in plots of 10 and 100 m²; saturated hydraulic conductivity is measured in at least 20 points for each surfaces class [11].

Hydrodynamic characteristics differ substantially between different types of soil surface features. Given the strong difference in runoff coefficients, land use /land cover evolution could probably explain the runoff increase and "re-greening" is evident in the Sahelian area (see section 9 below). However, what was rather observed within the AMMA experimental sites

**Figure 9.** Opposition between a "fallow on common sandy soil" (left) plot and a "fallow erosion (ERO) crusted soil" plot (right)

| Soil surface feature | Runoff coefficient % | saturated hydraulic conductivity mm.h-1 |
|---|---|---|
| Millet on common sandy soil | 4.0 +/- 1.4 | 172 +/- 79 (20)* |
| Fallow on common sandy soil | 10 +/- 4 | 79 +/- 41 (20) |
| Old fallow with bioderm | 25 +/- 7 | 18 +/- 12 (30) |
| Millet and fallow erosion (ERO) crusted soils | 60 +/- 8 | 10 +/- 5 (30) |

number of repetitions

**Table 2.** Comparison of the hydrodynamic properties of non-crusted and crusted soils.

| 1991-1994 | Eainfall | Runoff depth | Kr* | Rainfall/runoff | $r^2$ R = a P + b | Yearly runoff total duration in hours |
|---|---|---|---|---|---|---|
| TK amont | 513 | 180 | 0,36 | R = 0,56 P - 2,61 | 0,82 | 34,9 |
| TK aval | 513 | 133 | 0,26 | R = 0,43 P - 2,3 | 0,79 | 28,1 |
| TK bodo | 485 | 185 | 0,38 | R = 0,53 P - 2,14 | 0,68 | 62,7 |
| **2004-2009** | **rainfall** | **Runoff depth** | **Kr*** | **Rainfall/runoff** | **$r^2$ R = a P + b** | **Yearly runoff total duration in hours** |
| TK amont | 495 | 231 | 0,47 | R = 0,77 P – 4,9 | 0,85 | 34,2 |
| TK aval | 491 | 132 | 0,27 | R = 0,49 P – 3,5 | 0,74 | 18,2 |
| TK bodo (2007-2009) | 520 | 242 | 0,47 | R = 0,87 P - 7 | 0,81 | 25,9 |

Kr = runoff coefficient

**Table 3.** Runoff characteristics for the periods 1 (1991-1994) and 2 (2004-2009) on the three small Tondi Kiboro basins.

in Niger is a degradation of soils and vegetation during the last decades, without a noticeable recovery since the mid 1990s. In the Tondi Kiboro catchments, the area of degraded soils (mostly ERO crusted soils) in 2007 was twice that observed in 1993 (figure 10). As a matter of fact, the runoff coefficients were significantly increasing from the 1991-1994 to the 2004-2009 period (Table 3), as a consequence of the reduction of soil water holding capacity and the rise in crusted surfaces.

In the Wankama experimental catchment, the same evolution of land cover is observed; but there is no historical hydrological data available for comparison. However, figure 11 allows comparing the vegetation cover in 1950 (aerial photos) and in 2007 (pictures taken from a PIXY® drone; [15]).

In the lower part of this basin, the "ERO" crust areas are widespread and constitute a very active contributing area; 70% of the surface is covered by "ERO" crust (see Fig. 9 at right hand side). The ERO crust runoff coefficient is approximately 60% (Table 2) while it is only 4% in pearl millet crops and 10% in the fallow. An increase in "ERO" crusted area must then have strong hydrological consequences.

The "small koris" catchments (see Fig.3) are not gauged. However, taking the difference between the discharge at Niamey station, on the other hand, the sum of discharges of Niger at Kandadji and those of the Sirba and the Dargol (the two tributaries feeding Niger river between Kandadji and Niamey, Fig.3) [11] showed that a clear change in the behaviour of these "small koris" behaviour occurred after 1997. In general, between 1975 and 1996 the input volume of the Niger River at the Kandadji station was greater than the output volume at Niamey, presumably due to infiltration and evaporation losses. Since 1998, the opposite is observed, presumably due to new input from small tributaries, where crusted soils areas have increased in recent years. Some of these basins have shifted from endorheic prior to and during the 1990s to exorheic in recent years (see 6. section below), increasing the contribution to the Niger River from degraded areas with high runoff. These small tributaries have recently provided several billions cubic meters per year to the Niger River discharge [11] (Fig.12).

This cited study [11] was dedicated to the severe flooding of August and September 2010 in Niamey, were 300 ha in the river right bank were inundated. Twice, the level of Niger River reached a maximum during the rainy season (2030 $m^3.s^{-1}$ in early August, 2130 $m^3.s^{-1}$ in early September, the maximum previously registered value being 2000 $m^3.s^{-1}$). However, during August 2012, this maximum was widely exceeded. The discharge value reached 2473 $m^3.s^{-1}$ on 18th August, causing severe damage in the city of Niamey and extensive flooding downstream in Niger, Benin and Nigeria (Figure 13) [16].

Thus, previous studies show that from the measurement point scale to the meso-scale basin, the runoff increase is observable in the Sahelian area of the Middle Niger basin:

• at the point scale, it is shown that the new surface features created by land use change (crusted soils areas particularly) have a very low hydraulic conductivity and consequently high runoff coefficients (Table 2), the latter being measured in plots of 10 and 100 m²;

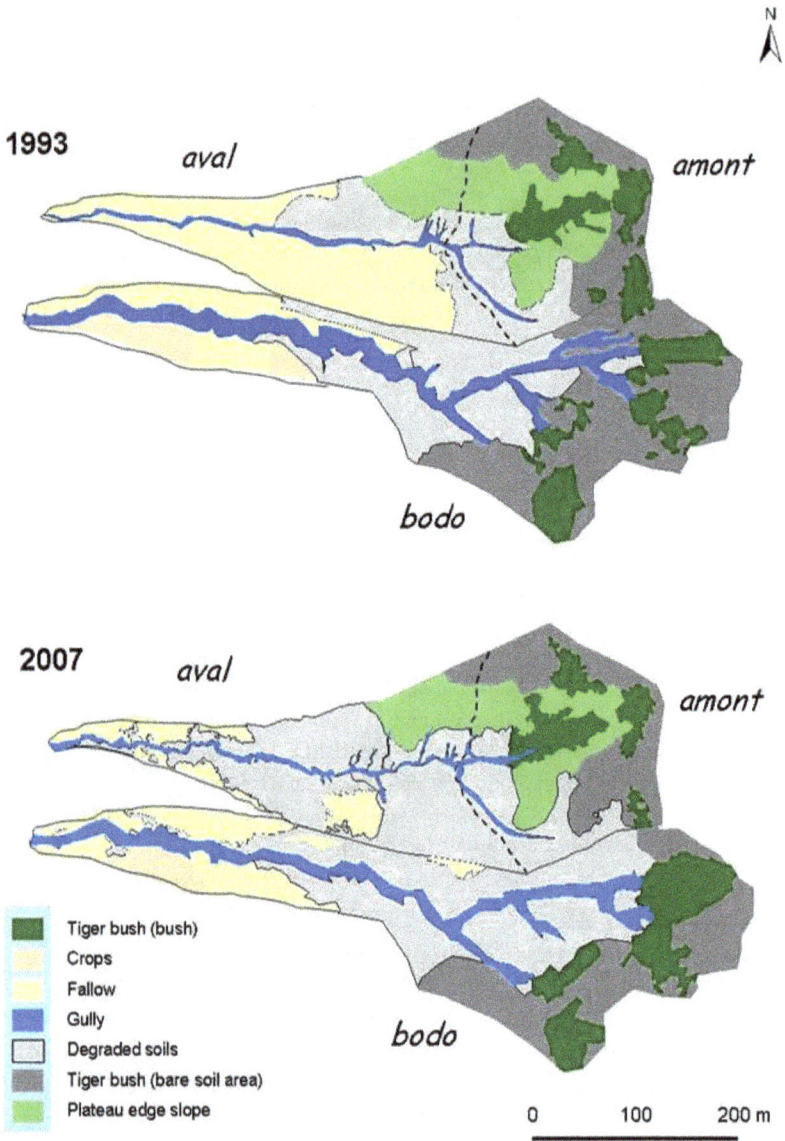

**Figure 10.** Land use over the Tondi Kiboro experimental catchments (Western Niger) in 1993 (up) and 2007 (down)

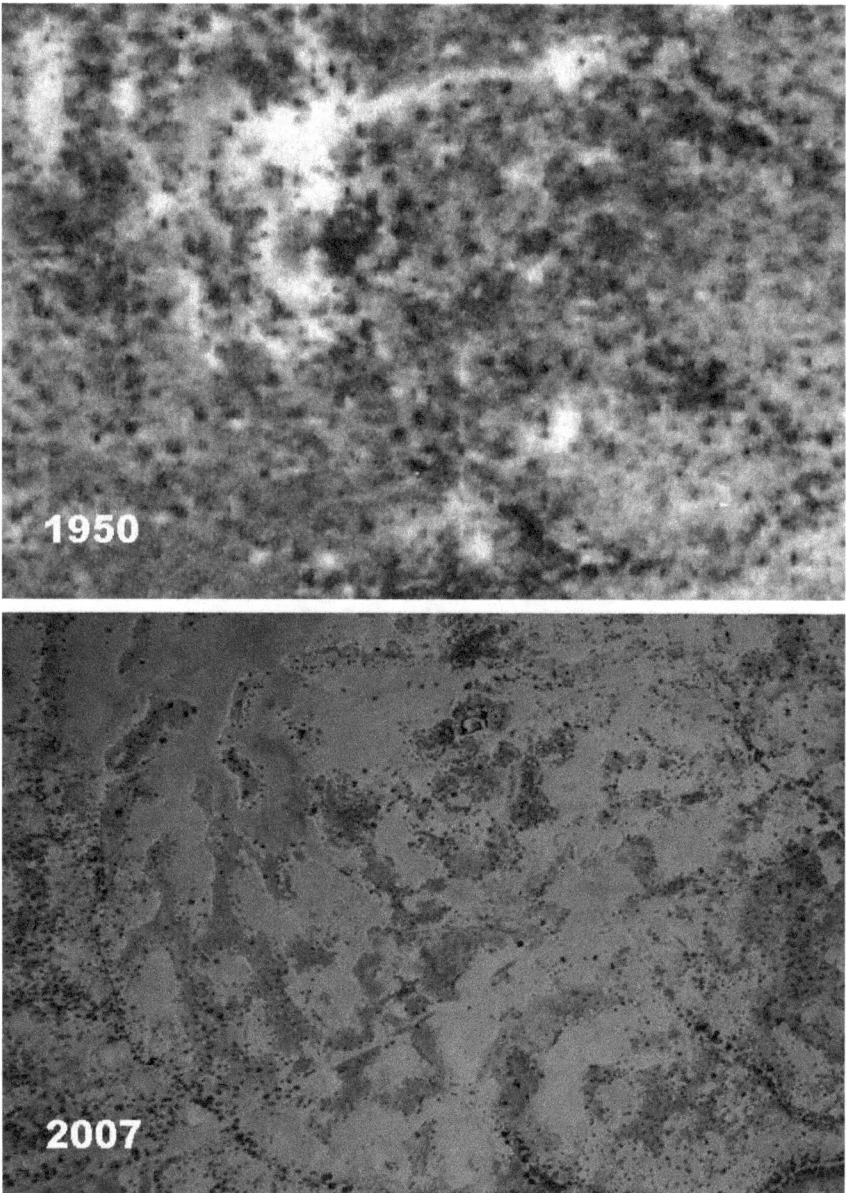

**Figure 11.** The lower part of Wankama experimental catchment in 1950 (up, aerial photo by IGN France) and in 2007 (photo taken from a PIXY drone); the generalisation of ERO crusted soils in a way of "hydro-aeolian depressions" is noticeable

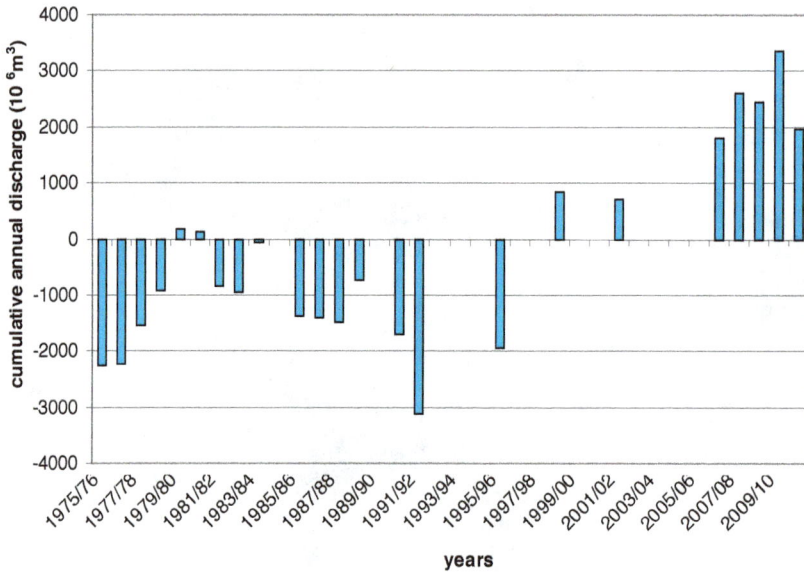

**Figure 12.** Evolution of the remaining water balance between Kandadji and Niamey from 1975 to 2010, after subtracting the discharge of the two main right bank tributaries (Dargol and Sirba); discharge of the Dargol and Sirba rivers were measured at the Kakassi and Garbey Korou stations (see Fig. 3).

- at the small basin scale, the experimental basins of Tondi Kiboro (12 ha) exhibited an increase in runoff and discharge probably due to the extension of crusted soil areas (Figure 10 and Table 3);

- the direct middle Niger river tributaries experienced a strong runoff increase after 1997 (Figure 12); the corresponding scale ranges from 10 to 2000 km²;

- the meso-scale Niger right bank tributaries basins range from 7000 km² (Dargol river basin) to 38,500 km² (Sirba) and 45,000 km² (Gorouol); they have shown a strong discharge increase since the beginning of the drought (figure 2); Amani and Nguetora [5] demonstrated that the flood was occurring during the 1980s almost one month earlier than during the 1960s, as a consequence of both decreasing vegetation cover and reduction in soil water holding capacity;

- the previous statements explain the significant rise in the first, red flood of the Niger river in its Middle basin (figure 13) and the earlier occurrence of this first flood (40 days earlier than during the 1970s as seen in figure 4).

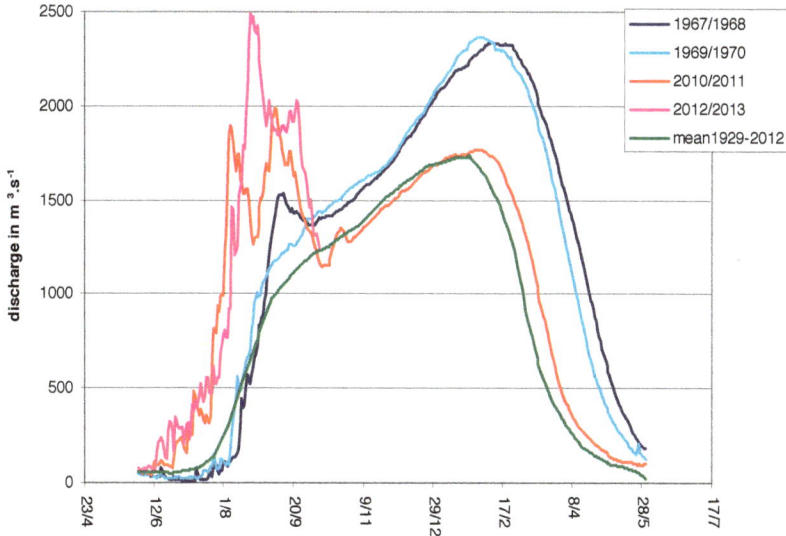

**Figure 13.** The 2012 red flood compared with previous remarkable years, included the two severe floods during the monsoon of 2010, and the pattern of the Guinean or black flood [16]

# 6. An increase in flood hazard

*"As it has been supposed from the end of the 1980's, the change in the hydrological functioning and the newly twin peak hydrograph" (of the Middle Niger river) "are linked to human factors, mostly to land use changes, particularly the land clearing and the extension of crops due to demographic pressure; this led to a soil baring and a fallow shortening which caused a soil crusting resulting in a severe decrease of soil infiltrability. This results in an increase in flood hazard" [11].*

The record high 2012 flood must remind policy makers that this hazard is becoming a big social and environmental concern to be accounted in land planning policy. Overall, environmental engineering must be performed and improved in order to offset and mitigate the effects of this trend: urban areas are firstly affected by flooding, but rural areas are those which have to be land managed in order to increase the soil water holding capacity. This should allow increasing land productivity and reducing the flood hazard downstream.

The immediate causes of the 2012 actual flood are not yet determined; however, rainfall amount was high, probably the highest measured since 1968 in the Middle Niger Basin. We focus here on the 2010 monsoon flood.

The first, small, mid July 2010 peak was due to all the sub-basins. It is worth noticing that the first of the two higher rainy season peaks (early August) was firstly due to discharges coming from Malian territory, upstream from Kandadji station and, then, to stream flows coming from the Gorouol basin. The second peak (early September) was mostly produced by flows coming from upstream from the Gorouol confluence, thus from the arid area of eastern Mali, and secondarily by the Sirba and Dargol rivers discharges. The small koris contribution is only estimated, as shown, by the difference between the discharge at Niamey and the sum of Niger at Kandadji, the Dargol and the Sirba. However it is obvious that these small koris had a large contribution to the first peak (figure 14).

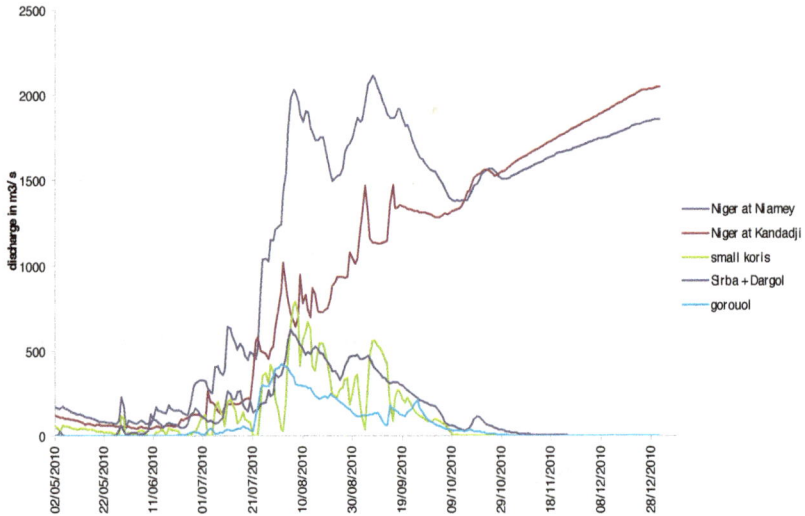

**Figure 14.** flood decomposition between different contributing areas.

The long term causes of these floods are the land use change and the increase in crusted soil areas. But recently, flood hazard was accentuated by three factors:

• the urbanisation

• the silting up

• the endorheic bursting

Urban population in Sub-Saharan Africa remains the lowest in the world. It is expected to increase strongly in the next years and decades. The percentage of urban population in the Sahel is only 30% in average (but 17% in Niger and 20% in Burkina Faso), compared with 50% in the southern West African countries on the shore of the Gulf of Guinea. The urban population in the Sahel is expected to reach 40% in 2030 and 50% in 2050. As the total population is increasing by 2.5-3% per year, it duplicates every 20-25 years. The urban population doubles every 14-15 years. Most of new resident come from rural areas and settle familial or informal housing. The latter is commonly settled on non-drained and non-buildable areas, which constitute the first areas being inundated in case of flooding. This was the case in 2003 at Saint Louis (Senegal), the 1st September 2009 at Ouagadougou (Burkina Faso) [17] or Agadès (Niger), in August 2010 and August 2012 at Niamey (Niger) [16], and at Dakar (Sénégal). Most of informal housing is built in adobe, straw or iron sheet, which is destroyed or severely damaged by flooding (Fig.15).

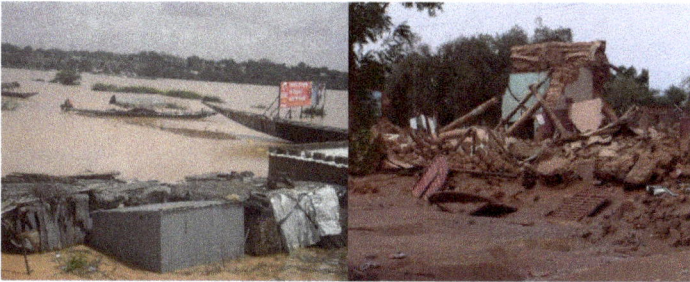

**Figure 15.** Flooded areas in Niamey (shore of Niger River) the 18th August 2012 (left) and in Agadès (right) the September 1st 2009 (*photo Ibrahim Noma and Baptiste Nay*)

In southern Sudanian areas [18] [19], the ongoing increase in the occurrence of inundations was highlighted during the last years. Flood related fatalities in Africa, as well as associated economic losses, have increased dramatically over the past half century [20]. This trend associated with urbanisation is expected to have dramatic consequences.

The second increasing flood hazard is the current silting up of river beds in the Sahelian region, linked to the observed erosion stage [21]. As runoff is rising, erosion is exacerbated, increasing sediment transport, leading to sandy deposits on the river beds, and contributing to flooding. This was the case for the last Niger River floods in Niamey, due to the invasion of its bed by alluvial fans coming from tributaries upstream from Niamey (Figure 16).

**Figure 16.** Evolution of the alluvial fan of the Kourtéré kori into the Niger river bed, just upstream from Niamey (Niger) [10]

A third increasing flood hazard in the Sahel is the increased exorheism. Some endorheic valleys became exorheic in recent years and decades, increasing the Niger river catchment area and the number of tributaries. Due to soil crusting, these new contributing areas have high runoff coefficients causing a rise in the available discharge for the Niger River (figure 17).

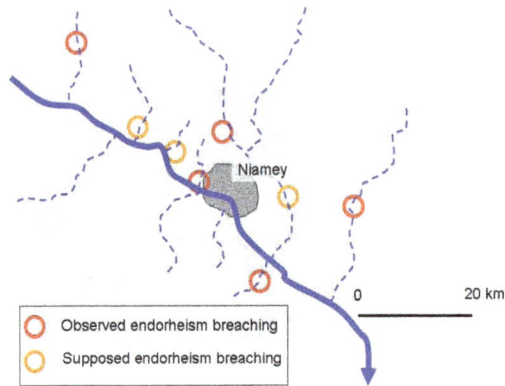

**Figure 17.** Observed and supposed endorheism bursting in the Niamey area of Niger River valley

## 7. Regionalisation

The runoff increase is not observed only in the Middle Niger river basin. The important contributions of Gil Mahé [4] [6] [22] [23] [24], before a first attempt of synthesis of Amogu *et al.*, [10], show the large extension of the increase in rivers discharge, justifying the termed "Sahelian Paradox".

We propose here a regionalisation of such mechanisms and an analysis of the respective role of natural (climatic) and Human (land use changes) factors in the appearance, evolution and geographical distribution of these hydrological behaviours.

Mahé and his colleagues highlighted the rise in the Nakambé River (Burkina Faso) discharge ([22]; figure 18), after documenting the same process on the Sahelian Niger river right bank tributaries [6]. More recently, they observed a similar evolution in the western part of the Sahel, in southern Mauritanian rivers ([24]; figure 19). In a newly published paper [23], the discharges in the Sokoto river in Nigeria showed an increasing trend, similar to those of the right bank tributaries.

The stream discharges of Sudanian rivers (Sudanian climate is characterized by annual rainfall amount exceeding 750 mm) were found to be decreasing [10], which was proposed by Mahé [25] to result from a drop in the water level in the aquifers sustaining the rivers during the low flow period. These areas have a "Hewlettian" hydrological behaviour instead of the "Hortonian" functioning typical for the Sahelian semiarid regions. Runoff only onset when the soil is saturated; in these areas, the reduction in rainfall affected firstly the part of rainy water previously dedicated to runoff, explaining the significant runoff and discharges decrease in these areas.

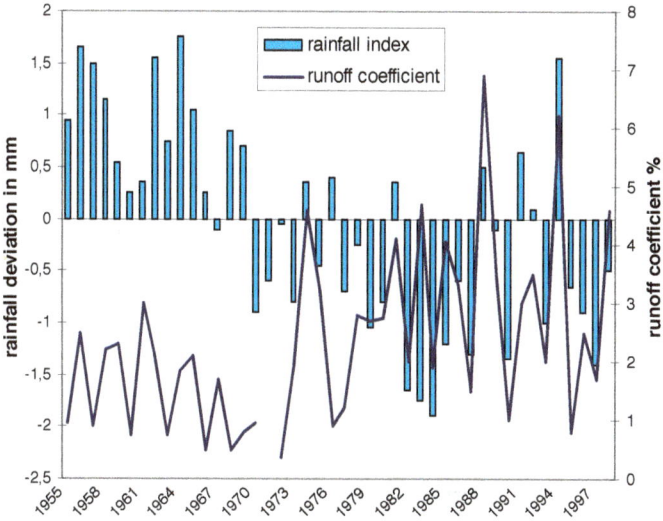

**Figure 18.** Runoff coefficients from observed measurements and rainfall indexes over the Nakambé river basin (Burkina Faso) [22]

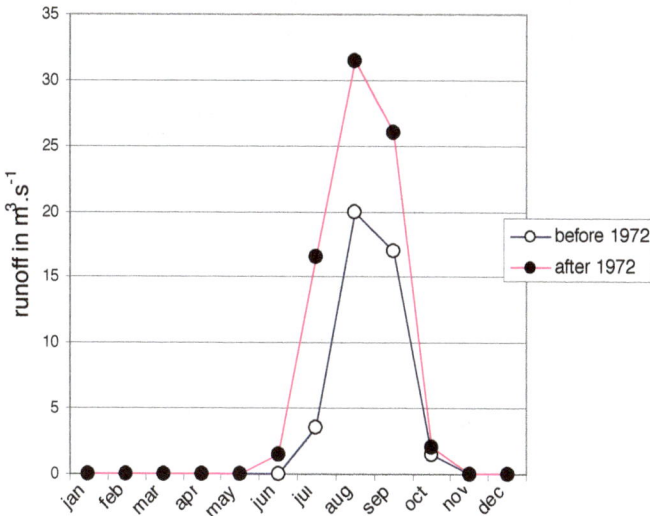

**Figure 19.** Monthly averaged discharges of seven Mauritanian tributaries of Senegal River, before and after 1972 [23]

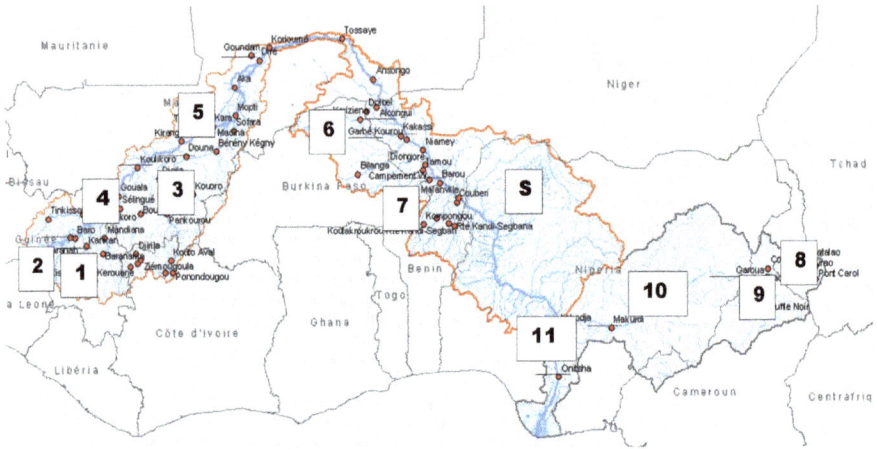

**Figure 20.** The Niger River basin [26]. Stations cited in Table 4 are underlined and the position of cited basins is indicated with its number in the table

The synthesis presented in fig.20 and Table 4 shows that, except in Sahelian areas, the discharge has been decreasing as expected, since the beginning of the drought in 1968. Figure 21 shows that there is a clear regional distribution of hydrological behaviour, with the Sahelian region showing a strong increase in runoff, the Sudanian ones exhibiting a significant decrease in river discharge, while Guinean areas (rainfall>1300mm) generally show a slight decrease in discharges.

| Basin | Niger | Niger | Niger | Niger | Niger | Niger | Niger | Benue | Benue | Benue | Niger |
|---|---|---|---|---|---|---|---|---|---|---|---|
| River | Milo | Niandan | Bani | Niger | Niger | Sahelian basin | Mekrou | Mayo Kébi | Benue | Benue | Niger |
| Station | Kankan | Baro | Douna | Koulikoro | Diré | Rive droite | Barou | Cossi | Garoua | Makurdi | Onitsha |
| Country | Guinea | Guinea | Mali | Mali | Mali | Burkina-Niger | Niger | Cameroon | Cameroon | Nigeria | Nigeria |
| Area km² | 9900 | 12600 | 101600 | 120000 | 366500 | 90500 | 10500 | 25100 | 64000 | 303600 | 1388300 |
| Numéro fig1 | 1 | 2 | 3 | 4 | 5 | 6 | 7 | 8 | 9 | 10 | 11 |
| years obs. | 1950-2000 | 1950-2000 | 1922-2000 | 1907-2000 | 1924-2000 | 1956-1995 | 1961-1999 | 1955-2000 | 1946-1991 | 1955-1995 | 1950-1987 |
| mean(-69) | 211 | 271 | 639 | 1552 | 2244 | 35,4 | 41 | 104 | 388 | 3549 | 6651 |
| mean(70-) | 155 | 186 | 235 | 1039 | 1349 | 38,0 | 23 | 74 | 244 | 2816 | 5016 |
| (70-)/(-69) % | -27 | -31 | -63 | -33 | -40 | +7 | -42 | -30 | -37 | -21 | -25 |

**Table 4.** mean discharges for two periods : before (until 1969) and after (since 1970) the Drought for 11 basins of Niger River [23]

## 8. And in endorheic basins?

As the "Sahelian Paradox" seems to apply in the whole Sahel, it must apply also in endorheic areas. In certain endorheic areas in the Sahel, the water table level has been found to be rising over the last several decades despite the strong reduction in rainfall observed after 1968. This phenomenon has been previously defined as the "Niamey Paradox" [27]. The excess in runoff has significantly increased the number of ponds. While ponds are the main zones of deep infiltration, their increase explains the rise of the water table level (figure 22).

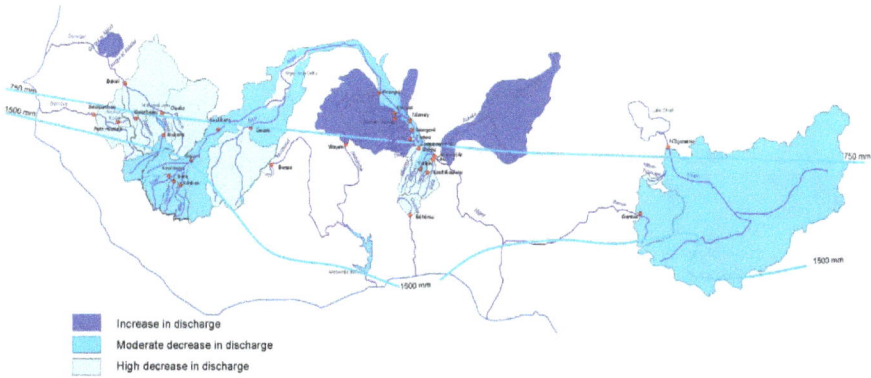

Increase in discharge
Moderate decrease in discharge
High decrease in discharge

**Figure 21.** The regional distribution of hydrological behaviour in West Africa; increasing discharges indicate a "horto-nian" functioning as well expected decrease in discharge ( [10] updated)

Indeed, the current active erosion processes are leading to the appearance of many new gullies and new spreading areas where sediments extracted by aeolian and hydric erosion are then transported in the gullies, and deposited. The gully beds are currently characterized by sand deposits ranging from 2 to 4 metres wide, several tens of centimetres deep (up to 1 to 2 m) and hundreds of metres long. Spreading areas are formed by these newly created streams when they reach gentler slopes, because their transport capacity becomes suddenly insufficient to carry such significant volumes of sand. They form sandy deposits of a magnitude of hundreds of square metres, up to several hectares, and, in some cases, tens of centimetres deep. These areas constitute new deep infiltration areas, accelerating the rise in water table [28].

In contrast, the water table level is falling in the Lake Chad area (figure 22). This lake and the Niamey square degree are located at the same latitude, but the behaviours of their respective groundwater systems are completely different. The area of the Niamey square degree is composed of small endorheic basins with only local water contribution. While Lake Chad is mostly water fed by tropical humid areas of the Upper Logone and Chari basins (95%). Recently it was shown that discharges are decreasing in Sudanian areas [10] and, thus, water supply to Lake Chad is decreasing. The Ari Koukouri well is located at some tens kilometres of the northern part of the Lake, dried repetitively since the 1980s. The difference in the water feeding patterns explains the opposite behaviour of groundwater under Lake Chad and under the Niamey square degree.

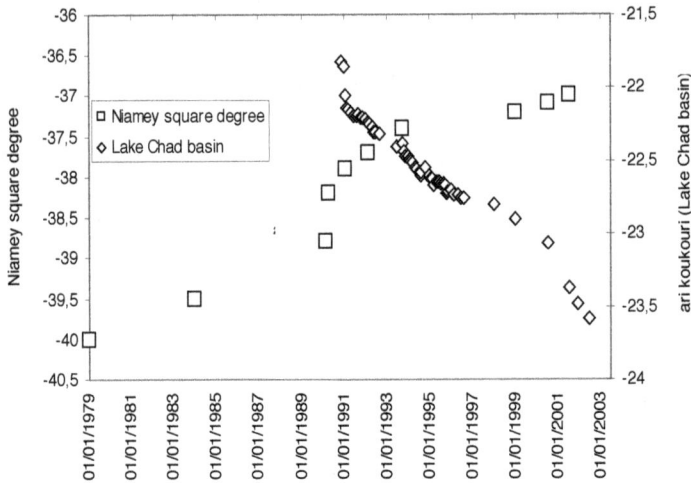

**Figure 22.** Evolution of water table level under the Niamey square degree area and under the Lake Chad [7] [28] [29], pers. Comm. of G. Favreau, 2007.

## 9. A persistent desertification?

The evolution of land use/land cover in the Sahel remains the subject of an ongoing debate.

At the Sahel regional scale, cultivated areas have been increasing significantly for more than 50 years due to population growth and very low crop yields (see Fig. 23 for the Niger Republic) [30].Some studies based on satellite data vegetation indices [31] [32] [33] have suggested a re-greening of the Sahel after 1994. However, other studies ([34] amongst others) have highlighted the limitations of vegetation indices for the determination of land use and land use changes when using remote sensing at coarse resolution. Land cover studies based on aerial photo-graph analyses [28] show on the contrary a decrease in vegetation cover in south western Niger.

Although the Sahel is probably re-greening, the western part of Niger and the eastern part of Burkina Faso are still suffering land degradation (see maps in [32] [33]; see figure 24).

North of our study area, in eastern Mali, an observed increase in runoff has been attributed to the removal of vegetation during the drought and its inability to recover during recent wetter years due to soil erosion and degradation [35]. In this case, the drought is the cause of the new hydrological behaviour of the area.

Therefore, there is a series of problems to account for in order to improve our knowledge of the Sahelian evolution towards regreening or desertification:

1.   The mapping methods to be used: before 1972, there was no satellite data. For this period only aerial photos can be used; therefore, there is no satellite product before the drought which begun in 1968. On the contrary nowadays, there is a large panel of satellite data available;

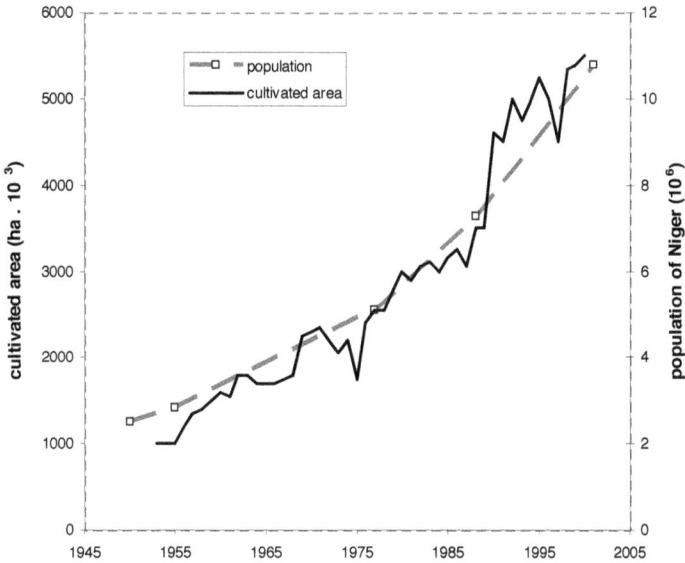

**Figure 23.** Evolution of population and cultivated area in Niger since 1950 [30]

2. It is not always easy to segregate either the respective role of the drought (climatic, thus "natural" trend) and of the societies (demographic pressure on the resources) within the soil and vegetation degradation (when observed), or the impact of reclamation actions;

3. Studies based on aerial photos are very time-consuming and limited in space: thus, the selected areas for the analyses are not always representative of the entire region.

4. The tools: the NDVI (normalised difference vegetation index) and the RUE (Rain Use Efficiency) are both indexes used to map the desertification trend.

5. Some elements have to be accounted for such as the adaptation of vegetation to drought, the evolution of resilience in a drought context, the weight of graminaceae and herbaceous species in the measurement of NDVI and RUE, particularly in northern Sahelian areas where they represent 99% of the biomass, and then of the NDVI signal; in cultivated areas, the fact that pearl millet crops have a high biomass thus making the NDVI exceed that of the natural vegetation;

6. Overall, there is a very high spatial and temporal variability in the evolution of land use/ land cover, making a short - term or small – basin based comparative analysis difficult to generalise.

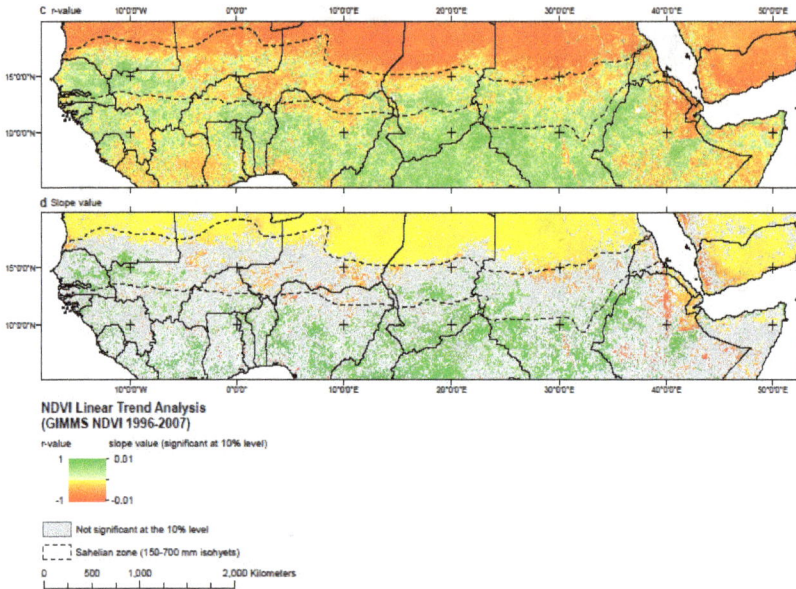

**Figure 24.** Evolution of NDVI values in the Sahel (area between the two doted lines) (up); down: mapping of the significant pixels (at 10% level) [33]

Finally, at the present time there is no certitude about the re-greening of the Sahel. However, hydrological studies showed that 5 to 10% of the land cover became crusted soils (ERO type, [14]) enough to significantly increase (twice) the runoff coefficient of a basin. Therefore, a general re-greening is consistent with the degradation of 5-10% of the total area, to explain the general increase in runoff in spite of an increase in vegetation cover.

## 10. Conclusion

A partial recovery of rainfall amount has been observed since the end of 20th century in the eastern and central parts of the Sahel. However, this cannot solely explain the increase in river discharges, since runoff coefficients have strongly increased (twice or more than those observed before the drought for basins of thousands of km$^2$). Since the other rainfall characteristics did not show any change which could explain the rise in runoff and the earlier occurrence of the yearly flood, this is likely due to the land use change. Instead of deforestation, the land cover evolution is due to the increase in crop areas; since almost all the cultivable areas are now cultivated, the demographic pressure leads to a shortening of fallows. This is the main cause of soil degradation and crusting. Degraded areas are characterised by low infiltration capacity and high runoff coefficients.

The increase in discharge obviously is leading to a rise in flood hazard in Sahelian areas. This is also observed in Sudanian areas as during the monsoon of 2007 when thousands of km² where flooded in Burkina Faso, Togo and Ghana [32] [18]. However, this evolution is much more marked in the Sahel because runoff coefficients are rather decreasing in Sudanian areas since the beginning of the drought.

Furthermore, although crops have been destroyed in some areas of the Sahel, the flood hazard is becoming a severe planning problem in urban areas, where extended zones are tarred, leading to a strong reduction in infiltration. Urbanisation in flooding areas (figure 25) is creating a new land management problem. The Sahelian hydrological paradox thus has negative consequences rather than the sole positive opportunity of getting more water to agriculture and grazing. Crops commonly increase infiltration in the Sahel; extension of cropping on unsuitable soils, shortening of fallows and no-use of fertilizers are some of the causes of soil crusting and degradation.

Policy makers should be alerted to the effects of intensive cropping and land clearing, in some areas, on the hydrological regimes of Sahelian rivers. They must consider the flood hazard in downstream urban areas as an emerging severe concern to sub-Saharan populations.

**Figure 25.** Inundation of a recent housing area in northern Dakar suburbs (Senegal) (left); a street became a temporary river in Ziguinchor (Casamance, Southern Senegal, Sudanian area) (right); pictures taken during the 2012 monsoon.

# Acknowledgements

We warmly acknowledge the colleagues of the Niger Basin Authority's "Projet Niger-Hycos" (NBA; http://nigerhycos.abn.ne/user-anon/htm/listStationByGroup.php) and the Niger Water Resources Department (DRE) for providing the Niger River discharge data. The Directors, and the colleagues of the Burkina Faso, the Mali and the Niger Meteorological Offices (Direction de la Météorologie Nationale) are also acknowledged. The SIEREM data were provided by A. l'Aour-Crès, N. Rouché and C. Dieulin (IRD-HSM), and the FRIEND animators provided rainfall data. This work was partly funded by French ANR projects ECLIS and ESCAPE, as well as by the AMMA project.

## Author details

Luc Descroix[1], Ibrahim Bouzou Moussa[2], Pierre Genthon[3], Daniel Sighomnou[4], Gil Mahé[3], Ibrahim Mamadou[5], Jean-Pierre Vandervaere[6], Emmanuèle Gautier[7], Oumarou Faran Maiga[2], Jean-Louis Rajot[8], Moussa Malam Abdou[1], Nadine Dessay[9], Aghali Ingatan[2], Ibrahim Noma[2], Kadidiatou Souley Yéro[1], Harouna Karambiri[10], Rasmus Fensholt[11], Jean Albergel[12] and Jean-Claude Olivry[13]

1 IRD / UJF, Grenoble, France

2 UAM University, Niamey, Niger

3 IRD-HSM, Montpellier, France

4 Niger Basin Authority, Niamey, Niger

5 University of Zinder, Niger

6 UJF-LTHE, Grenoble, France

7 Université Paris 8, France

8 IRD-BIOEMCO, Créteil, France

9 IRD-ESPACE-DEV, Montpellier, France

10 2iE International high School, Ouagadougou, Burkina Faso

11 University of Copenhague, Denmark

12 IRD-LISAH, Montpellier, France

13 IRD, France

## References

[1]  Ali, A.; Lebel, T.. The Sahelian standardized rainfall index revisited. Int. J. Climatol. 2009, 29, 1705-1714.

[2]  Olivry, J-C. Synthèse des connaissances hydrologiques et potential en resources en eau du Fleuve Niger. Internal report World bank- Niger Basin Authority, 2002, 156 p.

[3]  Albergel, J. Sécheresse, désertification et ressources en eau de surface : application aux petits bassins du Burkina Faso. In "The Influence of Climate Change and Climatic Variability on the Hydrologic Regime and Water Resources"; IAHS publication N° 168, Wallingford, UK, 1987,pp. 355-365.

[4] Mahe, G., Olivry, J.C. Assessment of freshwater yields to the ocean along the inter-tropical Atlantic coast of Africa". Comptes Rendus de l'Académie des Sciences, Series IIa 1999, 328, 621-626.

[5] Amani, A.; Nguetora, M. Evidence d'une modification du régime hydrologique du fleuve Niger à Niamey. In "FRIEND 2002 Regional Hydrology: Bridging the Gap between Research and Practice", Proceedings of the Friend Conference, Cape Town, South Africa, 18-22 March, 2002; Van Lannen, H., Demuth, S., Eds.; IAHS publication N°274, Wallingford, UK, 2002, pp. 449-456.

[6] Mahé, G.; Leduc, C.; Amani, A.; Paturel, J-E.; Girard, S.; Servat, E.; Dezetter, A. Augmentation récente du ruissellement de surface en région soudano sahélienne et impact sur les ressources en eau. In "Hydrology of the Mediterranean and Semi-Arid Regions, proceedings of an international symposium. Montpellier (France)", 2003/04/1-4, Servat E., Najem W., Leduc C., Shakeel A. (Ed.); Wallingford, UK, IAHS, 2003, publication n° 278, 2003, p. 215-222..

[7] Descroix, L., Mahé, G., Lebel, T., G., Favreau, G., Galle, S., Gautier, E., Olivry, J-C., Albergel, J., Amogu, O., Cappelaere, B., Dessouassi, R., Diedhiou, A., Le Breton, E., Mamadou, I. Sighomnou, D. Spatio-Temporal Variability of Hydrological Regimes Around the Boundaries between Sahelian and Sudanian Areas of West Africa: A Synthesis. Journal of Hydrology, AMMA special issue, 2009,375, 90-102. doi: 10.1016/j.jhydrol.2008.12.012

[8] Descroix, L., Laurent, J-P., Vauclin, M., Amogu, O., Boubkraoui, S., Ibrahim, B., Galle, S., Cappelaere, B., Bousquet, S., Mamadou, I., Le Breton, E., Lebel, T., Quantin, G., Ramier, D., Boulain, N. Experimental evidence of deep infiltration under sandy flats and gullies in the Sahel. Journal of Hydrology, 2012, 424-425, 1-15;. http://dx.doi.org/10.1016/j.jhydrol.2011.11.019

[9] Descroix, L., M. Esteves, K. Souley Yéro, J.-L. Rajot, M. Malam Abdou, S. Boubkraoui, J.-M. Lapetite, N. Dessay, I. Zin, O. Amogu, A. Bachir, I. Bouzou Moussa, E. Le Breton, and I. Mamadou. Runoff evolution according to land use change in a small Sahelian catchment. Hydrol. Earth Syst. Sci. Discuss., 2011, 8, 1569-1607, 2011www.hydrol-earth-syst-sci-discuss.net/8/1569/2011/doi:10.5194/hessd-8-1569-2011.

[10] Amogu O., Descroix L., Yéro K.S., Le Breton E., Mamadou I., Ali A., Vischel T., Bader J.-C., Moussa I.B., Gautier E., Boubkraoui S., Belleudy P. Increasing River Flows in the Sahel?. Water, 2010, 2(2):170-199.

[11] Descroix, L., Genthon, P. Amogu, O. Rajot, J-L., Sighomnou, D., Vauclin, M.. Change in Sahelian Rivers hydrograph: The case of recent red floods of the Niger River in the Niamey region. Global Planetary Change, 2012, 98-99, 18-30.

[12] Hulme, M. Climatic perspectives on Sahelian desiccation : 1973–1998. Global Environmental Change 2001, 11, 19–29 http://dx.doi.org/10.1016/S0959-3780(00)00042-X.

[13] Nicholson, S.E. On the question of the "recovery" of the rains in the West African Sahel. Journal of Arid Environments, 2005, 63, 615–641. doi:10.1016/j.jaridenv.2005.03.004

[14]  Casenave, A. & Valentin, C. A runoff capability classification system based on surface features criteria in semi-arid areas of West Africa. Journal of Hydrology, 1992, 130, 231–249 (1992).

[15]  Asseline, J., Noni, G.D., Chaume, R. Design and use of a low speed, remotely-controlled unmanned aerial vehicle (UAV) for remote sensing [French]. Photo Interprétation, 1999, 37, 3-9. http://www.documentation.ird.fr/hor/fdi:010026172.

[16]  Sighomnou, D. Evènements de crues du mois d'août 2012 sur le Niger. Projet GIRE 2, Autorité du bassin du Niger,, Niger Basin Authority, Niamey, 8 p. 2012.

[17]  Karambiri, H. Brève analyse fréquentielle de la pluie du 1$^{er}$ septembre 2009 à Ouagadougou (Burkina Faso). Note technique 2iE, 4 p. 2009.

[18]  Tschakert, P., Sagoe, R., Ofori-Darko, G. & Codjoe, S.M.. Floods in the Sahel: an analysis of anomalies, memory, and participatory learning. Climatic Change 2010, 103, 471-502. doi: 10.1007/s10584-009-9776-y.

[19]  Tarhule, A. Damaging rainfall and floodings: the other Sahel hazards. Climatic Change, 2005, 72, 355-377. doi: 10.1007/s10584-005-6792-4.

[20]  Di Baldassarre, G., A. Montanari, H. Lins, D. Koutsoyiannis, L. Brandimarte, and G. Blöschl. Flood fatalities in Africa: From diagnosis to mitigation, Geophys. Res. Lett., 2010, 37, L22402, doi:10.1029/2010GL045467.

[21]  Mamadou, I. la dynamique des koris et l'ensablement de leur lit et de celui du fleuve Niger dans la région de Niamey. PhD thesis, Paris 1 Panthéon Sorbonne University and Abdou Moumouni University of Niamey, 260 p. 2012.

[22]  Mahé, G.; Paturel, J.E.; Servat, E.; Conway, D.; Dezetter, A. Impact of land use change on soil water holding capacity and river modelling of the Nakambe River in Burkina-Faso. J. Hydrol. 2005, 300, 33-43.

[23]  Mahé, G., Lienou, G., Bamba, F., Paturel, J-E., Adeaga, O., Descroix, L., Mariko, A., Olivry, J-C., Sangaré, S., Ogilvie, A., Clanet, J-C. Le fleuve Niger et le changement climatique au cours des 100 dernières années. « Hydro-climatology variability and change (Proceedings of symposium held during IUGG 2011, Melbourne, Australia)"; IAHS pub. n° 344, 131-137. 2011.

[24]  Mahé, G.; Paturel, J-E. 1896-2006 Sahelian annual rainfall variability and runoff increase of Sahelian rivers. C.R. Geosciences, 2009, 341, 538-546.

[25]  Mahé, G. Surface/groundwater interactions in the Bani and Nakambe rivers, tributaries of the Niger and Volta basins. West Africa. Hydrol. Sci. J. 2009, 54, 704-712.

[26]  Boyer, J. F., Dieulin, C., Rouché, N., Crès, A. Servat, E. Paturel, J. E. & Mahé, G. SIEREM: an environmental information system for water resources. In: Climate Variability and Change – Hydrological Impacts (ed. by S. Demuth, A. Gustard, E. Planos, F. Scatena & E. Servat), 19–25. IAHS Publ. 308., 2006, IAHS Press, Wallingford, UK

[27]  Leduc, C., Favreau, G. and Shroeter, P. Long term rise in a Sahelian water-table: the Continental terminal in South West Niger. Journal of Hydrology 2001, 243, 43-54.

[28]  Leblanc, M., Favreau, G., Massuel, S., Tweed, S., Loireau, M., Cappelaere, B. Land clearance and hydrological change in the Sahel: SW Niger. Global and Planetary Change 2008, 61 (1-2), 49-62 (2008).

[29]  Zaïri, R. Etude géochimique et hydrodynamique de la nappe libre du Bassin du Lac Tchad dans les régions de Diffa (Niger oriental) et du Bornou (nord-est du Nigeria). PhD thesis, Montpellier 2 University, 210 p. 2008.

[30]  Guengant, J.-P., Banoin, M. Dynamique des populations, disponibilités en terres et adaptation des régimes fonciers: le Niger. FAO-CICRED, Publ., Roma, Paris, 142 p., 2003. http://www.cicred.org/Eng/Publications/pdf/MonoNiger.pdf

[31]  Anyamba, A. and Tucker, C.J. Analysis of Sahelian vegetation dynamics using NOAA-AVHRR NDVI data from 1981-2003. Journal of Arid Environments 2005, 63(3), 596-614.

[32]  Prince, S.D., Wessels, K.J., Tucker, C.J., Nicholson, S.E. Desertification in the Sahel: a reinterpretation of a reinterpretation. Global Change Biology 2007,13, 1308–1313. doi: 10.1111/j.1365-2486.2007.01356.x.

[33]  Fensholt, R., Rasmussen, K. Analysis of trends in the Sahelian 'rain-use efficiency' using GIMMS NDVI, RFE and GPCP rainfall data. Remote Sensing of Environment 2011, 115, 438-451. doi:10.1016/j.rse.2010.09.014.

[34]  Hein, L. & De Ritter, N. Desertification in the Sahel: a reinterpretation. Global Change Biology 2006, 12, 751-758.

[35]  Gardelle, J., Hiernaux, P., Kergoat, L. & Grippa, M. Less rain, more water in ponds: a remote sensing study of the dynamics of surface waters from 1950 to present in pastoral Sahel. (Gourma region, Mali) Hydro. Earth Syst. Sci. 2010, 14, 309–324. doi:10.5194/hess-14-309-2010.

# Chapter 10

# Changing Hydrology of the Himalayan Watershed

Arshad Ashraf

Additional information is available at the end of the chapter

## 1. Introduction

The Himalayan region is a source of ten major river systems that together provide irrigation, power and drinking water for 1.3 billion people i.e. over 20% of the world's population. The supply and quality of water in this region is under extreme threat, both from the effects of human activity and from natural processes and variation [1]. Population growth is already putting massive pressure on regional water resources, affecting water resource in terms of demand, water-use patterns and management practices. The change in hydrological cycle may affect river flows, agriculture, forests, biodiversity and health besides creating water related hazards [2]. The need for suitable strategies for climate resilient development has policy and governance implications [3]. Adaptation to climate change is the area that should be strengthened through policy advocacy supported by evidence through rigorous research and verified information.

Re-assessment of true catchments yields under existing and future scenarios of landuse and climate changes is very essential to devise watershed management strategies which can minimize adverse impacts both in terms of quantity and quality. Since trends are still unclear, the extent to which changes can be attributed to variable environmental changes is difficult to determine. It has become imperative to assess ongoing hydrological changes and changes that might occur in future to devise appropriate adaptation measures to foster resilience to future climate change, thereby enhancing water security.

In the present study, SWAT model developed by United States Department of Agriculture (USDA) [4] has been used to evaluate surface runoff generation, soil erosion and quantify the water balance of a Himalayan watershed in the Northern Pakistan. The response of watershed yield to historical landuse evolution and under variable landuse and climate change scenarios has been studied in order to mitigate the negative impacts of these changes and promote development activities in this region. The study would provide basis to recommend changes in the water management regimes so as to address future adaptation issues.

## 1.1. Modeling hydrological processes

Dealing with water management issues requires analyzing of different elements of hydrologic processes taking place in the area of interest. As such processes are taking place in a combine system that exists at a watershed level, thus the analysis must be carried out on a watershed basis. Understanding of relationship between various watershed characteristics such as morphology, landuse and soil, and hydrological components are very essential for water resources development in any area. Since the hydrologic processes are very complex, their proper comprehension is essential and for this watershed models are widely used. Most of the watershed models basically simulate the transformation of precipitation into runoff, sediment outflow and nutrient losses. Changes in landuse including urbanization and de(/re)forestation continue to affect the nature and magnitude of surface and subsurface water interactions and water availability influencing ecosystems and their services. One can formulate water conservation strategies only after understanding the spatial and temporal variations and the interaction of these hydrologic components. The alarming rate of soil erosion in context of changing landuse and climate in the Himalayan region calls for urgent attention for this problem. Assessment of erosion is a very difficult task when executed using conventional methods and requires to be done repetitively. The use of an appropriate watershed model is thus essential to deal with such problems.

Choice of watershed development model depends upon the hydrologic components to be incorporated in the water balance. The most important hydrologic elements from the water management point of view are surface runoff, lateral flow, baseflow and evapotranspiration. In presenting an appropriate view of reality, model must remain simple enough to understand and use. There are a number of integrated physically based distributed models, among which researchers have identified Soil and Water Assessment Tool (SWAT) as the most promising and computationally efficient [4]. The model is an integrated physically based distributed watershed model that has an ability to predict the impact of land management practices on water, sediment yield and agricultural chemical yield [5]. Distributed models also take the spatial variability of watershed properties into account.

## 1.2. Model description

The SWAT is a process-based continuous daily time-step model that offers distributed parameter, continuous time simulation, and flexible watershed configuration [6]. It has gained international acceptance as a robust interdisciplinary watershed modeling tool. Two methods are used for surface runoff estimation in SWAT i.e. the SCS curve number and Green-Ampt infiltration. This study is based on the use of curve number for surface runoff and hence stream flow simulation. A SWAT model can be built using the Arc-View interface called AVSWAT which provides suitable means to enter data into the SWAT code. Main processes include water balance calculations (i.e. surface runoff, return flow, percolation, evapotranspiration, and transmission losses), estimation of sediment yield, nutrient cycling and pesticide movement.

The spatial heterogeneity is represented by means of observable physical characteristics of the basin such as landuse, soils and topography etc. Model inputs include physical characteristics of the watershed and its sub-basins i.e. precipitation, temperature, soil type, land slope, Manning's n values, USLE K factor, and management inputs like crop rotations, planting and harvesting dates, tillage operations, irrigation, fertilizer use, and pesticide application rates. Model outputs include sub-basin and watershed values for surface flow, ground water and lateral flow, sediment, nutrient and pesticide yields. The main basin is divided into sub-basins which are further divided into hydrologic response units (HRU) composed of homogeneous landuse, soil types, relevant hydrological components and management practices. Sediment yield is estimated by the Modified Universal Soil Loss Equation (MUSLE; [7]. The model has been applied worldwide for the purpose of simulating sediment flow [8], modeling hydrologic balance [9], evaluation of the impact of landuse and landcover changes on the hydrology of catchments [10]. The model provides a flexible capability for creating climate change scenarios evaluating a wide range of "what if" questions about how weather and climate could affect our systems.

### 1.3. Equations of watershed hydrology

The hydrologic process in a watershed is simulated by the following water balance equation:

$$SW_t = SW + \sum_{i=1}^{t} \left( R_i - Q_i - ET_i - P_i - QR_i \right) \tag{1}$$

where: $SW_t$ is the final soil water content (mm), $SW$ is the initial soil water content minus the permanent wilting point water content (mm), $t$ is time in days, $R$ is rainfall (mm), $Q_i$ is surface runoff (mm), $ET_i$ is evapotranspiration (mm), $P_i$ is percolation (mm) and $QR_i$ is lateral flow (mm). The surface runoff is predicted by the following equation:

$$Q = \frac{(R - 0.2s)^2}{R + 0.8s} \text{ for } R > 0.2s \tag{2}$$

$Q = 0.0$     for $R < 0.2s$

$$s = 254\left(\frac{100}{CN} - 1\right) \tag{3}$$

Where, $Q$ = daily surface runoff (mm): $R$ = daily rainfall (mm), $S$ = retention parameter (mm); $CN$ = curve number.

Lateral flow is predicted by:

$$q_{lat} = 0.024 \frac{(2SSC\sin\alpha)}{\theta_d L}$$ (4)

Where, $q_{lat}$ = lateral flow (mm/ day); $S$ = drainable volume of soil water per unit area of saturated thickness (mm/day), $SC$ = saturated hydraulic conductivity (mm/h); $L$ = flow length (m); $\alpha$ = slope of the land: $\theta_d$ = drainable porosity

The base flow is estimated by:

$$Q_{gwj} = Q_{qwj-1} \cdot e^{\left(-\alpha_{gw}\cdot\Delta t\right)} + w_{rchrg} \cdot \left(1 - e^{\left(-\alpha_{gw}\cdot\Delta t\right)}\right)$$ (5)

Where, $Q_{gwj}$ = groundwater flow into the main channel on day j; $\alpha_{gw}$ = base flow recession constant; $\Delta t$ = time step. The computed runoff from each element is integrated using a finite difference form of the continuity equation relating moisture supply, storage and outflow.

### 1.4. Description of study area

Rawal watershed covers an area of 272 sq km within longitudes 73º 03′ - 73º 24′ E and latitudes 33° 41′ - 33° 54′ N comprising parts of Margalla hills and Murree mountains in the southern Himalayas of Pakistan (Figure 1). About 47% of the watershed area lies in the Islamabad Capital Territory while the rest in Punjab and Khyber Pakhtunkhwa (KPK) provinces, so it is well connected through a metalled road with other parts of the country. Korang is the main river flowing in the watershed that receives runoff from watershed via four major and 43 small streams [11]. Rawal dam is constructed on Korang river, which supplies 22 million gallons per day of water for drinking and other household needs to Rawalpindi city and a limited water for irrigation use to Islamabad area. The elevation ranges between 523 meters and 2145 meters above mean sea level (masl). Physiographically, the watershed comprises of 34% hilly area (Elev. <700 masl), 62% Middle mountains (Elev. within 700-2000 masl) and 4% High mountains (Elev. >2000 masl).

The Himalayas serve as a divide between Central Asia and South Asia. The Indo-Eurasian plates collision resulted in the formation of new relief and topography, which consists of series of mountain ranges located in the north and west of Pakistan, commonly known as the Himalayan Mountain System [12]. The principal uplift occurred during the middle or late Tertiary period, 12 to 65 million years ago. The study area lies in sub-humid to humid sub-tropical continental highlands. The hottest months are May, June and July. The mean maximum temperature ranges between 17.6ºC and 40.1ºC while mean minimum temperature between 2.1ºC and 21.6 ºC. The winter months are from October to March. The highest temperature was recorded as 46.6ºC in 2005 and the lowest as -3.9ºC during 1967 [13]. Mean annual rainfall of 1991-2010 period is about 1232 mm. The occurrence of rainfall is highly erratic both in space and time. Over 60 percent of the annual rainfall occurs during monsoon season i.e. from July to September. Most of the rainfall is drained out rapidly due to steep slopes and dissected nature of the terrain. Springs and streams are the main source of water for drinking and other domestic requirements. A prolonged dry season may cause water shortage in some parts of the area.

Underlying rocks consist of poorly compressed and highly folded and faulted Murree series that are moderately to severely eroded, shallow clayey loams of very low productivity [14]. The soils formed over shale are clayey while those developed on the sandstone are sandy loams to sandy clay loam in texture. The flora is mainly natural with xeric, broad-leaved deciduous, evergreen trees and diverse shrubs on the southern slopes. The dominating plant species are *Carissa spinarum* (Granda), *Dodonaea viscosa* (Sanatha) and *Olea ferruginea* (Wild Olive). Sub-tropical pine zone occupies steep and very steep mountain slopes [15]. Agriculture is practiced in small patches of land as terrace cultivation.

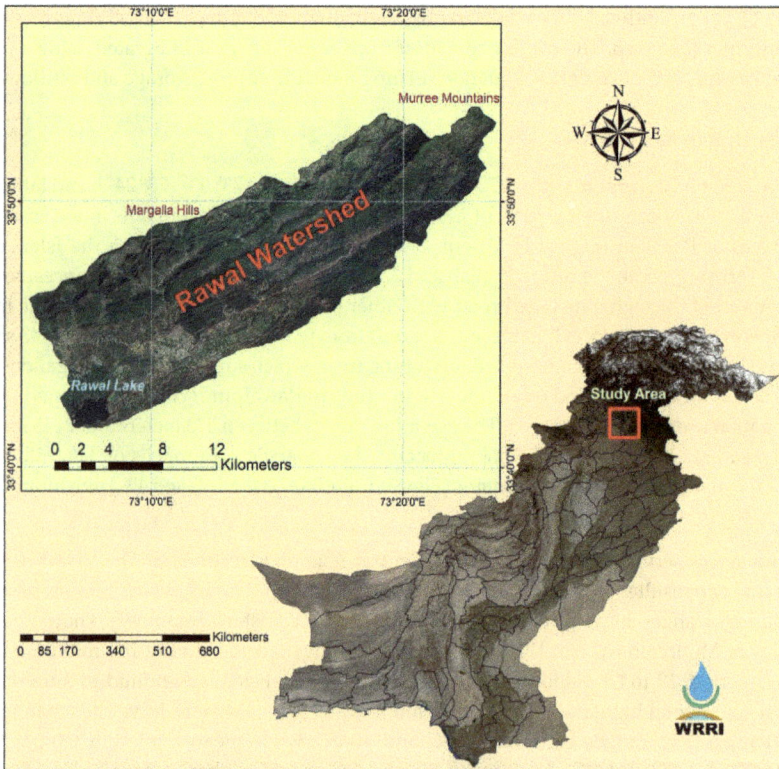

**Figure 1.** Location map of the study area

## 1.5. Main environmental issues

The watershed is confronting problems of rapid urban development and deforestation due to which its landuse is changing gradually. The population growth and addition of a number of housing colonies in the Rawal Lake catchment area are adversely affecting the regime of water coming into Rawal Lake. The activities like illegal cuttings due to high market value of forest wood and intensive use of forest wood for household needs (cooking, heating, timber etc.), ineffective forest management and forest disease etc. are accelerating the deforestation rate in the watershed area [16-17]. Destruction of aquatic habitat and a reduction of water quality are some of the negative impacts of deforestation. Extensive cattle grazing and fuel wood cutting by the local communities have deformed the plants to bushes [18]. The removal of a forest cover from steep slopes often leads to accelerated surface erosion and dramatically increases the frequency of land sliding and surface runoff. The storage capacity of the Rawal Lake which was 47,230 acre-ft when it was developed in 1960, has been reduced almost 34 percent due to sedimentation generating from natural and human induced factors in its catchment area [16]. The use of pesticides and herbicides in agriculture is a source of toxic pollution [19]. Many housing schemes, recreational pursuits e.g. Lake view point, Chatter and Valley parks etc. and farmhouses have been developed in the watershed. The construction of roads, pavements and other structures reduce the infiltration area that ultimately affect the recharging of the aquifer of the twin cities. No systematic study has been undertaken yet to document the landuse variability in the watershed.

# 2. Materials and methods

## 2.1. Data used

In the present study, the basic watershed data used to extract spatial input for SWAT model were hydrologic features, soil distribution, landuse information, and topography. The remote sensing technique has potential application in landuse monitoring and assessment at desired scales. RS images of LANDSAT-TM (Thematic Mapper) of period 1992 and LANDSAT-ETM + (Enhanced Thematic Mapper Plus) of 2000 and 2010 periods (Path-Row: 150-37) were used to delineate landuse/landcover of the watershed area on temporal basis. The LANDSAT ETM + sensor is a nadir-viewing, 7-band plus multi-spectral scanning radiometer (upgraded ver. of TM sensor) that detects spectrally filtered radiation from several portions of the electromagnetic spectrum. The spatial resolution (pixel sizes) of the image data includes 30 m each for the six visible, near-infrared, and short-wave infrared bands, 60 m for the thermal infrared band, and 15 m for the panchromatic band. The climatic parameters i.e. daily temperature (max& min) and precipitation data recorded at Satrameel observatory (73º 12′ 50″ E, 33° 45′ 57″ N & Elev: 610 m) maintained by Water Resources Research Institute had been collected for period 1991-2010. The discharge data of Korang river available on monthly basis from Small dams organization was acquired for the same period for model calibration and validation. The soil map developed by Soil Survey of Pakistan was utilized to extract soil data attributes for the study area.

## 2.2. Data preparation

The base map of the study area was prepared through generating and integrating thematic layers of elevation, physiography and infrastructure using ArcGIS 9.3 software. An integrated hydrological, spatial modeling and field investigations approach was adopted to achieve the study output. The boundaries of the watershed and sub-basins were delineated using Aster 30m DEM of the area in SWAT model 2005 software. Elevation map comprising of four classes i.e. >1600m, 1200-1600m, 800-1200m and <800m range, was prepared from Aster 30m DEM data (Figure 2). The image data was georeferenced using Universal Transverse Mercator (UTM) coordinate system (Zone 43 North). The satellite images were analyzed through visual and digital interpretation techniques to observe spatial variability of landuse. The visual interpretation was performed for qualitative analysis while digital interpretation for quantitative analysis of the image data. The false color composite of 5, 4, 2 (RGB) of LANDSAT image data was selected to extract signatures of representative landcover types from the image. In this bands combination, landcover is visible in true color i.e. vegetation in green, soil in pale to reddish brown and water in shades of blue color. The built-up area is shown in mixed pattern of white, brown, and purple colors due to variable types and density of constructed area, mixing of new and old settlements, presence of land features like lawns, parking sides, water ponds, roads/tracks etc. The signatures were evaluated using error matrix and an overall accuracy of more than 95 percent was achieved.

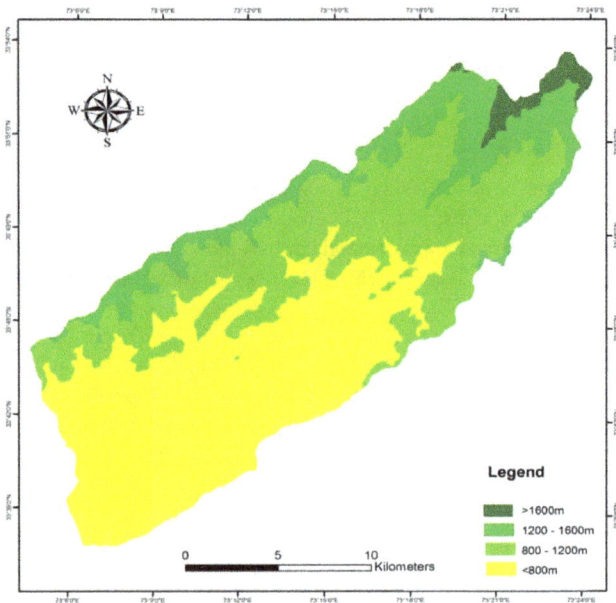

**Figure 2.** Elevation increases gradually towards northeast in the study area

The classification of the images was performed using supervised method following maximum likelihood rule mostly used to acquire reliable classification results. The classification output was supported with Normalized difference vegetation index - NDVI data that helps in segregating vegetative areas from non-vegetative [20]. The index which is based on the spectral characteristics of green vegetation cover in the area uses TM3 and TM4 bands of LANDSAT ETM+ image as given in the following equation:

$$NDVI = (TM4 - TM3) / (TM4 + TM3) \qquad (6)$$

The classification of the images was performed to obtain seven major landuse/landcover classes which include conifer forest, scrub forest, agriculture, rangeland, soil/rocks, settlement and water. The images were recoded and later filtering technique was applied to remove noisy/misclassified pixels from the recoded image data. The doubtful classes were modified after ground truthing i.e. performing field survey in the target areas. Finally change analysis of landuse/landcover classes was performed using spatial modeling functions of ERDAS Imagine 9.2 software.

### 2.3. Model baseline establishment

Main procedures in the model running includes: (a) development of streams and sub-basins databases, (b) landuse and soil data input within sub-basins, (c) Input variable parameters of climate and management options, (d) compilation of input data and running the model for generating output results. The entire watershed had been divided into 15 sub-basins by choosing a threshold area of 500 ha. A total of 73 HRUs were generated in those sub-basins. A threshold of 5% was defined landuse distribution and 15% for soil distribution over sub-basin area. The low percentage for landuse was used to accommodate conifer coverage distributed in patches over northeastern parts of the watershed area. The importance of land uses lies mainly in the computation of surface runoff with the help of SCS curve during the model operation [6]. Three soil classes were identified and mapped i.e. sandy clay loam over northwestern hilly terrain, sandy clay loam over valley area in the northeast and sandy loam over low plains in the southwestern part of the watershed area.

The subcomponents of the water balance identified for use in analyses are total flow (water yield) consisting of surface runoff, lateral and base flow, soil water recharge; and actual evapotranspiration. These components are expressed in terms of average annual depth of water in millimeters over the total watershed area. For estimation of sediment yield, C factor values were used on the basis of soil erosion study [21] carried out previously in Pothwar region. The C value of 0.176 was assigned to soil/rocks while 0.2 was assigned to agricultural land class. Higher C values indicate more risk of soil erosion. The conservation practice factor P was assigned value of 1 on account of no significant conservation practice present in the watershed area [22].

## 2.4. Model calibration

Calibration and validation of the SWAT model was performed using monthly river flows data of 1991-2010 period. Data pertaining to year 1991 to 2006 had been used for calibration and the rest for validation of the model. The purpose of model validation is to assess whether the model is able to predict field observations for time periods different from the calibration period [23]. Although the model was run for years 1991 to 2006, the first 6 years of the simulated output were disregarded in the calibration process, since these are required by the model as a warm-up period. This period is essential for the stabilization of parameters (e.g groundwater depth), as the results sometimes vary significantly from the observed values. Thus the final calibration period was from January 1997 to December 2006. The calibration accuracy was checked by calculating several indexes which include Nash & Sutcliffe coefficient (*NTD*), Root Mean Square Error (*RMSE*) and the correlation coefficient $R^2$ of the time series. The Nash & Sutcliffe coefficient [24] is an estimate of the variation of a time series from another as given by following equation:

$$NTD = 1 - \frac{\sum_{i=1}^{n}\left(Q_{obs,i} - Q_{sim,i}\right)^2}{\sum_{i=1}^{n}\left(Q_{obs,i} - \overline{Q_{sim,i}}\right)^2} \tag{7}$$

And root mean square error- *RMS* was computed using following equation:

$$RMSE = \sqrt{\frac{1}{n}\sum_{i=1}^{n}\left[Wi\left(Q_{sim,i} - Q_{obs,i}\right)\right]^2} \tag{8}$$

Where, $Q_{sim}$ = simulated time series, $Q_{obs}$ = observed time series, $\overline{Q_{sim}}$ = numerical mean for the simulated time series, $W$ = weight and $n$ = total number of measurements. A Nash & Sutcliffe coefficient approaching unity indicates that the estimated time series is almost identical to the observed one. The results of these tests are summarized in Table 1. The *NTD* index reached the value of 0.80, signifying a quite precise calibration. Later the model was validated using the same indexes, for the period of January 2007 to December 2010. The results of statistical analysis indicated a Nash Sutcliff efficiency of 0.80. The simulated river flows matched well with the observed values (Figure 3).

| Index | Calibrated period | Validated period |
|:---:|:---:|:---:|
| *NTD* | 0.80 | 0.80 |
| *RMSE (mm)* | 17.0 | 30.4 |
| $R^2$ | 0.81 | 0.91 |

**Table 1.** Criteria for examining the accuracy of calibration and validation processes

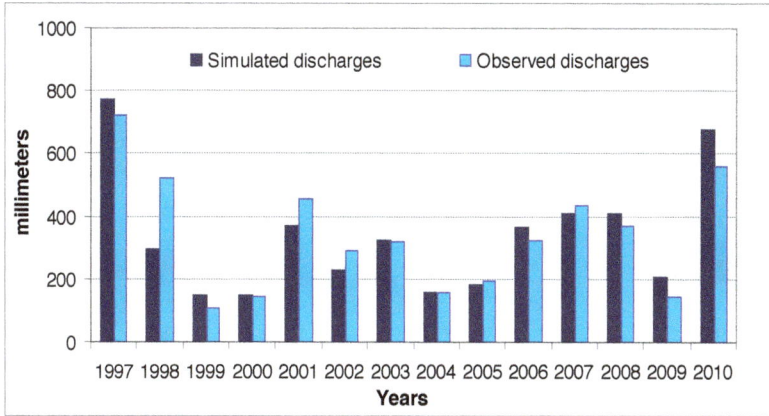

**Figure 3.** Time series of simulated and observed annual discharges for the Rawal watershed, period 1997-2010

# 3. Results and discussion

### 3.1. Assessment of changing landuse/landcover

Comprehensive information on the spatial and temporal distribution of landuse/landcover is essential for estimating hydrological changes at watershed level. The landuse/landcover condition of the watershed was estimated for three different periods i.e. 1992, 2000 and 2010 (Figure 4). Major landcover change was observed in the scrub class which indicated a reduction of about 4,515 ha during 1992-2010 period (Table 2). The rate of decrease in its coverage was about 1.5% per annum. The scrub wood is mostly used as fuel at local level due to non-availability of other energy sources. Major part of it had been converted into agriculture and built-up land, while in some areas it has changed into rangeland due to extensive wood cutting. These results are verified by the findings of [25] which highlighted maximum decrease in scrub forest during 30-year period i.e. 1977-2006 in Rawalpindi area. The settlement class had shown almost four times increase in coverage i.e. from 2.6% in 1992 to 8.7% in year 2010. The average rate of increase in this class was about 90 ha y-1. The rate was over 45 ha y-1 during 1992-2000 while it was about 125 ha y-1 during 2000-2010 period indicating rapid urbanization in the last decade (Figure 5). The conifer forest had shown a decline at a rate of about 2.1% y-1 within last two decades. Although FAO [26] had reported deforestation at a rate of about 1.5% annually in the country, but due to high urban development, the rate of forest decline was higher in the watershed area. The agriculture coverage indicated an average increase of about 26 ha annually during 1992-2010 period. The rate of increase was about 3.4% y-1 during 1992-2000 while it was 0.3% y-1 during 2000-2010 period. The situation indicates intense agriculture activity in the former decade that seems replaced by rapid growth in urban development in the later decade.

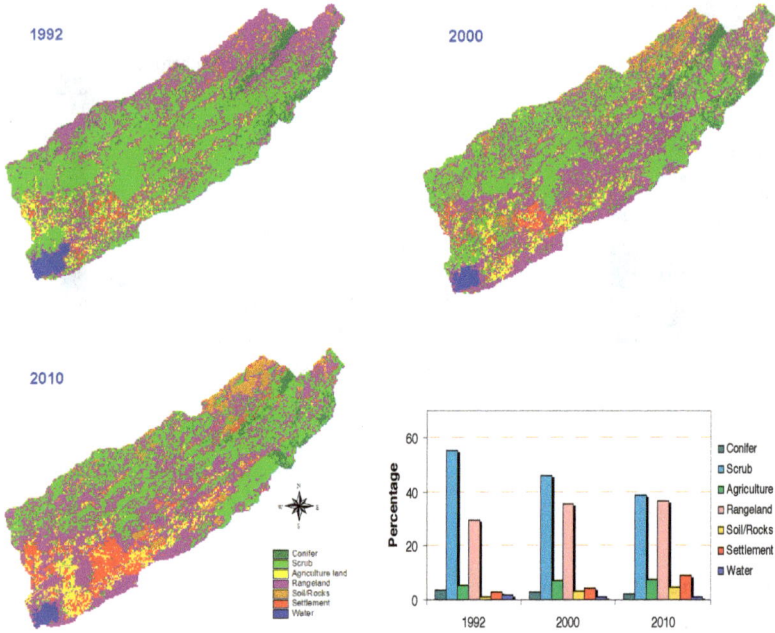

**Figure 4.** Spatio-temporal variations in landuse/landcover in Rawal watershed area during 1992-2010 period

| Landuse | 1992 | | 2000 | | 2010 | | 1992-2010 | |
| --- | --- | --- | --- | --- | --- | --- | --- | --- |
| | Area (ha) | % | Area (ha) | % | Area (ha) | % | Change (ha) | Change % |
| Conifer | 1006 | 3.7 | 762 | 2.8 | 626 | 2.3 | -381 | -1.4 |
| Scrub | 15069 | 55.4 | 12485 | 45.9 | 10554 | 38.8 | -4515 | -16.6 |
| Agriculture | 1496 | 5.5 | 1958 | 7.2 | 2013 | 7.4 | 517 | 1.9 |
| Rangeland | 8024 | 29.5 | 9629 | 35.4 | 9982 | 36.7 | 1958 | 7.2 |
| Soil/Rocks | 326 | 1.2 | 870 | 3.2 | 1306 | 4.8 | 979 | 3.6 |
| Settlement | 762 | 2.8 | 1170 | 4.3 | 2421 | 8.9 | 1659 | 6.1 |
| Water | 517 | 1.9 | 326 | 1.2 | 299 | 1.1 | -218 | -0.8 |
| Total | 27200 | 100 | 27200 | 100 | 27200 | 100 | - | - |

**Table 2.** Detail of landuse/landcover variations during 1992-2010 period

**Figure 5.** Growth of urbanization is causing rapid landuse change in the Rawal watershed area

The changes in landuse/landcover were variable on different elevation ranges during 1992-2010 period. The conifer forest has shown a decrease from 134 ha to 102 ha at greater than 1600m elevation range while this was from 343 ha to 238 ha within 1200-1600m elevation range during 1992-2010. The scrub class indicated a decrease of about 11 percent within 800-1200m range while 65% in less than 800m elevation range. In contrary to this, agriculture class had shown a increase of about 65% within 800-1200m range while 29% increase in less than 800m elevation range. About 86% settlement class was found below 800 m elevation during year 2010 indicating most of the urban development in the low lying areas of the watershed.

## 3.2. Model simulation

The model simulated an average water yield of about 378.6 mm/yr using base landuse of 2010 in the watershed area. About 49% of the yield was contributed by surface runoff and the rest by groundwater in the form of sub surface flows and springs etc. More than 70% of the annual yield was contributed during months of July, August and September. The surface runoff was found higher in the month of August i.e. over 83 mm while it was about 51 mm during July and 31 mm in September. The runoff was dominant over lower sub-basins likely due to higher impervious cover here than in the upper sub-basins of the watershed. The groundwater discharge to stream flows was maximum in the month of September and more than 70% of the discharge occurred during period from August to December. The long-term average soil loss in the watershed was estimated over 17 tons $ha^{-1} y^{-1}$ i.e. ranging from 0.4 to 36 tons $ha^{-1} y^{-1}$ in different sub-basins. These estimates of soil loss matched closely with the results of [22] which indicated soil loss ranged from 0.1 to 28 tons $ha^{-1} y^{-1}$ averaging 19.1 tons $ha^{-1} y^{-1}$ at Satrameel study site in this watershed.

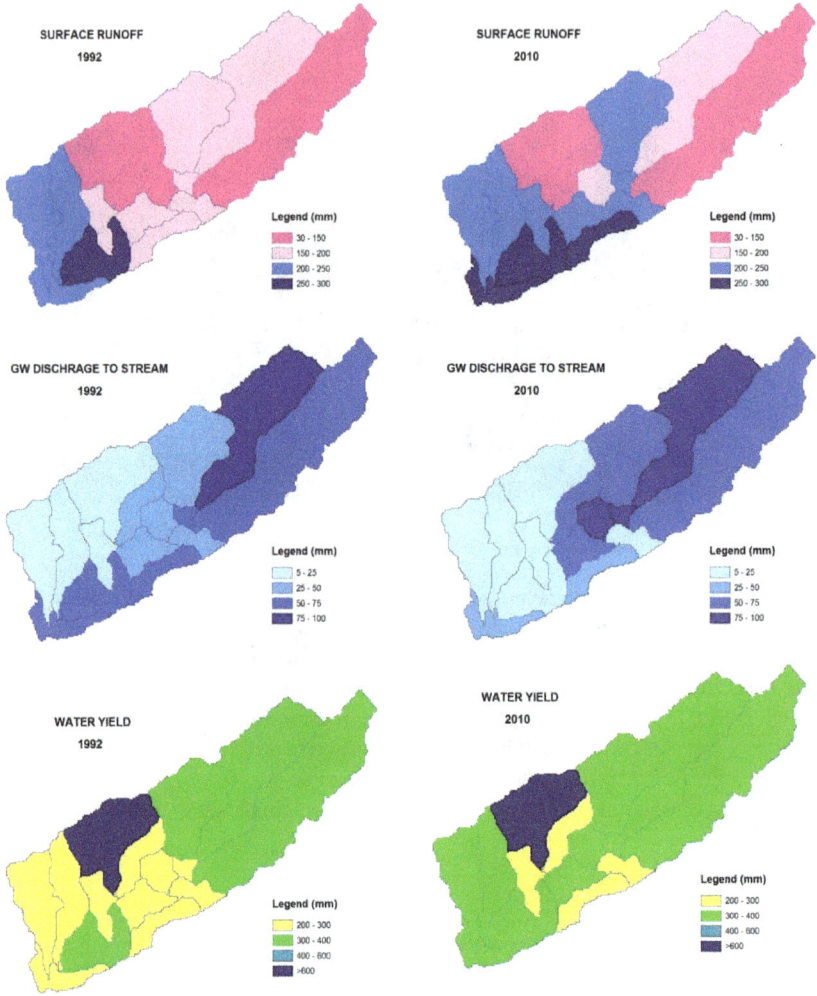

**Figure 6.** Comparison of the hydrological response of Rawal watershed to landuse conditions of 1992 and 2010 indicates dominant impact of landuse changes (i.e. urban development) in the southern low lying sub-basins on various hydrological parameters.

The model simulations showed a strong correlation between landuse evolution and the watershed runoff at the outlet. The change in landuse between years 1992 and 2010 indicated an increase of about 6% in the water yield and 14.3% in the surface runoff. The sub-basin wise hydrological response of the watershed during 1992-2010 period is shown in Figure 6. The sub-basins in the southern valley plains of the watershed indicate increase in surface runoff and water yield while decrease in groundwater contribution to the streams. The situation shows higher influence of urban landuse on hydrology of the low lying sub-basins as compared with sub-basins at higher elevations in the northeast of the watershed. Hydrologic changes due to increased impervious area and soil compaction generally lead to increased direct runoff, decreased groundwater recharge, and increased flooding, among other problems [27]. The combined effect of landuse and hydrological variations had exaggerated the problem of soil erosion resulting in an increase of about 17.4% in the sediment yield of watershed during 1992-2010 period. The increase in sediment yield can be attributed to the increase in surface runoff condition during this period (Figure 7). The zone of low sediment yield i.e. <5 tons ha$^{-1}$ y$^{-1}$ has shown a significant decrease while zones of medium sediment yield i.e. 5-10 tons ha$^{-1}$ y$^{-1}$ and high sediment yield i.e. >15 tons ha$^{-1}$ y$^{-1}$ a relative increase in the southeastern sub-basins of the Rawal watershed during 1992-2010 period (Figure 8).

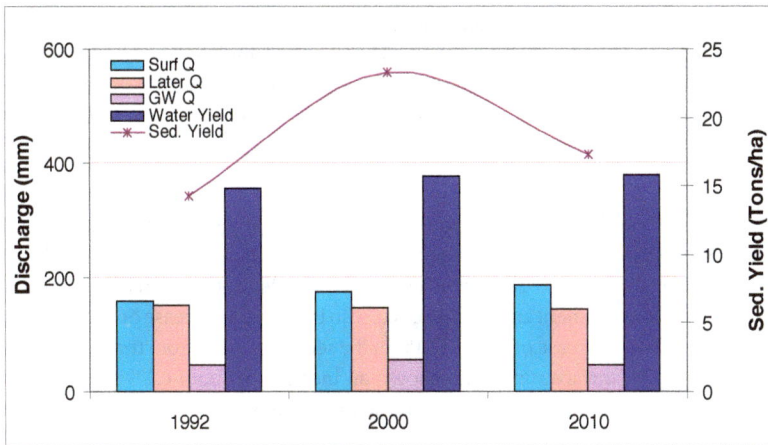

**Figure 7.** Hydrological parameters like surface runoff, water yield and sediment yield indicate an overall rise in values in reponse to landuse changes occurred within 1992-2010 period

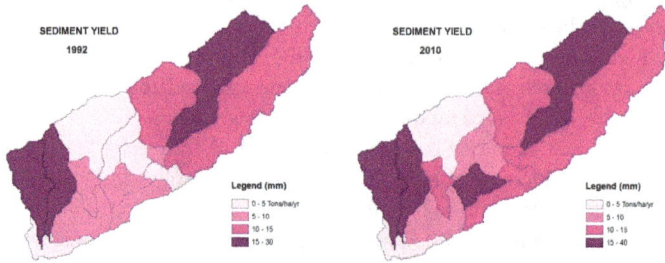

**Figure 8.** Temporal analysis of avarage annual sediment yield in the watershed during 1992-2010 peiod

## 3.3. Scenarios of extreme conditions

Different scenarios of landuse and climate change were developed to observe the response of water and sediment yields to the expected extreme conditions in future. The first three scenarios are related to probable changes in landuse/landcover of the watershed in future. As most of the landuse changes are taking place due to growth in urbanization i.e. development of built-up, agriculture land, deforestation/ afforestation in the area, so it formed the basis of these scenarios. The percentage coverage of landuse in the watershed under base line and three scenarios is shown in Table 3 and in map form in Figure 9. The other scenarios are based on the prediction scenarios of climate change for this region i.e. +0.9 ºC and +1.8 ºC change in temperature during 2020 and 2050 [28] and changes in precipitation. These were formulated in consultation with experts from the Intergovernmental Panel on Climate Change (IPCC) and are consistent with the scenarios generated using the Model for Assessment of Greenhouse gas Induced Climate Change (MAGICC) software. The analysis of different scenarios is given below:

- In the first scenario, all the rangeland below 800m elevation is assumed to be converted into built-up land (About 20% increase in the settlement class). It is based on our study findings that most of the urban development has been occurred in the low valley areas below 800m elevation during the last two decades. The runoff estimates in urban areas are required for comprehensive management analysis. The scenario indicates a decrease of about 0.1% in the water yield while an increase of about 12.1% in the sediment yield from that of the base year 2010 (Table 4). The surface runoff has shown an increase of about 0.2% while lateral discharge a decrease of about 2% due to increase in the impervious area during 2010 in the watershed.

- In the second scenario, all the scrub forest below 1200m elevation is assumed to be converted into agriculture land (About 31% increase in agriculture land) keeping other landuse conditions same as of base year 2010. This scenario is also based on our study findings that major agriculture development has occurred below 1200 m elevation during the last two decades in the watershed area. The scenario indicates an increase of about 3.6% in the water yield and about 73.6% in the sediment yield from the base year 2010. The surface runoff increases by 4.4% while lateral discharge decreases by 5.6% due to decrease in the scrub

forest coverage.

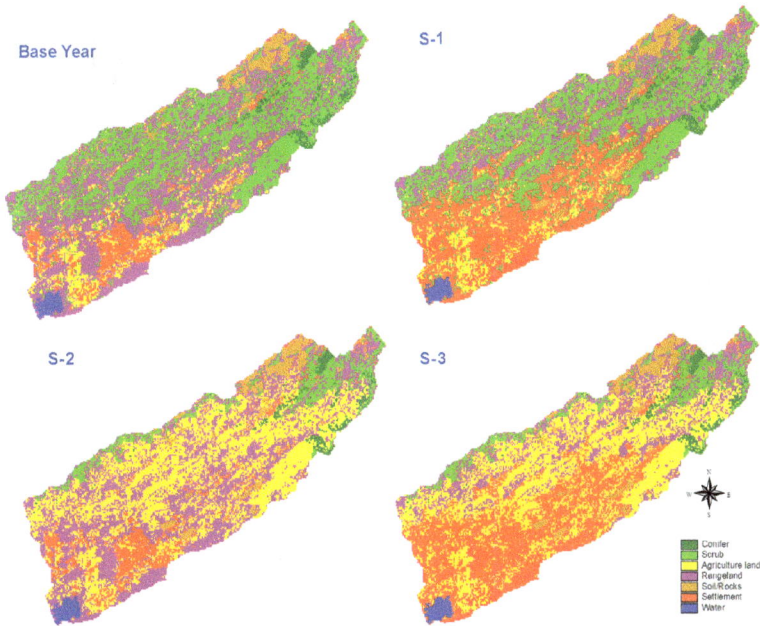

**Figure 9.** Landuse/landcover status during base year and three landuse change scenarios

| Landuse | Base year 2010 | Senario-1 | Senario-2 | Senario-3 |
|---|---|---|---|---|
| Conifer | 2.3 | 2.3 | 2.3 | 2.3 |
| Scrub | 38.8 | 38.8 | 7.7 | 7.7 |
| Agriculture | 7.4 | 7.4 | 38.5 | 38.5 |
| Rangeland | 36.7 | 16.3 | 36.7 | 16.3 |
| Soil/Rocks | 4.8 | 4.8 | 4.8 | 4.8 |
| Settlement | 8.9 | 29.3 | 8.9 | 29.3 |
| Water | 1.1 | 1.1 | 1.1 | 1.1 |
| Total | 100.0 | 100.0 | 100.0 | 100.0 |

**Table 3.** Percentage coverage of landuse in base year 2010 and under three landuse change scenarios

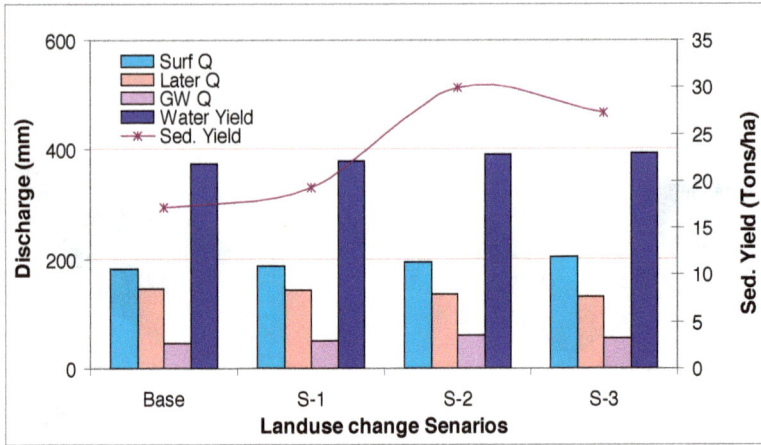

**Figure 10.** Hydrological response of the watershed under base year and three landuse change senarios

| No. | Scenarios | Surf. Runoff % | Water Yield % | Sediment Yield % |
|-----|-----------|----------------|---------------|------------------|
| S-1 | Urban develop. Below 800m elevation (converting rangeland into built-up land) | 0.2 | -0.1 | 12.1 |
| S-2 | Agriculture develop. below 1200m elevation (converting scrub forest into agriculture land) | 4.4 | 3.6 | 73.6 |
| S-3 | Combining scenarios 1 & 2 | 10.5 | 4.1 | 58.1 |
| S-4 | +0.9°C temperature in 2020 | -0.8 | -1.3 | 13.1 |
| S-5 | +1.8°C temperature in 2050 | -2.1 | -3.0 | 28.3 |
| S-6 | Increase of 10% rainfall in 2020 with no change in temperature | 24.5 | 19.1 | 26.1 |
| S-7 | Increase of 0.9°C temperature & 10% rainfall in 2020 | 23.6 | 17.8 | 41.8 |
| S-8 | Increase of 1.8°C temperature & 10% rainfall in 2050 | 22.1 | 15.8 | 58.1 |
| S-9 | Increase of 0.9°C temperature & decrease of 10% rainfall in 2020 | -22.7 | -19.0 | -13.6 |
| S-10 | Increase of 1.8°C temperature & decrease of 10% rainfall in 2050 | -23.8 | -20.6 | -1.5 |

**Table 4.** Percentage changes projected for surface runoff, water yield & sediment yield under different landuse and climate change scenarios using base conditions of 2010

- The third scenario is based on the combination of the 1st and 2nd scenarios i.e. increase in built-up and agriculture land below 800m and 1200m elevations, respectively. The scenario indicates an increase of about 4.1% in the water yield and 58.1% in the sediment yield of watershed. The surface runoff indicates an increase of about 10.5% while lateral discharge decrease of about 9% due to decline in the scrub forest cover and growth in the urban development. The scenarios 2&3 have also indicated an increase of 1.1% and 0.6% in actual evapotranspiration due to temperature variations. The hydrological response against different landuse change scenarios is shown graphically in Figure 10.

- The scenarios 4 to 10 are based on future climate changes in the watershed area and respective response of the water and sediment yields with the assumption that no change in landuse/landcover of base year 2010 will take place over the time. The rise of about 0.9°C temperature in year 2020 and 1.8°C in year 2050 indicates decrease of about 1.3% and 3.0% in the water yields of the watershed. Increase in temperature may result in higher evaporation rates that would affect the behavior of water yield.

- In scenario-6, increase of 10% in rainfall during 2020 keeping same temperature conditions as of base year 2010 has shown an increase of about 19% in the water yield and 26% in the sediment yield of the watershed. Similar increase in rainfall with same temperature conditions as of scenarios-4 and 5 i.e. +0.9°C in 2020 & +1.8°C in 2050, projects nearly 18% and 16% increase in the water yields and about 24% and 22% increase in the surface runoff as shown under scenarios-7 and 8 in Table 4. The sediment yield has also shown an increase ranging between 41% and 59% in these scenarios. The increase in rainfall usually causes increase in magnitude of floods which ultimately creates soil erosion and land degradation problems.

- On the other hand decrease of 10% in rainfall with same temperature conditions as of scenarios-4 and 5 projects decrease of about 19% and 21% in the water yields which ultimately lowered the sediment yield from that of base condition of 2010.

### 3.4. Risk mitigation of sediment yield

Appropriate strategies have to be defined separately for different landuse conditions for minimizing the risk of soil loss and sediment yield involved in scenarios of extreme conditions. In order to reduce high sediment yield from sub-basins with rapid urban development, the unplanned urbanization needs to be controlled by appropriate laws and means. In the non-urban areas, proper soil and water conservation measures can be adopted to mitigate the risk of soil erosion.

In high risk zone of sediment yield, mainly dissected gullies are more susceptible to soil erosion. The problem of gully erosion can be solved to great extent through restoration of vegetative cover for which proper structures could be placed to provide protection long enough to give vegetation a start. The structures may be of temporary or permanent in nature keeping in view the nature of problem. Conservation structures to reduce velocity of the runoff can also be developed to control the extent of gully erosion. In medium risk zone of soil erosion, modifying the cross section and grade of channel to limit the flow velocities can be performed to stabilize the gullies. The conservation measures like terracing, contour bunding and diversion channels can be adopted in the contributory watershed to control excessive surface runoff causing gully erosion. These practices will also provide additional moisture for growing crops and vegetative cover thus help reducing gully erosion. In low risk zone of soil erosion, contour benches having small bunds crossway the slope of the land on a contour may be established to reduce the erosion risk. High intensity rainfall during monsoon season invariably cause over saturation harmful for plants. The situation can be avoided through provision of water ways and grassy outlets to dispose off the excessive runoff. The risk of erosion can also be minimized through adopting practices like strip farming, terracing, contour farming besides modifying bunds and minor land leveling in the cultivated area of watershed.

## 4. Conclusions

The recent changes in landuse/landcover conditions have brought significant impact on water flows, sediments and threat to eco-hydrology of the Himalayan area. The rapid growth in urbanization has increased the demand for land for development purposes consequently forest and water resources are coming under enormous pressure. The general trends of landuse change are gradual decline in coverage of scrub and coniferous forest, increase in urban development and somewhere in agriculture area. The increase in built-up land in the valleys has reduced the recharge source of groundwater which needs to be protected through controlling unplanned growth of urbanization. The rise in global warming accompanied with high variability in precipitation projects extreme changes in water balance and ultimately deterioration of the land quality. It is essential to regulate the urban development properly, affordable substitute-fuels for household use should be made available and an extensive community reforestation programme is undertaken to improve the fragile eco-system of the region. An integrated adaptation strategy needs to be developed at national and regional levels to cope with future implications of hydrological changes through focusing key policy areas and improving adaptive capacities of the communities at risk. Existing knowledge and data gaps need to be filled by systematic observations and enhanced capacities for research since these will be fundamental for developing climate change adaptation and mitigation programmes for the Himalayan region in future.

## Author details

Arshad Ashraf

Water Resources Research Institute, National Agricultural Research Center, Islamabad, Pakistan

# References

[1] Behrman, N. The Waters of The Third Pole: Sources of Threat, Sources of Survival: 2010; 48. www.chinadialogue.net/UserFiles/File/third_pole_full_report.pdf (accessed 17 October, 2012).

[2] ICIMOD. Climate Change and the Himalayas: More Vulnerable Livelihoods, Erratic Climate Shifts for the Region and the World. Newsletter, Sustainable Mountain Development 2007; No.53: p55. ISSN 1013-7386 2007.

[3] Singh, SP; Bassignana-Khadka, I; Karky, BS and Sharma. Climate change in the Hindu Kush-Himalayas: The state of current knowledge, Kathmandu: ICIMOD; 2011.

[4] Neitsch, S.L.; Arnold, J.G.; Kiniry, J.R.; and Williams, J.R. Soil and water assessment tool theoretical documentation. available at: http://swatmodel.tamu.edu/media/1292/SWAT2005theory.pdf; 2005.

[5] Neitsch S.L.; Arnold J.G. and Williams J.R. Soil and Water Assessment Tool, User's Manual; 1999.

[6] Arnold, J.G.; J.R. Williams; R. Srinivasan; K.W. King and R.H. Griggs. SWAT - Soil and Water Assessment Tool, USDA, Agricultural Research Service, Grassland, Soil and Water Research Laboratory, 808 East Blackland Road, Temple, TX 76502, revised 10/25/94; 1994.

[7] Williams, J.R. Chapter 25. The EPIC Model. p. 909-1000. In Computer Models of Watershed Hydrology. Water Resources Publications. Highlands Ranch, CO.; 1995.

[8] Ndomba, P.M. Modelling of Sediment Upstream of Nyumba Ya Mungu Reservoir in Pangani River Basin, Nile Basin Water Science & Engineering Journal 2010; Vol. 3, No. 2: 25-38.

[9] Setegen, S.G.; R. Srinivasan and B. Dargahi. Hydro-logical Modelling in Lake Tana Basin, Ethiopia Using SWAT Model, The Open Hydrology Journal 2008; Vol. 2: 49-62.

[10] Odira, P.M.A.; M.O. Nyadawa; B. Okello; N.A. Juma and J.P.O. Obiero. Impact of land use/cover dynamics on stream flow: A case study of Nzoia River Catchment, Kenya, Nile Water Science and Engineering Journal 2010; Vol. 3, No. 2: 64-78.

[11] Aftab, N. Haphazard colonies polluting Rawal Lake, Daily Times Monday, March 01. http://www.dailytimes.com.pk/default.asp...009_pg11_1. 2010.

[12] Roohi, R.; Ashraf, A.; Naz, R.; Hussain, S.A. and Chaudhry, M.H. Inventory of glaciers and glacial lake outburst floods (GLOFs) affected by global warming in the mountains of Himalayan region, Indus Basin, Pakistan Himalaya. Report prepared for ICIMOD, Kathmandu, Nepal; 2005.

[13] Ghumman, A.R. Assessment of water quality of Rawal Lake by long term monitoring. Environmental Monitoring Assessment Journal. DOI 10.1007/s10661-010-1776-x.; 2010.

[14] Khan, M.I.R. Pakistan Journal of Forestry 1962; 12: 185.

[15] Soil Survey Report. Reconnaissance soil survey of Haro Basin 1976, Preliminary Ed. Soil Survey of Pakistan, Lahore; 1978.

[16] IUCN. Rapid environmental appraisal of developments in and around Murree Hills, IUCN Pakistan; 2005.

[17] Tanvir, A.; B. Shahbaz and A. Suleri (2006), Analysis of myths and realities of deforestation in northwest Pakistan: implications for forestry extension. International Journal of Agriculture & Biology.1560–8530/2006/08–1–107–110. http://www.fspublishers.org.

[18] Shafiq, M.; S. Ahmed; A. Nasir; M.Z. Ikram; M. Aslam and M. Khan. Surface runoff from degraded scrub forest watershed under high rainfall zone. Journal of Engineering and Applied Sciences 1997; 16(1): 7-12.

[19] EPA (2004), Report on Rawal lake catchment area monitoring operation, Pakistan Environmental Protection Agency, Ministry of Environment, Islamabad: pp. 19.

[20] Roohi, R.; A. Ashraf and S. Ahmed. Identification of landuse and vegetation types in Fatehjang area, using LANDSAT-TM data. Quarterly Science Vision 2004; 9(1): 81-88.

[21] Oweis, T. and M. Ashraf (eds). Assessment and options for improved productivity and sustainability of natural resources in Dhrabi watershed Pakistan, ICARDA, Aleppo, Syria; 2012.

[22] Nasir A.; K. Uchida and M. Ashraf. Estimation of soil erosion by using RUSLE and GIS for small mountainous watersheds. Pakistan Journal of Water Resources 2006; 10(1): 11-21.

[23] Donigian Jr., A. S. Watershed model calibration and validation: The HSPF experience. National TMDL Science and Policy, Phoenix, AZ. November 13–16, 2002: 44-73.

[24] Nash, J.E. and Sutcliffe V. River flow forecasting through conceptual models, I. A discussion of principles. Journal of Hydrology 1970; 10: 282-290.

[25] Arfan, M. Spatio-temporal Modeling of Urbanization and its Effects on Periurban Land Use System. M.Sc Thesis. Department of Plant Sciences. Quaid-e-Azam University, Islamabad; 2008.

[26] FAO. State of the world's forests – 2005. Food and Agricultural Organization (FAO), Rome, Italy; 2005.

[27] Booth, D. Urbanization and the natural drainage system - impacts, solutions and prognoses. Northwest Environmental Journal 1991; 7: 93-118.

[28] INC Report. Pakistan's Initial National Communication on Climate Change to UNFCCC, Ministry of Environment, Islamabad; 2003.

# Chapter 11

# A Review of the Effects of Hydrologic Alteration on Fisheries and Biodiversity and the Management and Conservation of Natural Resources in Regulated River Systems

Peter C. Sakaris

Additional information is available at the end of the chapter

## 1. Introduction

Hydrologic alterations resulting from dam construction and other human activities have negatively impacted the biodiversity and ecological integrity of rivers worldwide (Dudgeon 2000, Pringle et al. 2000). These alterations have included habitat fragmentation, conversion of lotic to lentic habitat, variable flow and thermal regimes, degraded water quality, altered sediment transport processes, and changes in timing and duration of floodplain inundation (Cushman 1985, Pringle 2000). The negative impacts of altered hydrologic regimes on aquatic organisms are well documented. For example, dam construction has blocked the migratory routes of diadromous and potamodromous species (e.g., salmonids and white sturgeon, *Acipenser transmontanus*), which has severely reduced their spawning and overall reproductive success (Wunderlich et al. 1994, Beamesderfer et al. 1995). In the Alabama River system (USA), flow-modification in regulated reaches has resulted in losses of river-dependent ("fluvial") fish species, and distributions of federally listed species have been restricted by main stem impoundment (Freeman et al. 2004). Several researchers have documented major changes in fish assemblage structure following dam construction (Paragamian 2002, Quinn and Kwak 2003; Gillete et al. 2005). For example, Quinn and Kwak (2003) reported that long-term changes in the fish assemblage after dam construction on the White River (Arkansas, USA) included a shift from warmwater to coldwater species, a substantial decrease in fluvial specialists, and dramatic reductions in species richness. The negative effects of dam construction are not only limited to fishes. Freshwater mussel diversity has declined substantially, particularly in the southeast USA, as a consequence of hydrologic alteration (Watters 2000).

This chapter begins with an examination of the hydrologic alterations that may be caused by dam construction. Several examples are presented for different continents, emphasizing that hydrologic alteration is an issue of global concern. Next, the effects of altered hydrologic regimes on the growth, recruitment, and survival of organisms and on the overall biodiversity and community structure in regulated river systems are reviewed. Subsequently, tools and strategies to manage and conserve aquatic fauna in regulated river systems are discussed. In the past several decades, a wealth of information has been published on these topics. This chapter provides a general overview of the impacts of hydrologic alteration and presents several management approaches, which have been developed to address it. More detailed information may be found in the review papers referenced in this chapter.

## 2. Alteration of the flow regime

In their highly impactful paper, Poff et al. (1997) outlined five important characteristics of a flow regime: *magnitude, frequency, duration, timing* (or predictability), and the *rate of change* (or flashiness). These major components of the flow regime are ecologically relevant to the system. For example, the magnitude of flow (e.g., mean monthly discharge) may define habitat characteristics such as wetted area or habitat volume in a stream or river (Richter et al. 1996). The frequency of episodic flows (e.g., high or low pulse frequencies) may lend insight on how often drought or flood conditions occur within a system (Richter et al. 1996). Each of these flow attributes may be altered by dam construction and hydropeaking operations. For example, flows have rapidly fluctuated between extremely low and high discharges as a result of hydropeaking operations downstream of Harris Dam on the Tallapoosa River, USA (Irwin and Freeman 2002). Extreme discharge fluctuations during a period of only four to six hours have generated a highly variable flow regime that has potentially threatened the persistence of several native fishes (i.e., fluvial specialists) below the dam (Irwin and Freeman 2002). Irwin and Freeman (2002) reported significant changes in hydrology after construction of Harris Dam in 1982, which included increases in high-pulse frequency, low-pulse frequency, fall rate, and the number of flow reversals. Irwin and Freeman (2002) documented release-driven, diel temperature fluctuations as high as 10°C, producing highly stressful conditions for resident organisms.

Hydrologic alterations that occur as a result of dam construction have been well documented (Galat and Lipkin 2000, Maingi and Marsh 2002, Yang et al. 2008). For example, Yang et al. (2008) applied the Range of Variability Approach (RVA discussed later; Richter et al. 1997) to evaluate the effects of dam construction on the hydrologic regimes of middle and lower river networks in Yellow River, China. The authors stressed that assessments of hydrologic alteration are extremely complex, particularly in systems that are impounded by more than one dam (Yang et al. 2008). In addition, both pre- and post-impact discharge data must be sufficient to effectively assess the effects of dams on hydrologic processes. The Yellow River in China was impounded by the Sanmenxia and Xiaolangdi dams to meet several objectives: flood control and electricity generation, to reduce downstream sediment deposition, and to provide water for irrigation. Unfortunately, the natural flow regime has become significantly altered in the system, with the lower Yellow River recently experiencing zero flow conditions as a result of increased water consumption (Yang et al. 2008). Significantly reduced flows will have negative effects on biodiversity and the persistence of viable wetlands and fisheries in the Yellow River Delta (Yang et al. 2008). The analysis of Yang et al. (2008) indicated that Xiaolangdi dam significantly altered the natural flow regime of the lower Yellow River in the following ways: decreased median of monthly flow, decreased medians of annual 1-, 3-, 7-, 30-, and 90-day minimum and maximum flows, higher low pulse and high pulse counts, and decreased medians of fall rate, rise rate, and number of reversals in the post-impact period.

Maingi and Marsh (2002) studied the effects of dam construction on hydrologic conditions in the Tana River, Kenya. Kenya has a growing population, and water needs have increased as populations are forced to expand into semi-arid regions (Maingi and Marsh 2002). Five dams were constructed from 1968 to 1988 along the upper Tana, the largest river in Kenya (Maingi and Marsh 2002). The largest dam (Masinga Dam) was built to provide hydropower, to increase irrigation potential in the lower basin, and to increase use of dry season flows in the upper Tana (Maingi and Marsh 2002). Of special concern is a tract of riverine tropical forest along the mid- to lower Tana River. This forest extends 0.5 to 3 km from the bank of the river, and the forest largely depends on regular flooding and sufficient groundwater. With decreased peak flows and a declining water table due to river regulation, preservation and regeneration of this riverine forest has become a challenge (Maingi and Marsh 2002). Analyses revealed that major changes in the flow regime occurred after Masinga Dam was constructed, including a significant reduction in May flows, reduced variability in monthly discharges, reduced 7-d, 30-d, and 90-d maximum annual discharges, decreased mean low pulse duration from 14.6 to 7.9 days, increased annual rises and annual falls of the river, and increased mean fall rates from 15.1 m/s to 21.6 m/s (Maingi and Marsh 2002). Experiments with vegetation sample plots indicated that vegetation located above 1.80 m of dry season river level has experienced an average 67.7% reduction in days flooded after construction of Masinga Dam. Experiments also revealed that flood pulse duration declined significantly for all vegetation plots by an average of 87.6% (Maingi and Marsh 2002). These reductions in flood frequency and duration have negative implications for the preservation of riverine forest in the Tana River Basin.

Along the Missouri River (USA), a series dams were constructed for improved navigation, irrigation, and flood control, as well as hydropower generation (Galat and Lipkin 2000). Galat and Lipkin (2000) attributed the listing of the Missouri River as North America's most endangered river in 1997 (American Rivers, 1997) to the numerous alterations that have significantly impacted the ecosystem. In their study, Galat and Lipkin (2000) divided the Missouri River into three sections and classified them as 1) an upper, least-altered section with four dams, 2) a middle highly impacted section with six large mainstem dams, and 3) a regulated and channelized lower section. The authors also used Richter's RVA approach (Richter et al. 1997) to compare pre- and post-impoundment hydrologic conditions at the upper (least-altered), middle, and lower (channelized) sections (Galat and Lipkin 2000). Their analyses indicated that numerous hydrologic changes occurred after mainstem impoundment of the Missouri River including: 1) increased mean annual discharges (i.e., 30 to 38% higher at channelized locations), 2) a stabilization of mean monthly discharges with higher flows from August through February and a reduction in June and July high flows, 3) loss of a natural bimodal flood pulse, 4) lower flow variability at most stations, 5) higher 1-, 7-, 30-day annual minimum flows at all stations, 6) altered timing of annual peak and minimum flows (particularly at middle, inter-reservoir sites), 7) increased frequency of high pulses at two middle, inter-reservoir locations and three of four lower-basin channelized stations, 8) decreased frequency of high pulses at one middle, inter-reservoir and two channelized stations, 9) changes in mean duration of high-flow pulses, 10) decrease in the number of low-flow pulses, and 11) reduction in mean rise and fall rates. Some changes, such as stabilization of mean monthly discharges and variation in the number of high pulses, were absent or mild at least-altered locations and more pronounced at middle and channelized locations (Galat and Lipkin 2000). Pegg et al. (2003) used an alternative, time-series approach for assessing the impacts of hydrologic alteration on the Missouri River, and their findings generally corroborated those of Galat and Lipkin (2000).

Modification of the landscape, or watershed, through land-use activities has also influenced hydrologic processes and has complicated our understanding of how hydrologic alteration affects the ecological integrity of freshwater ecosystems. The ability to return to some semblance of a natural flow regime will require knowledge of how land-use (e.g., residential, industrial, agriculture, etc.) also impacts hydrologic conditions in a regulated river system (Poff et al. 1997, Poff et al. 2006). Streams and rivers are four-dimensional systems (i.e., including their temporal dimension) that are intimately linked with the groundwater and landscape, and their lateral interactions with the landscape are vital to maintaining the integrity and function of these ecosystems (Fausch et al. 2002). Stream and riverine ecosystems exchange sediments, nutrients, and energy with the landscape, and, therefore, are part of a larger "riverscape" (Fausch et al. 2002), open to and affected by external processes.

Poff et al. (2006) sought to understand how geomorphological alterations due to land-use changes interact with hydrologic alterations from dam construction to influence the hydro-geomorphic integrity of streams in the USA. The interaction between the natural flow regime and geomorphology of a stream is important in defining the habitat that is available to organisms in a system (Poff et al. 2006). Therefore, alterations of the flow and/or geomorphic properties of a system will likely induce changes in resident fauna. Poff et al. (2006) also emphasized the importance of understanding how the underlying natural variation in physiography across major regions can influence the impact of land-use changes on hydrogeomorphic processes in a stream. The authors explained that the natural topographical, geological, and climatic features of a region will influence how deforestation and agricultural activities, for example, might affect the rates of sediment and nutrient input and water flow in local streams (Poff et al. 2006). In their study, four U.S. regions, distinct in their natural vegetation, climate, geology, and physiography, were identified and examined: 1) Pacific Northwest, with Pacific Lowland Mixed Forest and Cascade Mixed Forest-Coniferous Forest-Alpine Meadow provinces, 2) Southwest, with the Colorado Semi-Arid, American Semi-desert and Desert provinces, 3) Central, consisting of the Prairie Parkland Temperate Forest, and 4) Southwest Region, consisting of South Eastern Mixed Forest (Poff et al. 2006). Within each region, areas were classified as *Least Disturbed*, *Agriculture*, or *Urban*, and the percent of each class was calculated (Poff et al. 2006). Poff et al. (2006) discovered numerous regional differences in hydrologic responses to flow alteration along gradients of increasing urban and

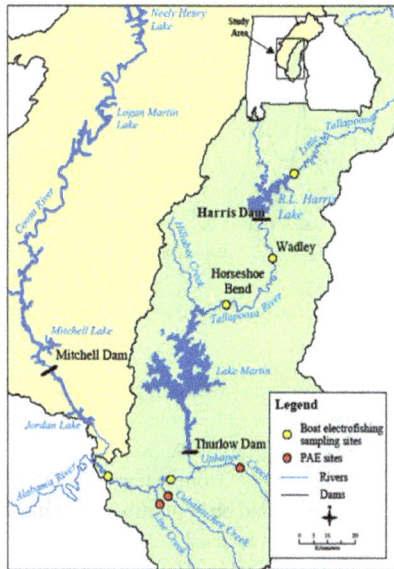

**Figure 1.** Hydrologic data were retrieved from USGS stream gauge stations at two regulated sites downstream of Harris Dam, Wadley and Horseshoe Bend, and at one unregulated site, Hillabee Creek (from Sakaris 2006).

agricultural land cover. For example, with increased urban cover, peak flows increased in the Southeast and Northwest regions, minimum flows increased in the Central Region and decreased in the Northwest, duration of near-bankfull flows declined in the Southeast and the Northwest, and flow variability increased in three regions (Southeast, Central, and Northwest; Poff et al. 2006). Poff et al.'s (2006) study highlighted the importance of accounting for regionally specific, landscape-level effects in the assessment of local hydrologic conditions in stream ecosystems.

## 3. Case study

In the Alabama River system (USA), four hydropower dams were constructed on the main stem of the Tallapoosa River (Boschung and Mayden 2004). In the Northern Piedmont, flows have rapidly fluctuated between extremely low and high flows as a result of hydropeaking operations downstream of Harris Dam on the Tallapoosa River (Irwin and Freeman 2002). Hydrologic data were retrieved from United States Geological Survey (USGS) stream gauge stations at three locations downstream of Harris Dam: 1) Wadley (USGS 02414500), a regulated site downstream of Harris Dam, 2) Horseshoe Bend (USGS 02414715), a regulated site downstream of the Wadley location, and 3) Hillabee Creek (USGS 02415000), an unregulated tributary of the Tallapoosa River (Figure 1, website: waterdata.usgs.gov/al/nwis).

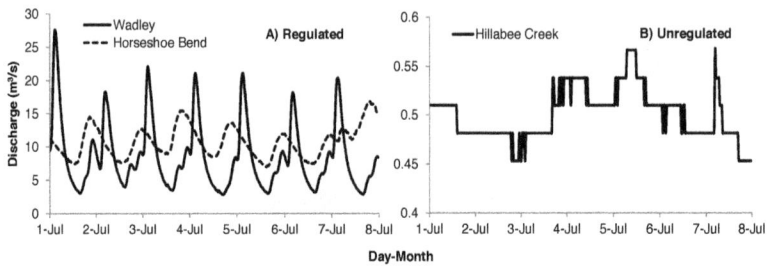

**Figure 2.** Daily variation in river discharge at two regulated locations (A) and at an unregulated site (B) in the Talla-poosa River Watershed.

Hourly variation in stream discharge (m³/s) was compared among the three sites during the first week of July 2012. Ecologically relevant hydrologic variables were also calculated in the Indicators of Hydrologic Alteration Program (IHA, Sustainable Waters Program, The Nature Conservancy, Boulder, CO) for each location from water years 1987 to 2012. Water years were started on October 1 of each year and ended on September 30 of the following year (e.g., water year 1987 = 10/01/86 – 9/30/87). Analyses focused on annual high and low pulse frequencies, number of reversals, and rise rates. Annual hydrologic conditions were compared between the two regulated sites and the unregulated site, which was treated as a "reference site."

Hydrologic regimes were markedly different between the regulated locations and the un-regulated site (Figure 2). In early July, daily hydropeaking operations produced unnatural flow variation below Harris Dam, while a more natural flow regime persisted in the local unregulated tributary (Figure 2). Daily variation in discharge was substantially dampened at the Horseshoe Bend site (Figure 2), which is located farther downstream of Harris Dam (Figure 1). This reduced variation in flow indicated that the effects of hydropeaking operations may not be as severe at more downstream locations. However, unnatural and rapidly fluctuating flows, such as those observed below Harris Dam, generally produce a stressful environment for the river fauna that reside there.

As expected, high pulse and low pulse frequencies, the number of reversals, and rise rates were similar between the two regulated sites, as well as the overall annual variation in these hydrologic parameters (Figure 3). All four hydrologic parameters were substantially lower at the unregulated site (Figure 3). Fewer high pulses, low pulses, and reversals at Hillabee Creek indicated that the flow regime was much less variable and may be more representative of natural flow conditions in this region. Higher rise rates at the regulated sites are likely due, in part, to the rapidly increasing flows during hydropeaking events.

## 4. Hydrologic effects on recruitment, growth, survival

Altered flow regimes below dams have typically produced unfavorable conditions for the recruitment of fishes (Fraley et al. 1986; Brouder 2001, Freeman et al. 2001, Wildhaber et al.

A Review of the Effects of Hydrologic Alteration on Fisheries and Biodiversity and the Management
and Conservation of Natural Resources in Regulated River Systems

273

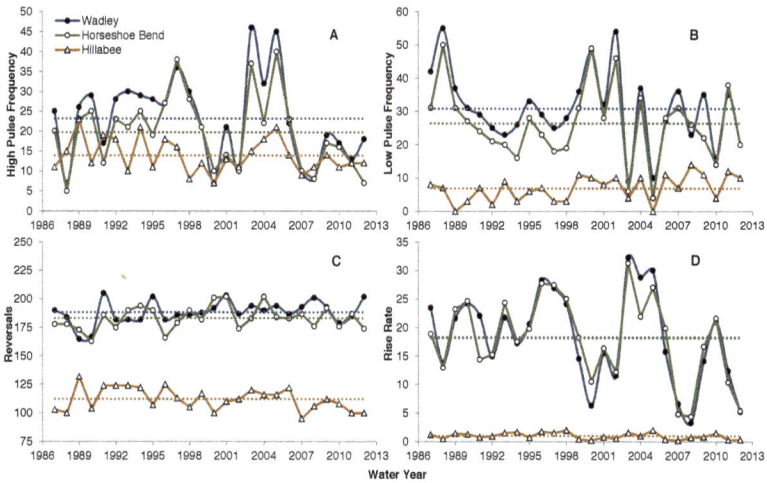

**Figure 3.** Annual variation in high pulse frequencies (A), low pulse frequencies (B), number of reversals (C), and rise rates (D) at two regulated locations and at an unregulated site in the Tallapoosa River Watershed. Dotted lines represent the statistical mean for each location. Mean rise rate was nearly equal at the two unregulated sites.

2000, Propst and Gido 2004). Freeman et al. (2001) reported that juvenile fish abundances were strongly related to the persistence of shallow habitats in a regulated reach of the Tallapoosa River, Alabama. However, the persistence of these habitats was severely reduced by rapid flow fluctuations resulting from hydropeaking operations (Freeman et al. 2001). In a regulated section of the Neosho River (Kansas), the reduction of minimum flows below John Redmond Dam reduced the availability of riffle habitats that were suitable for Neosho madtoms (*Noturus placidus*, Wildhaber et al. 2000). Moreover, hypolimnial-release of coldwater from dams will generally slow growth and development and alter physiology of fish during early life stages; whereas, the release of warm water from small, surface release dams may result in reduced densities of coldwater fish species (Clarkson and Childs 2000, Lessard and Hayes 2003).

Recruitment of fishes has been related to hydrology in freshwater ecosystems; however, most studies have been conducted in reservoirs (see, for example: Maceina and Stimpert 1998, Buynak et al. 1999, Sammons and Bettoli 2000, Schultz et al. 2002). Few studies have directly examined relations between hydrology and recruitment of fishes in regulated river sections. In a regulated section of the Roanoke River (North Carolina), Rulifson and Manooch (1990) reported that striped bass *Morone saxatilis* recruitment was highest when river flows were low to moderate (142 – 283 m³/s) during the spawning season. During the years when recruitment was highest, *flows typically resembled pre-impoundment flow conditions* (Rulifson and Manooch 1990). Striped bass require a specific flow regime for successful transport of eggs and larvae to nursery habitats (Rulifson and Manooch 1990). Irwin et al. (1999) reported that riffle habitats (i.e., shallow-fast and shallow coarse) were utilized by juvenile channel catfish *Ictalurus punctatus* and flathead catfish *Pylodictis olivaris*. However, persistence of these habitats may decrease in highly regulated systems (Bowen at al. 1998), thereby negatively influencing the recruitment of catfishes. Furthermore, Holland-Bartels and Duval (1988) suggested that variation in channel catfish productivity was related to river discharges. A decrease in age-0 channel catfish abundance was attributed to a sharp increase in river discharge that likely disrupted spawning activity and flushed young from nests (Holland-Bartels and Duval 1988). Therefore, one would suspect that highly variable flows (i.e., high rise and fall rates) during hydropeaking operations would negatively affect the spawning success and recruitment of channel catfish. In middle reaches of the regulated Missouri River, Pegg et al. (2003) identified a significant reduction in spring spawning flows as a major impairment of fish spawning and recruitment.

Studies have indicated that reduced flooding, or a diminished flood pulse, has contributed to low fish recruitment in river systems. Bonvechio and Allen (2005) studied recruitment of sunfishes *Lepomis* spp. and black basses *Micropterus* spp. in relation to hydrology in four Florida rivers. The authors suggested that high flows in the fall would increase access to floodplain habitats, thereby increasing prey availability (i.e., invertebrates) for adult sunfishes before the spawning season (Bonvechio and Allen 2005). Sunfishes would likely consume more prey and allocate more energy towards reproduction (i.e., fecundity), producing a stronger year class. In the inter-reservoir and lower channelized sections of the Missouri River, changes in the magnitude, frequency, timing, and duration of the annual flood pulse (i.e., inundation of the floodplain) was indicated as the likely cause of reduced recruitment and production of flood-plain fishes (Galat and Lipkin 2000). The elimination of a fall flood pulse in the lower Missouri River has limited fish and wildlife access to floodplain habitats (Galat and Lipkin 2000). Brouder (2001) found a strong positive relationship between maximum mean daily discharge and recruitment of the roundtail chub, *Gila robusta*, in the upper Verde River, Arizona. Brouder (2001) explained that a reduction or elimination of flooding through hydrologic alteration would be deleterious to the recruitment of native roundtail chub. Flooding helps to prepare spawning substrate by clearing interstitial spaces for eggs, potentially reduces population sizes of nonnative species (thereby reducing competition), and possibly dilutes contaminants that would negatively affect reproductive success (Brouder 2001).

Alteration of the thermal regime as a consequence of dam construction can also have negative effects on fish recruitment (Clarkson and Childs 2000, Horne et al. 2004). Hydroelectric dams on the Manistee River, a tributary of Lake Michigan, negatively impact steelhead (anadromous rainbow trout, *Oncoryhnchus mykiss*) recruitment, by preventing steelhead access to potential upstream spawning habitats (Horne et al. 2004). Furthermore, the release of warm surface water from the reservoir results in increased summer temperatures that reduce the survival of age-0 steelhead in the river (Horne et al. 2004). Clarkson and Childs (2000) proposed that declines of native big-river fishes of the Colorado River Basin were partly due to the release of cold, hypolimnial water from dams. In laboratory experiments, growth rates of four species (razorback sucker, flannelmouth sucker, humpback chub, and Colorado squawfish) were slower and their development was delayed at colder temperatures (Clarkson and Childs 2000). Larval fish also lost equilibrium when transferred from 20°C to 10°C (Clarkson and Childs 2000). Slow growth, delayed development, and loss of equilibrium at early life history stages all likely contribute to reduced recruitment in a system.

Growth of fishes may also be related to hydrology in river systems. Quist and Guy (1998) suggested that increased growth of channel catfish in the Kansas River (USA) resulted from floodplain inundation. Inundation of the floodplain typically provides shallower, prey-rich habitats for fishes (Welcomme 1979). Mayo and Schramm (1999) hypothesized that growth of flathead catfish was also influenced by water temperature during the growing season, in addition to the number of flood days in the lower Mississippi River system. Unfortunately, hydrologic alterations may include changes to the timing and duration of floodplain inundation as well as thermal regimes (Cushman 1985, Pringle 2000). Rutherford et al. (1995) determined that growth of age-0 channel catfish in the Mississippi River was also related to the length of the growing season, which could theoretically be shortened with the release of cold, hypolimnetic water from a dam. Coldwater from hypolimnial-release dams may dramatically lower spring and summer tailwater temperatures, which may slow the growth and development of fishes (Clarkson and Childs 2000).

Hydrologic alteration has also strongly impacted the growth and recruitment of riparian and wetland vegetation (Young et al. 1995, Burke et al. 2008). For example, the recruitment of cottonwoods (*Populus* spp.) is closely linked to hydrologic and geomorphic processes (Scott et al. 1997, Burke et al. 2008). Burke et al. (2008) applied a hierarchical approach to studying the impacts of dam operations on the Kootenai River Ecosystem, focusing on cottonwood recruitment as their ecological response to river regulation. This hierarchical approach assessed *first order* impacts (changes in hydrologic conditions) that led to *second order* impacts, such as altered sediment transport and channel morphology (Burke et al. 2008). Burke et al. (2008) then described *third-level* impacts as biological functions that are influenced by first and second-level impacts. *Fourth-level* impacts are those involving feedbacks between biological and physical conditions (Burke et al. 2008). Overall, this hierarchical approach studied the effects of hydrologic alteration at the ecosystem level, assessing the physical, chemical, and biological changes that occur in a system.

The Kootenai River system, located in parts of Idaho, Montana, and British Columbia, has been modified in various ways. Levees were constructed in lower floodplain sections of the river to convert floodplain for agriculture, and Corral Linn Dam was constructed in the 1930s for hydropower and flood control (Burke et al. 2008). In 1974, Libby Dam was constructed upstream of Corral Linn Dam to provide additional hydropower and flood control, impounding 145 km of the river upstream of Libby Dam (Burke et al. 2008). As a result, natural flow conditions were altered and inundation of the floodplain was limited in the system (Burke et al. 2008). Specifically, higher flows are now maintained during naturally low-flow periods (fall-winter), while much lower flows are maintained during naturally high-flow periods (spring-summer; Burke et al. 2008). Burke et al. (2008) analyzed three time periods in their study: 1) *historic*, before Corral Linn Dam was operational, 2) *pre-Libby Dam*, and 3) *post-Libby Dam*. First, second, and third-order impacts as a result of dam construction and other activities were examined in a 233-km study reach between Libby Dam and the downstream Corral Linn Dam. To examine first-order impacts, Burke et al. (2008) utilized *Indicators of Hydrologic Alteration* (IHA) software (Richter et al. 1996) to compare hydrologic regimes across the time periods. The authors used a one-dimensional hydrodynamic flow model to evaluate second-order changes in hydraulics and bed mobility. Third-order impacts were assessed by changes in the recruitment potential of black cottonwoods. Due to the significant reduction of naturally high flows during the spring snowmelt, cottonwood seed germination sites are no longer prepared through the mobilization and redistribution of sediments. In addition, germination sites no longer experience the slow and gradual recession of these natural spring flows that would help maintain adequate soil moisture for the establishment of seedling roots (Burke et al. 2008). Burke et al.'s (2008) analyses revealed major changes in the hydrologic regime below Libby Dam, which included significantly greater median monthly flows during winter and the near elimination of a spring snowmelt peak (i.e., an "inverted annual hydrograph"). Notable second order, temporal and spatial alterations were also detected in stage fluctuation, stream power, shear stress, and bed mobility. The authors determined that the activities of both dams have contributed to lower cottonwood recruitment, by increasing stage recession rates during the seedling establishment period in the lower study reach and changing the timing, magnitude, and duration of flow in the upper and middle sections of the study reach (Burke et al. 2008).

Young et al. (1995) conducted a study examining how the growth of the Baldcypress, a tree common to wetlands in the southeastern United States, may respond to an altered flow regime. This tree is often found in wetlands or swamps that are subject to frequent or permanent flooding (Young et al. 1995). Before their study site was impounded by road construction in 1973, it existed as a floodplain swamp with permanent shallow flooding. The impoundment resulted in increased water levels on the upstream side of the road and apparently no effects on water levels on the downstream side (Young et al. 1995). Young et al. (1995) discovered that trees at the impacted, upstream site initially exhibited a significant growth surge due to increased flooding, but an overall decline in growth followed for 16 years. The initial surge in growth was possibly due to a pulse of sediment nutrient deposition with flooding, while the later decline in growth may have been a result of increased anoxic conditions in the rooting zone (Young et al. 1995).

The recruitment and survival of native riparian vegetation may not only depend on the restoration of more natural flow conditions, but also the removal of invasive riparian species that have better success in altered flow conditions (Merritt and Poff 2010). In western North America, the cottonwood, *Populus deltoides*, has declined substantially, while the invasive saltcedar, *Tamarix*, has become well established (Merritt and Poff 2010). Merritt and Poff (2010) determined that recruitment potential of *Tamarix* was highest along unregulated reaches in the Southwestern United States, *but remained high across a gradient of regulated flows*. In contrast, recruitment of cottonwoods was highest under natural flow conditions and declined abruptly with even slight flow modification (Merrit and Poff 2010). In addition, *Tamarix* was most dominant along the most altered river reaches, whereas *Tamarix* and *Populus* were equally dominant at the least regulated sites. The authors concluded that altered flow regimes have further enhanced the dominance of *Tamarix* over native plant species (Merritt and Poff 2010).

River regulation can also affect the reproductive success of marine fishes and invertebrates (Drinkwater and Frank 1994). For example, the recruitment of marine fish and invertebrates appears to be highly correlated with freshwater input (Drinkwater and Frank 1994). In most cases, increased river runoff into coastal oceans positively influences fish and invertebrate production, by increasing nutrient inputs that enhance primary production (Drinkwater and Frank 1994). The impoundment of rivers by dams, the diversion of water for agricultural purposes (irrigation), and regulated release of water from dams can modify the amount and/ or timing of freshwater released to coastal estuaries. See Drinkwater and Frank (1994) for a thorough review of the effects of river regulation on marine fish and invertebrates.

## 5. Hydrologic effects on community structure and biodiversity

The effects of hydrologic alteration on community structure of aquatic organisms have been well documented (Dudgeon 2000, Pringle et al. 2000, Marchetti and Moyle 2001, Humphries et al. 2008). Marchetti and Moyle (2001) reported that most native fish species of the regulated Putah Creek, a tributary of the Sacramento River, were more often found in habitats that were characteristic of the natural flow regime (i.e., increased canopy, higher streamflow, decreased conductivity, cooler temperatures, and fewer pools). In contrast, most nonnative species appeared to be adapted to conditions opposite to those of the natural flow regime (Marchetti and Moyle 2001). Therefore, restoration of a native-dominated fish assemblage would require a return to natural flow conditions. In the regulated Campaspe River (Australia), only four of ten native species were consistently documented from 1995 to 2003, while historically an estimated 18 to 20 native fishes once inhabited the river (Humphries et al. 2008). The authors also documented the presence of six exotic fishes, with common carp and European perch being the most abundant fishes (i.e., 36% of the overall fish abundance, Humphries et al. 2008).

Dam construction and river regulation has threatened aquatic and terrestrial biodiversity worldwide (Dudgeon 2000, Pringle et al. 2000). Pringle et al. (2000) provided a thorough review of the effects of hydrologic alterations on riverine biota in temperate and neotropical regions. The authors explain that, although construction of new dams in the United States has declined, large dam construction has occurred more recently in tropical regions of South America (Pringle et al. 2000). In temperate regions, migratory diadromous fishes, such as salmon, sturgeon, American shad, and American eel, have been extirpated from much of their native ranges due to dams that block their spawning migrations (Drinkwater and Frank 1994, Pringle 2000). Movements of potamodromous fishes have also been impeded, restricting their reproductive success and overall distributions (Pringle et al. 2000). Habitat fragmentation and the conversion of lotic, free-flowing habitat to more lentic conditions, in the form of impoundments, reservoirs, etc., have resulted in the imperilment of small-bodied fishes, particularly fluvial-dependent species (Pringle et al. 2000). This increase in availability of lentic habitat has also allowed for the expansion of lentic fishes and the introduction of lentic fishes into systems beyond their native range (Pringle et al. 2000). As mentioned in the previous section, Pringle (2000) also identified a reduction or alteration in the timing and/or duration of floodplain inundation as a major factor contributing to the decline of flood-dependent taxa. Reduced freshwater flows into estuarine habitats have also threatened species, such as the delta smelt in San Francisco Bay, California (Pringle et al. 2000). In Pringle et al.'s (2000) case study of the

Mobile River Basin in the southeast USA, the authors report from the literature that at least 16 endemic mussels and 38 gastropods are thought to be extinct.

Although the negative impacts of hydrologic alterations on biota are well documented in North America, less is known about the effects of river regulation on South American tropical systems (Pringle et al. 2000). Pringle et al. (2000) explained that biota of Neotropical rivers are highly vulnerable to hydrologic alterations for several reasons. The habitat heterogeneity of tropical ecosystems has produced highly diverse communities with high rates of endemism. Many South American fishes are highly migratory and have complex life cycles and depend on seasonal floodplain inundation that provides food, refuge, and nursery habitat for young fishes. The accumulation of organic material in reservoirs of low-gradient Amazonian streams has led to undesirable water quality conditions (e.g., hypoxia) that can result in fish kills. Reduced freshwater input to estuaries has also led to an increased presence of marine fishes in lower sections of rivers, replacing native freshwater species (Pringle et al. 2000). Agostinho et al. (2008) provided a more recent, extensive review of the effects of dams on fish fauna in the Neotropical Region. The authors focused on the highly regulated Paraná River, which flows through the most highly populated region in Brazil. Data are presented illustrating the negative impacts of dams on fish diversity and fisheries in the region. Agostinho et al. (2008) expressed the need for improved management approaches in Brazil, such as taking a more ecosystem-level rather than reductionist approach. The authors also mention that little information exists regarding the effects of hydrologic alteration (dams), fishery exploitation, and other impacts on aquatic resources in Brazil. These data needs must be addressed so that managers can formulate and inform effective management decisions (Agostinho et al. 2008).

Asia possesses a proportionately high number of dams, with the number of large dams increasing from 1,541 in 1950 to a staggering 22,701 in 1982 (Dudgeon 2000). China has constructed the greatest number of dams in tropical Asia (Dudgeon 2000). The climate of this region alternates between a wet and dry season, with many organisms depending on the wet season and the associated flood pulse for sustenance and access to floodplain habitats (Dudgeon 2000). Dam construction, however, has focused on flood control during wet periods and storing water during dry periods, resulting in significantly altered hydrologic regimes in most major rivers (Dudgeon 2000). Dudgeon (2000) also mentions that other factors, such as pollution, deforestation, overharvesting and rapidly growing human populations in the landscape, have further exacerbated conditions in these systems. Hydrologic alteration has negatively impacted a wide diversity of taxa in this region, including crocodiles, terrestrial mammals, fishes, and river dolphins (Dudgeon 2000). The Mekong River Basin supports a high diversity of over 500 fishes. Unfortunately, the construction of large dams on the Mekong River has threatened the persistence of many species (Dudgeon 2000). See Dudgeon (2000) for a thorough overview of the ecological consequences of large dam construction on the Mekong River.

Negative impacts of hydrologic alteration are not only limited to aquatic organisms. Riverine forest along the regulated Tana River in Kenya serves as habitat for two endemic primates, the rare Tana River Red Colobus and the critically endangered Tana River Mangabey (Maingi and Marsh 2002). Regulation of the Tana River threatens the persistence of riverine forest along mid and lower sections of Tana River and, therefore, further endangers these rare primates

(Maingi and Marsh 2002). Hill et al. (1998) examined the effects of dams on the shoreline vegetation of lakes and reservoirs in southern Nova Scotia, Canada. Hill et al.'s (1998) study included 37 unregulated and 13 regulated lakes, for which plant species inventories were conducted. Plant communities of regulated lakes were less diverse, contained more exotic species, and typically lacked rare shoreline herbs. The authors attributed this reduction in diversity and introduction of nonnative species to the altered hydrologic regimes of reservoirs that produce extreme fluctuations in water levels.

A significant reduction in *hydrologic connectivity*, as a result of dam construction and other anthropogenic activities in the landscape, has also threatened aquatic biodiversity in riverine systems (Pringle 2003). Hydrologic connectivity refers to "the water-mediated transfer of matter, energy, and/or organisms within or between elements of the hydrologic cycle (Pringle 2001, Pringle 2003)." Pringle's (2001, 2003) definition of hydrologic connectivity emphasizes its importance at a regional or global scale, whereas *river connectivity* refers to the continuity or linkages of a river ecosystem as it operates across its four dimensions (i.e., temporal, and longitudinal, lateral, and vertical spatial dimensions, Freeman et al. 2007). Dam construction (i.e., reduced hydrologic connectivity) has impeded the spawning migrations of anadromous fishes, preventing these fishes from returning to their natal sites. Substantial reductions in the distribution and abundance of freshwater mussel species have been attributed to reduced habitat connectivity. Fragmentation of habitats isolates local populations from others, limiting or eliminating the exchange of individuals and the potential for recolonization of habitat patches when a local extinction occurs. Reduced hydrologic connectivity also has negative impacts on broader-scale functions, such as biogeochemical cycling in ecosystems (Pringle 2003). For example, dams act as barriers to the transport of silica to coastal oceans, limiting primary production (i.e., diatom production) and the integrity of coastal food webs (Pringle 2003, Freeman et al. 2007).

# 6. Flow management and modeling

In 1997, Poff et al. succinctly explained that "current management approaches often fail to recognize the fundamental scientific principle that the integrity of flowing water systems depends largely on their *natural dynamic character*." Although in today's society returning riverine systems to their "natural dynamic character" is nearly impossible, the authors indicated that conservation and management strategies should attempt to restore the ecological integrity of these regulated systems by enhancing their "natural" flow variability (Poff et al 1997). Although regulated rivers can never be fully restored to natural conditions, flows below dams should be managed to best represent natural flow conditions (Poff et al. 1997). Previous management strategies focused on improving water quality and simply implementing minimum flow requirements (Poff et al. 1997, Richter et al. 1997, Arthington et al. 2006). Richter et al. (1996) also mentioned that past management strategies focused on the flow requirements of only a few selected aquatic species and neglected the flows needed to maintain aquatic-riparian systems and broader ecosystem functions. Management of freshwater resources was also conducted in a compartmentalized or "fragmented" fashion (Poff et al.

1997, Karr 1991). Management approaches today have evolved to incorporate the prescription and implementation of natural aspects of the flow regime. In addition, a more concerted effort is applied to coordinate management activities among various resource agencies. Furthermore, current strategies attempt to apply a more holistic, ecosystem-level (rather than reductionist) approach to the management of regulated rivers and conservation of freshwater resources.

Various techniques and modeling approaches have been developed to enhance our under-standing of how hydrologic alteration affects aquatic ecosystems, as well as improve our management of regulated rivers (Richter et al. 1996, Richter et al. 1997, Irwin and Freeman 2002, Olden and Poff 2003, Arthington et al. 2006, Poff et al. 2009, Merrit and Poff 2010, Sakaris and Irwin 2010). Richter et al. (1996) emphasized the importance of selecting hydrologic parameters that are most "biologically relevant" when assessing hydrologic alteration in a regulated system. In other words, we should focus on the parameters that most influence the ecological integrity of a system. Richter et al. (1996) presented a well-structured approach for hydrologic assessment, *Indicators of Hydrologic Alteration* (IHA), which accounts for the most biologically relevant parameters. This approach defines and calculates a series of hydrologic attributes and then compares the hydrologic regime of a system before and after impact (e.g., impoundment). A total of 32 biologically relevant hydrologic parameters are calculated for each year from these five IHA statistics groups: 1) magnitude of monthly water conditions, 2) magnitude and duration of annual extreme water conditions, 3) timing of annual extreme water conditions, 4) frequency and duration of high and low pulses, and 5) rate and frequency of water condition changes (Richter et al. 1996). These parameters account for the five impor-tant and ecologically relevant characteristics of a flow regime: magnitude, frequency, duration, timing, and the rate of change (Poff et al. 1997). The four steps of Richter et al.'s (1996) approach are: 1) define the data series for pre- and post-impact periods (usually collected from USGS flow gauges), 2) calculate values of hydrologic attributes for each year in each data series (i.e., pre-impact and post-impact data series), 3) compute inter-annual statistics for the 32 param-eters in each data series, specifically 32 measures of central tendency and 32 measures of dispersion, and 4) calculate values of the IHA. The fourth step involves comparing the 64 inter-annual statistics between pre- and post-impact periods, as a percent deviation of one time period to the other (Richter et al. 1996). The IHA approach can also be used to compare hydrologic regimes between regulated and "reference" sites.

Richter et al. (1997) further improved the approach to river management with the develop-ment of the "Range of Variability Approach (RVA)." Richter et al. (1997) mention that previ-ous approaches did not provide specific flow targets to be met, focused on a limited number of features of the hydrologic regime, and/or focused on only a few target species and a limited number of their habitat requirements. In addition, research studies examining relationships between hydrologic conditions and ecological responses in a system are typically time-consum-ing (often taking several years) and are usually not completed within the timeframe during which flow management decisions are typically made (Richter et al. 1997). Richter et al.'s (1997) RVA assists river managers in the identification of flow-based management targets that should enhance the overall ecological integrity of a system. For systems with highly altered hydrolog-ic regimes, the main idea is *to restore hydrologic conditions within the historical or "natural" range of variation*, particularly for streamflow characteristics that are well outside the historical range

(Richter et al. 1997). Richter et al. (1997) recommended that the RVA be applied in the preliminary stages of *adaptive flow management programs* (see below), providing initial flow management targets that can be modified as more ecological information is gathered for a specific ecosystem. The RVA approach has six steps, which are briefly described here. For a more in-depth overview, see Richter et al. (1997). The six steps are as follows: 1) Characterize the natural range of streamflow variation using the IHA approach (Richter et al. 1996) described above. 2) Select management targets, one for each of the 32 hydrologic parameters, with the idea that each management target should fall within the natural range of variation. Each target may have upper and lower bounds (e.g., ± 1 standard deviation). 3) The river management team formulates a management "system" or plan, using the RVA targets as design guidelines. 4) Scientists conduct routine ecological monitoring and/or river research program to evaluate ecological effects of the management system as it is implemented. 5) Characterize actual streamflow variation using the IHA method at the end of each year and compare the values of hydrologic parameters with the RVA target values. 6) Revise either the management system or RVA targets based on new information that is collected (Richter et al. 1997).

An adaptive approach, termed *adaptive-flow management*, has been recommended for the management of regulated river systems (Irwin and Freeman 2002). In adaptive-flow management, managers attempt to restore rivers to near-natural flow regimes while accounting for societal needs (Irwin and Freeman 2002). The main goal of adaptive-flow management is to continually improve management as uncertainty about a river system is reduced. This management approach requires the cooperation and long-term commitment of natural resource personnel, private industry, landowners, and other stakeholders. Adaptive-flow management can be best described as an iterative process with a series of steps that include 1) prescription of a flow/management regime that satisfies all stakeholders, 2) monitoring and evaluation of the flow regime's effect on habitat and biota, and 3) the recommendation of a new and improved management regime. By quantifying relationships between features of the flow regime and responses in the biota and overall ecosystem, models can be developed to predict how populations, communities or the ecosystem may respond to the prescription of flow regimes, or an "environmental flow standard (Arthington et al. 2006)." These models can be continually improved as more is learned about the ecological responses to hydrologic alteration in the managed river system.

Olden and Poff (2003) addressed a major issue confronting managers in determining which of the many published approaches and hydrologic (and "ecologically relevant") parameters should be used in river management. The authors recognized that many of the hydrologic variables proposed for use in the characterization of a flow regime (e.g., 32 hydrologic parameters, Richter et al. 1996) were inter-correlated, and little guidance was provided for the selection of appropriate parameters. Olden and Poff's (2003) main goal was to provide a standardized framework for the selection of a reduced set of hydrologic indices and to minimize redundancy among the selected parameters. This reduced set of indices would still account for the majority of the statistical variation in the complete set of hydrologic indices, minimize multicollinearity among the selected hydrologic variables, and adequately represent the critical attributes of a system's flow regime. The authors also examined the effectiveness of IHA and the overall transferability of indices to facilitate comparisons across systems that differ in their streamflow characteristics (Olden and Poff 2003). See Olden and Poff (2003) for a detailed overview of the approach and a review of the 171 hydrologic indices published in the literature.

In 2006, Arthington et al. proposed a mechanism for developing regional environmental flow "standards." Their rationale was that hydrologic and ecological data are often lacking for specific streams or rivers in a region, which makes it quite difficult to prescribe system-specific flow regimes in the management of regulated rivers. Arthington et al. (2006) recommended classifying rivers and streams that share important flow attributes into "ecologically meaningful groups." Within a region, the logical assumption is that rivers that are similar in their flow variability and geomorphic properties would exhibit similar ecological responses to management regimes. Arthington et al. (2006) described their approach as grouping the systems into "practical management units." See Arthington et al. (2006) for a complete overview of their management approach, which shares common features with the ELOHA approach described below.

Poff et al. (2010) explained that a strong need exists to develop ecological goals and management standards for streams and rivers at a regional or even global scale. Water resource and environmental flow management has become highly complex, because management must account for diverse societal needs while attempting to restore the ecological integrity of degraded ecosystems. Meanwhile, rapidly growing human populations will further increase water consumption and energy demands and require increased food production. As a result, restoring systems with highly altered flow regimes to "natural" flow conditions will become even more difficult. The authors, consisting of a group of international scientists, presented a framework for evaluating environmental flow needs that could potentially form the basis for implementing flow standards at a regional scale (Poff et al. 2009). Poff et al. (2009) refer to this framework as the *Ecological limits of hydrologic alteration* (ELOHA), with the goal of presenting *"a logical approach that flexibly allows scientists, water resource managers and other stakeholders to analyze and synthesize available scientific information into coherent, ecologically based and socially acceptable goals and standards for management of environmental flows."* Poff et al. (2010) recognize that water resource managers from different regions are often confronted with unique challenges, may operate in different social and political environments, and may be at different stages of water-resource development. The necessary scientific foundations of the ELOHA framework exist and consist of: 1) essentially years of research has been conducted examining the effects of altered hydrologic regimes on population dynamics, community structure, and ecosystem-level functions, 2) the previous application of various methods for managing environment flows, which the authors refer to as a "rich toolbox" from which methods or tools can be applied by water resource managers, 3) a conceptual foundation that facilitates regional flow assessments, 4) the development of hydrologic models, and 5) an understanding that river management is complex and adaptive and must meet both ecological and societal goals (Poff et al. 2010).

The ELOHA framework consists of four major steps that can be flexibly applied by managers from different regions (Poff et al. 2009). 1) *Building a "hydrologic foundation" for the region* involves collecting hydrologic time-series data and constructing hydrographs that represent "baseline" (minimally altered) and "developed" (altered) hydrologic conditions throughout the region, particularly for all locations that require environmental flow management and protection. 2) *Classifying rivers according to their hydrology and geomorphology* assumes that rivers with similar hydrologic regimes (e.g., snowmelt driven rivers) and geomorphic characteristics would likely respond similarly to hydrologic alteration and other disturbances, whereas rivers that are dissimilar in type (e.g., snowmelt vs. desert rivers) would likely respond differently when altered. When classifying rivers based on their hydrologic regimes, chosen hydrologic features should collectively characterize the flow regime of the system and avoid redundancy in the parameters used (Olden and Poff 2003). The selected hydrologic metrics should also be ecologically relevant and be applicable in management. River classification is important, because flow management decisions will likely vary based on river type. Furthermore, if the "hydrologic foundation" is not fully built for a region, the hydrologic models and management targets developed for one river may be extrapolated to similar systems until more system-specific data are collected. 3) *Computing flow alteration* involves estimation of the degree of hydrologic alteration for each system, for which hydrologic data are available. Any deviation in the hydrologic regime from "natural" (baseline) conditions may have an ecological impact, and this ecological impact generally becomes more severe as the disparity between developed and baseline conditions widens. Programs, such as IHA (Richter et al. 1996; Mathews and Richter 2007), can be used to calculate a set of hydrologic alteration values as a percent or absolute deviation from baseline condition for each developed site. 4) *Conducting research and monitoring programs to assess ecological responses to altered hydrologic regimes* addresses the critical need for improved understanding of biotic and ecosystem responses to flow alteration in the ELOHA framework. Flow alteration-ecological response relationships guide river managers in establishing flow management targets, or "standards," and in developing flow management plans that will most likely enhance the ecological integrity of an altered system. It is important to note that the ELOHA approach is an adaptive process. Scientists play an important role in this process, by conducting research programs that attempt to reduce uncertainty and build our understanding of ecological responses to hydrologic alteration. With new information, management flow standards can be updated and implemented over time. See Poff et al. (2010) for a detailed overview of the application of ELOHA, the various models and tools that can be used in each step of the process, and the potential challenges that may confront river managers and scientists that adopt this approach.

The ELOHA framework (Poff et al. 2010) requires the assessment of "ecologically significant" differences between baseline and developed hydrologic regimes in a region. Merritt and Poff (2010) recently developed an *index of flow modification* (IFM), which is a composite metric of the most biologically relevant hydrologic variables that essentially measures how modified an altered flow regime is compared to unregulated (or baseline) conditions. Pre-dam and post-dam flow data are collected for each location (or study reach), typically from USGS (United States Geological Survey) gauges. Biologically relevant hydrologic variables are then obtained for pre-dam and post-dam periods using IHA (Indicators of Hydrologic Alteration) software (Richter et al. 1996), and then the absolute or percent change in each variable is calculated from pre-dam to post-dam periods. In their study, Merritt and Poff (2010) calculated the percentage change in spring flow (mean of April through June), summer flow (mean of July through September), low flow (mean of October through February flows), and 2-, 10-, and 25-year recurrence interval peak flows. The change in the number of days of minimum flow and the change in maximum flow were also calculated (Merrit and Poff 2010). Principle Components Analysis (PCA) is then conducted using these calculated metrics (i.e., hydrologic metrics) of all study sites, and only significant principle component axes are used in the calculation of IFM. Merrit and Poff (2010) developed the IFM *"by calculating the Euclidean distance of each observation (study reach) from the centroid of the significant PCA axis scores of relatively unregulated rivers for the hydrologic metrics."* The IFM can then be used, for example, to examine relationships between the recruitment, abundance, and/or growth of organisms and the degree of flow modification (IFM) across sites ranging from relatively unregulated to regulated conditions.

Population matrix models have also been developed by scientists to evaluate and predict how riparian and aquatic populations respond to hydrologic variation in systems with altered flow regimes (Lytle and Merritt 2004, Sakaris and Irwin 2010). These models may be useful in step 4 of the ELOHA approach. Lytle and Merritt (2004) developed a stochastic, density-dependent model to predict how annual hydrologic variation affects the mortality, recruitment, and population dynamics of the riparian cottonwood and to project how altered flow regimes might affect cottonwood populations. Lytle and Merritt (2004) simulated the effects of channelization and damming in the Yampa River, Colorado, and their model suggested that the observed natural flow regime would likely produce the most abundant mature cottonwood forest. Sakaris and Irwin (2010) developed matrix models for predicting the effects of altered flow regimes on the dynamics of a flathead catfish population in a regulated section of the lower Coosa River, Alabama. Matrix construction required the collection of fertility, survival, and body growth data (for size-classified matrices) for the fish population. The authors conducted multiple regression analyses to assess the influence of hydrologic features of the altered flow regime on annual recruitment of the flathead catfish in the system. Using this information, the effects of environmental stochasticity (hydrologic variation) on the long-term growth dynamics of the catfish populations was projected. Sakaris and Irwin (2010) also used their model to predict the effects of prescribed flow regimes on fish population dynamics in the river. Sakaris and Irwin (2010) presented their model as a potential tool that could be used in the adaptive flow management of regulated rivers (e.g., below Harris Dam on the Tallapoosa River, Alabama, Irwin and Freeman 2002).

Arthington et al. (2006) mentioned that general agreement exists among scientists and most river managers that to maintain the ecological integrity and biodiversity of a system, we must attempt to restore, or "mimic," natural flow conditions. That is, all general features of the natural flow regime (magnitude, frequency, duration, timing, and the rate of change, Poff et al. 1997), to some degree, should be accounted for when prescribing a flow-management regime in a regulated system. As mentioned earlier, previous management strategies typically focused on implementing a single environmental flow standard, such as maintaining a minimum flow requirement below a dam. For example, on the Tallapoosa River in the East Gulf Coastal Plain in Alabama (USA), a minimum continuous flow (34 $m^3/s$) was established below Thurlow Dam as part of a re-licensing agreement in 1991. Although diversity of fishes increased approximately 3 km downstream of the dam (Travnichek et al. 1995), Thurlow Dam has still exhibited high annual variability in discharge that often exceeds dam capacity, which has typically resulted in prolonged periods of high flow (> 283 m$^3$/s). Wildhaber et al. (2000) evaluated relations between Neosho madtom densities and flows in the Neosho River Basin below John Redmond Dam and suggested that higher minimum flows be prescribed in the river to improve densities of Neosho madtoms and other ictalurids. Studies have evolved since to prescribe or, at least, model flow regimes that are more natural in character. For example, Propst and Gido (2004) attempted to partially mimic the natural flow regime in a regulated reach of the San Juan River (Colorado), by increasing reservoir releases to mimic timing and only partially mimicking the amplitude, volume, and duration of spring snowmelt discharge. Densities of native fishes typically increased in years with high spring discharges (Propst and Gido 2004). Horne et al. (2004) modeled the effects of two management scenarios, bottom withdrawal and actual dam removal, on the recruitment of steelhead in the Manistee River. The authors' models predicted that bottom withdrawal (of hypolimnetic water) would slightly cool summer water temperatures and modestly enhance steelhead recruitment in the river (Horne et al. 2004). Horne et al. (2004) mention, however, that their model for dam removal did not account for the added benefit of increased steelhead access to upstream spawning habitat, which would likely improve recruitment in the system. In the lower Kootenai River, Burke et al. (2008) discovered that recruitment potential of the back cottonwood improved in 1997 and 1999, partly due to experimental flow releases from Libby Dam. These water releases helped to mimic pre-dam hydrologic conditions during the spring snowmelt, with a sustained peak in flow during the early growing season and a subsequent gradual recession of flow (Burke et al. 2008). These experimental flow releases have been implemented since 1993 to enhance the spawning success of white sturgeon (Burke et al. 2008).

River management and the conservation of natural resources in regulated rivers will become increasingly difficult, as the ecological needs of an ecosystem must be delicately balanced with societal needs. Furthermore, the management of altered flow regimes has become quite complex, as we must also account for and understand how the interaction of local climate, land use, and the unique geological and topographical features of a region influence hydrologic processes in a river system. Future management approaches will require the involvement and cooperation of governmental agencies, scientists, non-profit organizations, and the public to develop solutions that attempt to restore features of the natural flow regime, conserving and enhancing biodiversity, while providing for the needs of society.

## Author details

Peter C. Sakaris

Address all correspondence to: psakaris@spsu.edu

Department of Biology and Chemistry, Southern Polytechnic State University, Marietta, GA, USA

# References

[1] Agostinho, A. A, Pelicice, F. M, & Gomes, L. C. (2008). Dams and the fish fauna of the Neotropical Region: impacts and management related to diversity and fisheries. *Brazilian Journal of Biology* , 68, 1119-1132.

[2] American Rivers (1997). North America's most endangered and threatened rivers of 1997. American Rivers, Washington D. C.

[3] Arthington, A. H, Bunn, S. E, Poff, N. L, & Naiman, R. (2006). The challenge of providing environmental flow rules to sustain river esosystems. *Ecological Applications* , 16, 1311-1318.

[4] Beamesderfer, R. C. P, Rien, T. A, & Nigro, A. A. (1995). Differences in the dynamics and potential production of impounded and unimpounded white sturgeon populations in the lower Columbia River. *Transactions of the American Fisheries Society* , 124, 857-872.

[5] Bonvechio, T. F, & Allen, M. S. (2005). Relations between hydrological variables and year-class strength of sportfish in eight Florida waterbodies. *Hydrobiologia* , 532, 193-207.

[6] Boschung, H. T, & Mayden, R. L. (2004). Fishes of Alabama. Smithsonian Books, Washington D. C.

[7] Bowen, Z. H, Freeman, M. C, & Bovee, K. D. (1998). Evaluation of generalized habitat criteria for assessing impacts of altered flow regimes on warmwater fishes. *Transactions of the American Fisheries Society* , 127, 455-468.

[8] Brouder, M. J. (2001). Effects of flooding on recruitment of roundtail chub, *Gila robusta*, in a Southewestern River. *The Southwestern Naturalist* , 46, 302-310.

[9] Burke, M, Jorde, K, & Buffington, J. M. (2008). Application of a hierarchical framework for assessing environmental impacts of dam operation: changes in streamflow, bed mobility, and recruitment of riparian trees in a western North American river. *Journal of Environmental Management* , 90, 224-236.

[10] Buynak, G. L, Mitchell, B, Bunnell, D, Mclemore, B, & Rister, P. (1999). Management of largemouth bass at Kentucky and Barkley Lakes, Kentucky. *North American Journal of Fisheries Management* , 19, 59-66.

[11] Clarkson, R. W, & Childs, M. R. (2000). Temperature effects of hypolimnial-release dams on early life stages of Colorado River Basin big-river fishes. *Copeia* , 2, 402-412.

[12] Cushman, R. M. (1985). Review of ecological effects of rapidly varying flows downstream from hydroelectric facilities. *North American Journal of Fisheries Management* , 5, 330-339.

[13] Drinkwater, K. F, & Frank, K. T. (1994). Effects of river regulation and diversion on marine fish and invertebrates. *Aquatic Conservation: Freshwater and Marine Ecosystems* , 4, 135-151.

[14] Dudgeon, D. (2000). Large-scale hydrological changes in tropical Asia: prospects for riverine diversity. *BioScience* , 50, 793-806.

[15] Fausch, K. D, Torgersen, C. E, Baxter, C, & Li, H. (2002). Landscapes to riverscapes: bridging the gap between research and conservation of stream fishes. *Bioscience* , 52, 1-16.

[16] Fraley, J. J, Mcmullin, S. L, & Graham, P. J. (1986). Effects of hydroelectric operations in the Kokanee population in the Flathead River System, Montana. *North American Journal of Fisheries Management* , 6, 560-568.

[17] Freeman, M. C, Bowen, Z. H, Bovee, K. D, & Irwin, E. R. (2001). Flow and habitat effects on juvenile fish abundance in natural and altered flow regimes. *Ecological Applications* , 11, 179-190.

[18] Freeman, M. C, Irwin, E. R, Burkhead, N. M, Freeman, B. J, & Bart, H. L. Jr. (2004). Status and conservation of the fish fauna of the Alabama River system. *American Fisheries Society Symposium* , 45, 557-585.

[19] Freeman, M. C, Pringle, C. M, & Jackson, C. R. (2007). Hydrologic connectivity and the contribution of stream headwaters to ecological integrity at regional scales. *Journal of the American Water Resources Association* , 43, 5-14.

[20] Galat, D. L, & Lipkin, R. (2000). Restoring ecological integrity of great rivers: historical hydrographs aid in defining reference conditions for the Missouri River. *Hydrobiologia* 422/, 423, 29-48.

[21] Gillette, D. P, Tiemann, J. S, Edds, D. R, & Wildhaber, M. L. (2005). Spatiotemporal patterns of fish assemblage structure in a river impounded by low-head dams. *Copeia* , 2005(3), 539-549.

[22] Hill, N. M, Keddy, P. A, & Wisheu, I. C. (1998). A hydrological model for predicting the effects of dams on the shoreline vegetation of lakes and reservoirs. *Environmental Management* , 22, 723-736.

[23] Holland-bartels, L. E, & Duval, M. C. (1988). Variations in abundance of young-of-the-year channel catfish in a navigation pool of the Upper Mississippi River. *Transactions of the American Fisheries Society* , 117, 202-208.

[24] Horne, B. D, Rutherford, E. S, & Wehrly, K. E. (2004). Simulating effects of hydrodam alteration on thermal regime and wild steelhead recruitment in a stable-flow Lake Michigan tributary. *River Research and Applications* , 20, 185-203.

[25] Humphries, P, Brown, P, Douglas, J, Pickworth, A, Strongman, R, Hall, K, & Serafini, L. (2008). Flow-related patterns in abundance and composition of the fish fauna of a degraded Australian lowland river. *Freshwater Biology* , 53, 789-813.

[26] Irwin, E. R, Freeman, M. C, & Costley, K. M. (1999). Habitat use by juvenile channel catfish and flathead catfish in lotic systems in Alabama. *in* E. R. Irwin, W. A. Hubert, C. F. Rabeni, H. L. Schramm, Jr., and T. Coon, editors. Catfish 2000: proceedings of the international ictalurid symposium. *American Fisheries Society, Symposium 24*, Bethesda, Maryland., 223-230.

[27] Irwin, E. R, & Freeman, M. C. (2002). Proposal for adaptive management to conserve biotic integrity in a regulated segment of the Tallapoosa River, Alabama, U.S.A. *Conservation Biology* , 16, 1212-1222.

[28] Karr, J. R. (1991). Biological integrity: a long-neglected aspect of water resource management. *Ecological Applications* , 1, 66-84.

[29] Lessard, J. L, & Hayes, D. B. (2003). Effects of elevated water temperature on fish and macroinvertebrate communities below small dams. *River Research and Applications* , 19, 721-732.

[30] Lytle, D. A, & Merritt, D. M. (2004). Hydrologic regimes and riparian forests: a structured population model for cottonwood. *Ecology* , 85, 2493-2503.

[31] Maceina, M. J, & Stimpert, M. C. (1998). Relations between reservoir hydrology and crappie recruitment in Alabama. *North American Journal of Fisheries Management* , 18, 104-113.

[32] Maingi, J. K, & Marsh, S. E. (2002). Quantifying hydrologic impacts following dam construction along the Tana River, Kenya. *Journal of Arid Environments* , 50, 53-79.

[33] Marchetti, M. P, & Moyle, P. B. (2001). Effects of flow regime on fish assemblages in a regulated California stream. *Ecological Applications* , 11, 530-539.

[34] Mayo, R. M. and H. L. Schramm Jr. (1999). Growth of flathead catfish in the lower Mississippi River. *in* E. R. Irwin, W. A. Hubert, C. F. Rabeni, H. L. Schramm, Jr., and T. Coon, editors. Catfish 2000: proceedings of the international ictalurid symposium. *American Fisheries Society, Symposium 24*, Bethesda, Maryland., 121-124.

[35] Matthews, R, & Richter, B. (2007). Application of the Indicators of Hydrologic Alteration software in environmental flow-setting. Journal of the American Water Resources Association. , 43, 1-4.

[36] Merritt, D. M, & Poff, N. L. (2010). Shifting dominance of riparian *Populus* and *Tamarix* along gradients of flow alteration in western North American rivers. *Ecological Applications* , 20, 135-152.

[37] Olden, J. D, & Poff, N. L. (2003). Redundancy and the choice of hydrologic indices for characterizing streamflow regimes. *River Research and Applications* , 19, 101-121.

[38] Paragamian, V. L. (2002). Changes in the species composition of the fish community in a reach of the Kootenai River, Idaho, after construction of Libby Dam. *Journal of Freshwater Ecology* , 17, 375-383.

[39] Pegg, M. A, Pierce, C. L, & Roy, A. (2003). Hydrological alteration along the Missouri River Basin: a time-series approach. *Aquatic Sciences* , 65, 63-72.

[40] Poff, N. L, Allan, J. D, Bain, M. B, Karr, J. R, Prestegaard, K. L, Richter, B. D, Sparks, R. E, & Stromberg, J. C. (1997). The natural flow regime: a paradigm for river conservation and restoration. *BioScience* , 47, 769-784.

[41] Poff, N. L, Bledsoe, B. P, & Cuhaciyan, C. O. (2006). Hydrologic variation with land use across the contiguous United States: geomorphic and ecological consequences for stream ecosystems. *Geomorphology* , 79, 264-285.

[42] Poff, N. L, Richter, B, Arthington, A. H, Bunn, S. E, Naiman, R. J, Kendy, E, Acreman, M, Apse, C, Bledsoe, B. P, Freeman, M, Henriksen, J, Jacobson, R. B, Kennen, J, Merritt, D. M, Keeffe, J. O, Olden, J. D, Rogers, K, Tharme, R. E, & Warner, A. (2010). The Ecological Limits of Hydrologic Alteration (ELOHA): a new framework for developing regional environmental flow standards. *Freshwater Biology* , 55, 147-170.

[43] Pringle, C. M, Freeman, M. C, & Freeman, B. J. (2000). Regional effects of hydrologic alterations on riverine macrobiota in the new world: tropical-temperate comparisons. *BioScience* , 50, 807-823.

[44] Pringle, C. M. (2001). Hydrologic connectivity and the management of biological reserves: a global perspective. *Ecological Applications* , 11, 981-998.

[45] Pringle, C. M. (2003). What is hydrologic connectivity and why is it ecologically important? *Hydrologic Processes* , 17, 2685-2689.

[46] Propst, D. L, & Gido, K. B. (2004). Responses of native and nonnative fishes to natural flow regime mimicry in the San Juan River. *Transactions of the American Fisheries Society* , 133, 922-931.

[47] Quinn, J. W, & Kwak, T. J. (2003). Fish assemblage changes in an Ozark river after impoundment: a long-term perspective. *Transactions of the American Fisheries Society* , 132, 110-119.

[48] Quist, M. C, & Guy, C. S. (1998). Population characteristics of channel catfish from the Kansas River, Kansas. *Journal of Freshwater Ecology* , 13, 351-359.

[49] Richter, B. D, Baumgartner, J. V, Powell, J, & Braun, D. P. (1996). A method for assessing hydrologic alteration within ecosystems. *Conservation Biology* , 10, 1163-1174.

[50] Richter, B. D, Baumgartner, J. V, Powell, J, & Wigington, R. (1997). How much water does a river need? *Freshwater Biology 37*, 231-249.

[51] Rulifson, R. A. and C. S. Manooch III. (1990). Recruitment of juvenile striped bass in the Roanoke River, North Carolina, as related to reservoir discharge. *North American Journal of Fisheries Management 10*, 397-407.

[52] Rutherford, D. A, Kelso, W. E, Bryan, C. F, & Constant, G. C. (1995). Influence of physicochemical characteristics on annual growth increments of four fishes from the lower Mississippi River. *Transactions of the American Fisheries Society* , 124, 687-697.

[53] Sakaris, P. C. (2006). Effects of hydrologic variation on dynamics of channel catfish and flathead catfish populations in regulated and unregulated rivers in the southeast USA. Ph.D. Dissertation, Auburn University, Alabama, USA.

[54] Sakaris, P. C, & Irwin, E. R. (2010). Tuning stochastic matrix models with hydrologic data to predict the population dynamics of a riverine fish. *Ecological Applications* , 20, 483-496.

[55] Sammons, S. M, & Bettoli, P. W. (2000). Population dynamics of a reservoir sport fish community in response to hydrology. *North American Journal of Fisheries Management* , 20, 791-800.

[56] Schultz, R. D, Guy, C. S, & Robinson, D. A. Jr. (2002). Comparative influences of gizzard shad catch rates and reservoir hydrology on recruitment of white bass in Kansas reservoirs. *North American Journal of Fisheries Management* , 22, 671-676.

[57] Scott, M. L, Auble, G. T, & Friedman, J. M. (1997). Flood dependency of cottonwood establishment along the Missouri River, Montana, USA. *Ecological Applications* , 7, 677-690.

[58] Travnichek, V. H, Bain, M. B, & Maceina, M. J. (1995). Recovery of a warmwater fish assemblage after the initiation of a minimum-flow release downstream from a hydroelectric dam. *Transactions of the American Fisheries Society* , 124, 836-844.

[59] Watters, G. T. (2000). Freshwater mussels and water quality: a review of the effects of hydrologic and instream habitat alterations. *Proceedings of the First Freshwater Mollusk Conservation Society Symposium* , 1999, 261-274.

[60] Welcomme, R. L. (1979). Fisheries ecology in floodplain rivers. Longman, New York, New York.

[61] Wildhaber, M. L, Tabor, V. M, Whitaker, J. E, Allert, A. L, Mulhern, D. W, Lamberson, P. J, & Powell, K. L. (2000). Ictalurid populations in relation to the presence of a main-stem reservoir in a Midwestern warmwater stream with emphasis on the threatened Neosho madtom. *Transactions of the American Fisheries Society* , 129, 1264-1280.

[62] Wunderlich, R. C, Winter, B. D, & Meyer, J. H. (1994). Restoration of the Elwha River ecosystem. *Fisheries* , 19, 11-19.

[63] Yang, T, Zhang, Q, Chen, Y. D, Tao, X, Xu, C, & Chen, X. (2008). A spatial assessment of hydrologic alteration caused by dam construction in the middle and lower Yellow River, China. *Hydrological Processes* , 22, 3829-3843.

[64] Young, P. J, Keeland, B. D, & Sharitz, R. R. (1995). Growth response of baldcypress [Taxodium distichum (L.) Rich.] to an altered flow regime. American Midland Naturalist , 133, 206-212.

# Chapter 12

# Current Challenges in Experimental Watershed Hydrology

Wei-Zu Gu, Jiu-Fu Liu, Jia-Ju Lu and Jay Frentress

Additional information is available at the end of the chapter

## 1. Introduction

The river basin, watershed or catchment is central to many of concepts in hydrology [1] including contaminant hydrology. In fact, two different studies in the Seine River basin during the end of seventeenth century, independently made by Pierre Perraut and Edme Mariotte have been identified by A. K. Biswas as the beginning of quantitative hydrology [2], Along with Perraut and Mariotte, Biswas also suggested Edmond Halley as co-founder of experimental hydrology. It was later accepted that scientific hydrology was founded on these two basin studies [3].

However, basin studies developed slowly until the end of nineteenth century when public demands accelerated. The first modern basin studies commenced in Emmental in Switzerland during the 1890s and were focused on hydrological differences between two small, ca 60 ha, basins [1]. A multitude of basin studies have appeared since the early twentieth century from many parts of the world. These include: (1) Wagon Wheel Gap experiment of USA begun in 1910 in two forested basins [4]; (2) Valday Hydrological Laboratory of USSR begun in 1933 and focused on field investigations of multiple hydrological parameters in watersheds with different scales [5]; (3) Coweeta Hydrologic Laboratory of USA set up since 1934 for forest hydrology and, ecological research in two main basins with 32 sub-watersheds under various treatments [6]; (4) Harz Mountains experiment of Germany with a pair of catchments, Wintertal and Large Bramke, begun in 1948 and focusing on hydrological effects of land use [7]; (5) Alrance experiment of France started in 1950 and focused on streamflow patterns[8]; and (6) Bluebrook Runoff Experiment of China established on 1953 and focused on drainage problems of the vast agricultural plain of the North Huai He River [9].

A period of rapid worldwide development of hydrological basin studies resulted from the International Hydrology Decade (IHD) Representative and Experimental Basin Programme 1965- 1974, with an estimated 3000 basin studies conducted during the Decade [1]. "Representative basins (RBs), which are selected as representative of a region of presumed hydrological similarity, are used for intensive investigations of specific problems of the hydrological cycle (or part thereof) under relatively stable, natural conditions" [10]. Experimental basins (EBs) are relatively homogeneous in soil and vegetation, have uniform physical characteristics, and are deliberately modified for study [10]. However, many EBs are just well-instrumented basins of small area [11]. Sometimes both RB and EB are included, as in the fruitful Plynlimon experiment in the UK [12].

If we designate the first phase of basin study (until ca the middle twentieth century) as foundational and the second phase (during/after the IHD) as developmental, it would appear that experimental watershed hydrology is inevitably going into a third phase of transition and innovation. Most field experiments in watershed science to date remain largely descriptive, with results that are difficult to generalize [13]. In effect, "... catchment hydrology is trapped in a dead-end track, a theoretical impasse" [14]. Experimental watershed studies, the core of watershed hydrology, are now confronting tremendous and even more complicated challenges, which are raised mainly from changing environment conditions and partly from the weakness of current watershed studies. This chapter provides an overview of the main challenges facing experimental watershed hydrology. Addressing these challenges will hopefully lead to substantial innovation in the field.

An important and instructive paper addressing a new vision of watershed hydrology [13] asked "What's wrong with the status quo?" If we posed this same question to the experimental hydrologist, the limitations of most existing RBs/EBs might be characterized as mainly twofold. First, study basin designs have been limited by the black box concept and many misconceptions (e.g., the linearity, non-heterogeneity, additivity of hydrologic systems etc.). Second, operation has been substantially bounded by the hydraulic conception of these watersheds as isolated hydrological systems. All of these watershed studies monitor only total runoff at the stream-outlet and the subsurface responses of the watershed are only estimated by hydrograph separation, etc. These characteristics undermine the formulation of a unified theory of watershed hydrology [14] and the development of watershed models [13,15].

There is a clear need to move beyond the status quo and expand from this narrow hydrological perspective to generate hypotheses governing general behavior across places and scales [13], with the ultimate aim "to advance the science of hydrology" [15]. For the third phase, the stage of transition and innovation, instead of classic RBs as described above, another kind of experimental watershed, the Critical Zone Experimental Block (CZEB) is suggested, with its concept and infrastructures very different from that of current EB. Following the concept of Critical Zone defined by the National Research Council [16], the CZEB will have practical boundaries for different geomorphologic regions.

An ongoing trial application of the CZEB approach is presented in this chapter.

## 2. Challenges

Numerous challenges face researchers as they move forward to advance hydrology science using experimental watershed studies.

### 2.1. Demands of advancing hydrology science

More than two decades ago, V. Klemes proclaimed that "hydrology is as yet lacking a solid scientific foundation needed for its development as a natural science" [17] and this sentiment has been reiterated by others. Bras and Eagleson said "in the modern science establishment, this niche is vacant" in their paper titled "Hydrology, the fogotten earth science" [18]. Jim Dooge asked "Is hydrology now an established science? Is hydrologic practice now firmly based on scientific principles?"[19]. McDonnell et al concluded that "the critical ideas and positive vision presented in that (Dooge) paper remain just as fresh, relevant and, unfortunately, very much unfulfilled" [13]. Sivapalan also suggested several steps toward a new unified hydrologic theory [14]. Substantial progress in hydrologic science "ultimately depends on new experimental work, new field observations, and new data collection networks" as summarized by Kirchner [15]. But, what shape will these new experiments and networks take?

*Coupling processes.* A watershed is an inherently dynamic ecological system composed of a variety of biotic and abiotic processes. For such systems, it is important not only to focus on hydrological process but also on linked ecological, biogeochemical, pedological and geological processes. Misunderstandings of the interrelatedness of these processes are commonplace (e.g., summarizing the net effect of all biological processes as evapotranspiration). The vegetation root system forms a dynamically complex net for preferential flowpaths. It changes runoff generation mechanisms not only for surface runoff but the subsurface flows, and the root system itself is effected via feedback from the various, changing flow patterns. Vegetation growth also adapts according to climate, infiltration, soil water and even evapotranspiration while feedback mechanisms reinforce the interconnectedness of the watershed system [20].

*Innovative measurements.* More than twenty years ago, JE Nash suggested that discovery in hydrology sciences has been limited by "a deficiency in our empiricism" and, "our tolerance of poor methods of observations" [21]. Since then, methods of observations have improved significantly. However, our observation techniques still inhibit hydrology and a fundamental change to systematic measurement programs is needed. The current observation programs are aimed mostly at classic watershed hydrological process. Instead we should focus on a holistic description of heterogeneity of not only the watershed hydrological process but all related processes as described above, as well as landscape properties and climate inputs.

*Supporting and testing hydrological model.* Physically based hydrological models are one of the most promising directions for advancing hydrologic science. Grayson et al concluded that "the models are enabling us to ask more questions, many of which are fundamental to our understanding of the natural systems that are the subject of our models" [22]. However, vital weaknesses in current watershed models exist. First, they are generally based on "well known small-scale theories such as Darcy's law and the Richards equation for coupled balance of mass and momentum" [13] and do not accurately capture the spatial heterogeneity, inherent non-linearity, and non-additive properties of natural watershed systems. Second, a large number of models are heavily over-parameterized, leading to equi-finality, wherein multiple combinations of parameters can yield equivalent results [13]. Grayson et al showed that comparable fits to a hydrograph could be achieved using a saturated overland flow model or Hortonian flow assumptions [22]. Kirchner argued that "parameter-rich models may succeed as mathematical marionettes dancing to match the calibration data even if their underlying premises are unrealistic" [15]. Kirchner concluded that if "the present trend away from physical processes and toward mathematistry ('blackboard hydrology') continues in hydrologic education and practice, hydrology will end up in a dead end as a science and become useless for applications" [17].

To achieve substantive improvements, models should be developed "in conjunction with, and as an integral part of, carefully planned and executed field studies established for the purpose of advancing our knowledge of the natural system"[22]. Key experimental basin requirements for model-generated hypothesis testing include: use of innovative basin designs and inclusion of sufficient variation in physiographic settings, hydroclimates, and watershed scales.

## 2.2. Hydrologic replumbing and natural climate oscillations

The National Research Council of the US identified anthropogenic perturbation and replumbing of the hydrologic cycle [23] as a fundamental challenge [16] in watershed science. Human influences dominate "the natural cycle of freshwater causing environmental changes that have been argued to move the planet to a new geologic era termed the 'Anthropocene'" [23,24].

Natural climate oscillations with observed large, abrupt events, the widespread millennial scale climate changes of the last glaciation (consisting of Dansgaard-Oeschger oscillations, Heinrich events etc.) are hypothesized to have been forced by North Atlantic atmosphere-ocean-ice interactions [25]. It has been argued that these oscillations are ultimately driven by variations in eccentricity, axial tilt and precession of the Earth's orbit as described in the Milankovitch theory. The decadal and multidecadal climate variations are generally linked with recognized dynamics, such as the El Niño Southern Oscillation (ENSO) [23].

Anthropogenic modification of the water cycle could "push the climate into new regimes" because "climate change has taken the climate system out of the repeated cycle of glacial-interglacial episodes" [23]. These alterations may accelerate the arrival of millennial scale events and trigger a "tipping point" transition resulting in "major climatic perturbations on time scales of decades to centuries"[26]. How might experimentalists, a small proportion of total hydrologists, address these challenges with progressive basin studies?

*Long term monitoring.* Well-established representative basins and benchmark experimental basins distributed across different physiographic and hydroclimatic conditions are critical for assessing hydrologic changes due to replumbing. Basins instrumented during or before the 1950's may also be used to determine patterns of rainfall redistribution, resultant stream flow and related biological issues.

*Consequences of replumbed hydrology.* Monitoring and unraveling the hydrological and ecological responses and the feedbacks due to the replumbed hydrological cycle from water conservancy projects, especially those in arid and semi-arid regions, is needed. Dams are perhaps the most dramatic examples of human capacity to transform nature in the name of development [27], but development programs and projects create both winners and losers [28]. Dams in arid and semi-arid basins, especially in endorheic basins, break the natural cycle, looting downstream groundwater and accelerating desertification [29]. Long distance water diversion may reasonably be viewed as an ecological "planning disaster"[30]. There are concerns that development is outpacing scientific understanding, a circumstance that may prove disastrous for the environment for the generations to come.

*Fundamental studies.* Various hydrological and biogeochemical fluxes through catchments are important to understanding the effects of hydrologic replumbing, but many of these are poorly understood and rarely measured. For example, various subsurface ecological and hydrological components cannot be measured directly and must instead be estimated from a small number of point measurements. Particular emphasis has previously been placed on the downward flux of catchment processes components (i.e., rainfall and discharge) rather than upward flux- land evaporation, transpiration, and upward recharge of groundwater flux, etc – even though "these fluxes serve as important regulators of the dynamics of the cycle"[23]. There are various unknown or poorly understood mechanisms regarding natural catchment behavior that are affected by the hydrologic replumbing occurring at local, regional and global scales. These include: (1) the old water paradox [31]; (2) network-like preferential flow[13], natural symmetry between water, landscape, soil and vegetation and the underlying organizing principles between soil, vegetation and other biotic elements [14]; (3) the combined mechanisms of the geological water-cycle and hydrologic water cycle in the shallow hydrosphere, as well as the deep circulation of groundwater, which go far beyond current hydrogeological boundaries [32]; (4) the puzzle of the missing sink for the remaining 40% of watershed carbon balance, which after years effort a consensus for it still has yet to emerge [33]; (5) the mechanisms involved in low flow hydrology and its relation to shallow and deep groundwater, regolith, land use, water quality and the biodiversity of aquatic ecosystems [34]; and (6) the coupling of basin geomorphologic processes with hydrologic and ecologic processes, including the poorly understood processes and mechanisms of environmental release, transport, and biological transformation of various contaminants in the unsaturated zone, shallow aquifers and deep aquifers.

*Historical data on water replumbling.* Historical data from sources other than hydrometric stage records will be very helpful to understand large-scale environmental changes, serving as a reference for the natural variation and anthropogenic impact. Figure 1 illustrates desertification in an endorheic basin as the result of human impacts. Four stages of man-earth relationship are illustrated, from the 'stage of natural harmony' until the stage of the environmental disasters, the punishment of nature (see Figure 1 caption).

*Design systematic experimental facilities.* The dynamic systems currently facing experimental hydrologists exhibit a wide range of heterogeneity and process complexity across large spatiotemporal scales. There are multiple ways to overcome such scaling problems [13,14], but hydrometric measurement systems in natural basins still face challenges (e.g., the ability of current rain gauges with conventional deployment techniques to obtain precise precipitation inputs for natural basins remains controversial). It is necessary to have systematic experimental facilities consisting of laboratory physical models, hillslope "catchment", small-size natural watershed with uniformly dominant vegetation, small-size natural watershed with more complex vegetation, unchannelized catchments, nested sub-watersheds, and watersheds of large size. Such experimental facilities need to be situated in zones with different physiographic and hydroclimatic conditions. Obviously, this would require collaborative efforts at an international scale.

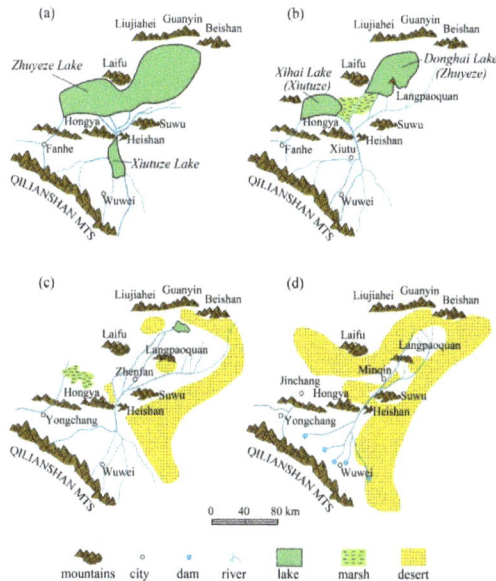

**Figure 1.** Desertification of the Rocksheep River (Shiyang River) Basin [35], an endorheic basin with total drainage area of 40,690 km². (a) prior to 121 BC, the "stage of natural harmony", the area of its terminal lake around 540 km²; (b) 420- 589 AD, the "stage of relying"; (c) ca 1900AD, the "stage of impact", the cultivated area during Qing Dynasty was developed quickly, which reached to an area about 4/5 of that of modern time. (d) ca 1990's, the "stage of damage", there are 4 dams since 1959, 22 dams until 1972 [35].

# 3. The Critical Zone Experimental Block (CZEB)

For the future of hydrologic basin study, another kind of experimental basin is suggested: the Critical Zone Experimental *Block* (CZEB) customarily known as Critical Zone Experimental *Basin*.

## 3.1. Deficiencies in the design and operation of current experimental basins

*Watershed surface laterality.* Ignoring the inherent interconnectedness of surface and subsurface watershed components reflects the fact that watershed study is driven by practical rather than scientific requirements. Hydrologists still lack a comprehensive understanding of surface responses to nonvisible and difficult to measure subsurface constraints. A physiological analogy might be studying the skin while ignoring the functions beneath. Focusing on surface responses using surface monitoring will quickly trap the observer with misconceptions. For instance, hydrologists delineate watersheds with surface topography, though actual watershed boundaries may cross topographically determined boundaries depending on subsurface features. Even our most fundamental measurements of the watershed – its shape and size – are strongly influenced by our inability to peer beneath the surface of the catchment.

*Hydrologic process laterality.* The hydrologic cycle is tightly coupled with other cycles [36]. The hydrologic cycle is also dynamic, not static, in nature. Studying the hydrologic cycle while ignoring the interconnectedness of other cycles is equivalent to studying blood circulation while only looking at blood vessels.

*Downward components laterality.* There is a tendency to emphasize measurements of downward components (e.g. rainfall, discharge and infiltration) rather than upward components (e.g. evaporation and transpiration aforementioned), even though upward components can be the dominant terms in water balance [37].

## 3.2. Lessons learned from the Chinese experimental basin studies over the last fifty years

The Chinese experimental watershed studies experienced a saddle-backed fluctuation with two peaks [9] during their fifty-year history. The lessons learned were summarized in cooperation with J.J. McDonnell, C. Kendall and N.E. Peters, as following [9]:

1.  The research facilities need not only the RB and EB of natural conditions, but also those with controlled boundary conditions. All EBs of recent decades referred only to the natural boundary of the surface watershed, ignoring that of the bedrock. This frequently led to controversial results.

2.  The EB should be treated as an integrated system, from surface to bedrock.

3.  The runoff response of an EB, including surface and subsurface components, should be monitored hydrometrically. For all EBs in past decades, these components were monitored incompletely or incorrectly, insufficient even for 'black box' evaluations. As a consequence, most of the results can't be explained physically.

4.  It is necessary to use multiple tracers to look inside the physical processes of the hydrological cycle and the anthropogenic impacts at the catchment scale.

5.  Comparative EBs should be located in different climate and morphological regions and should have comparable monitoring strategies.

It was also suggested that hydrological experimentation should not address "only to the hydrosphere but the interactions between atmosphere, lithosphere, biosphere and the intelligence-sphere." To this end, there is a need to "develop approaches using physical, chemical, isotopic, biogeochemical and hydrometeorological methods for basin studies as a 'hybrid' basin research"[38].

### 3.3. What is CZEB?

The term "critical zone" has been used in many publications to refer to wide-ranging things in areas of geosciences and mineral, etc (e.g., the geological formation, the rhizosphere, the transitional zones in alluvial coastal plain rivers, etc) [36]. In 2001, the National Research Council of US recommended the integrated study of the "Critical Zone"   (CZ) [36], defined as "the heterogeneous, near surface environment in which complex interactions involving rock, soil, water, air and living organisms regulate the natural habitat and determine availability of life sustaining resources" [16]. The CZ is one of the most compelling research areas in Earth sciences in the 21st century [36]. This zone was further defined as extending from "the vegetation canopy to the zone of groundwater" in 2005 [39], or "the bedrock to the atmosphere boundary layer" [40] and, "top of the vegetation down to the bottom of the aquifers" [36] by 2010.

The Critical Zone Experimental Block (CZEB) geologically is a monolith-block with its surface, the watershed, bounded by topographical water divides, which define the surface boundary, and which lacks defined subsurface boundaries. A drainage basin conceptually is only its surface part, the visible face (Figure 2). The CZEB is actually an Experimental "Block" within the Critical Zone with a surface drainage basin (the watershed). It is a dynamic ecosystem coupled with various supporting systems but using hydrological processes as the unifying theme. It is a living, breathing, evolving boundary layer where rock, soil, water, air and living organisms interact [41]. The CZEB is the experimental hydrological watershed study in the CZ framework. It follows that the components within the CZEB (including monolith, hillslope, sub-watershed, etc) will be monitored according to the "downward and upward approaches to theory development in catchment hydrology" [14, 42]. "The Critical Zone Observatory network is the only type to integrate biological and geological sciences so tightly"[40], and following this, the CZEB network will perhaps be the only network to integrate the hydro-logical, biological and geological sciences so tightly as well.

### 3.4. Boundaries of CZEB

*Top.* The evaporation surface of the canopy in general is not the mean surface of the canopy but slightly lower. If the mean evaporation surface of the canopy is $h$ $(m)$ above the mean ground surface, then the top boundary of a CZEB is defined as $H(m)$, where $H=(1.5$ to $2.0)\times h$ with coefficients of $h$ varying according to the vegetation. This is mainly for the purpose of energy budget and eddy covariance flux observations. $H$ is just the lower part of the atmos-phere boundary layer.

*Bottom.* There are three cases:

Case I: Bedrock is situated at a relatively shallow depth from the ground surface while the regolith is shallow, too. The bottom boundary is defined as the geological boundary (Figure 3a).

Case II: Bedrock is deep. The bottom boundary is defined as the plane where the tritium content of groundwater approaches zero or the detection limit of ±0.7 TU (the "tritium naught line" (TNL)[43]), which is same as that of Case III (Figure 3b).

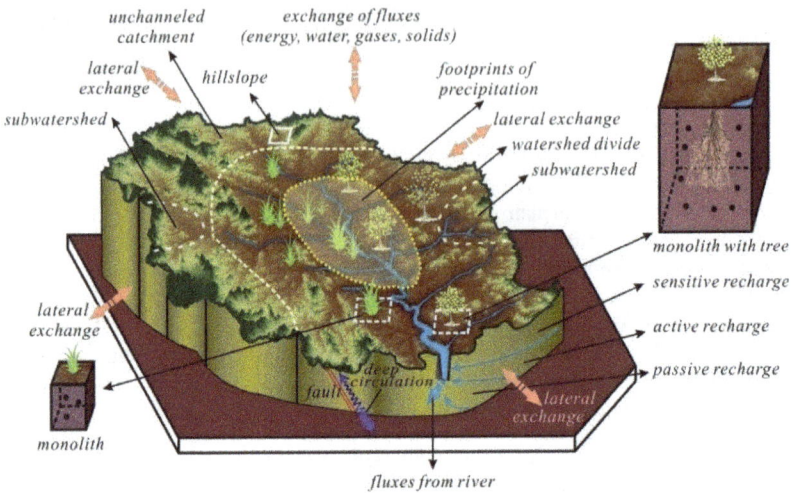

**Figure 2.** A conceptualized Critical Zone Experimental Block (CZEB)

Case III: There are stratified alluvia, potentially with multiple aquifers with aquicludes and aquitards, common in flat areas with thick deposit and deep bedrock, up to hundreds of meters or more. Groundwater recharge to the river can be separated into sensitive, active and passive zones. In this case, the bottom boundary of CZEB is defined as the bottom of the active recharge zone, the plane of TNL (Figure 3c).

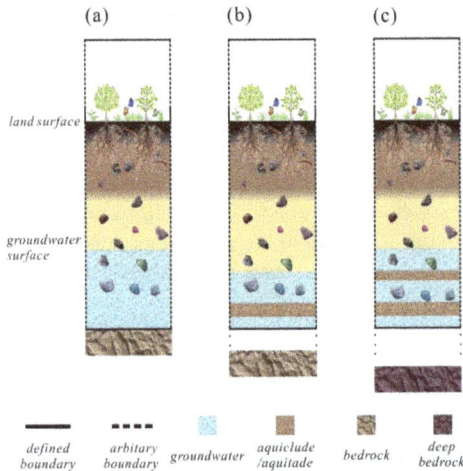

**Figure 3.** Boundaries of CZEB

*Lateral sides*

Part I: Above the ground surface up to height *H* as described above, delineated according to the surface topographic watershed boundary (Figure 3).

Part II: Below the ground surface, it is in general defined arbitrarily, except in the case of existing geological boundaries.

### 3.5. Functioning of CZEB

CZEB is an ecological dynamic and evolving system. It also is a natural open system, exchanging mass and energy with its surrounding environment across 'arbitrary' boundaries. It is a dissipative, complex system, however, with some degree of self-organization [36].

*Interfaces.* CZEB encompasses "the near-surface biosphere and atmosphere, the entire pedosphere, and the surface and near-surface portion of the hydrosphere and lithosphere" [36]. Within the CZEB, various processes including hydrologic, atmospheric, lithospheric, geomorphic and geochemical processes are coupled and dynamically interrelated. To simplify the organization of various interfacing processes throughout the CZEB, compartment zones can be broadly separated as follows:

1. The zone above ground surface, the "aboveground vegetation zone" [36].

2. The unsaturated zone, "the belowground root zone and the deeper vadose zone "[36].

3. The saturated zone, the "saturated aquifer zone" [36].

This layering has a general trend of increasing density with depth, has a dampening effect on state variables with depth, and an increase in distance to energy input at the soil surface [36]. There is also "an overall trend of increasing response time" [36, 44].

*Mass – the material aspect of CZEB.* The material base of the CZEB is fixed and limited, involving rock, soil, water, air and living organisms [16]. On this point it happens to coincide with the ancient Chinese philosophy, the so-called *"Five-xing"*, which holds that five fundamental elements form the universe and the Earth: "Jin"(metal), corresponding with the term "rock"; "Mu" (wood), with that of "living organisms"; "Shui"(water) with that of "water"; "Huo"(fire) with that of "air"; and "Tu"(soil).

*Energy and force – the driven aspect of CZEB.* Continuous energy fluxes and inherent forces drive the CZEB system. External solar energy is certainly the vital component of external energy source. Various inherent forces include: gravity, surface tension, intermolecular forces, capillary force, etc which control its dynamic situation.

*Organization and entropy – the philosophical aspect of CZEB.* This open system of structural dissipative processes and irreversible evolution tends to increase its entropy spontaneously and go towards disorder [36]. Even the exchange of entropy fluxes across its boundaries is continuous. The second law of thermodynamics, from which the concept of entropy was derived, is one of the fundamental natural laws. However, the role of feedbacks of this nonlinear system will promote "self-organization" as energy dissipates, providing opportunities for dissipating energy to act again within the system towards the direction of order. "Conservation without evolution is death. Evolution without conservation is madness" [45,

46]. Conservation of energy appears in the CZEB at all scales, with observable 'behaviors' of organization, symmetry, genes, etc. Lin suggested including "information" as one of the four general factors (i.e., conservation of energy and mass and, the accumulation of entropy and information, which dictates the evolutionary outcome and the functioning of the CZ ) [36]. This is equally applicable to CZEB.

### 3.6. The reactors of CZEB

The CZEB can be separated into three interdependent zones as mentioned above with each zone consisting of the same base materials, the "*five-xing*" rock, soil, water, air and living organisms as well as non-material bases of energy, force and information. This coincides well with the '*Bagua*' ('*eight-gua*') of the earliest Chinese philosophical work "*I –Jing*" ("The Book of Chang"). According to I-Jing, the origin of the universe was just a singularity of chaos ('*Tai-ji*'), consisting of '*Yang*' and '*Yin*', the unity of opposites, which produced '*eight-gua*,' the general principles governing development in the material world. Each "*gua*" is a combination of three whole lines ('*Yang*') and broken lines ('*Yin*'). As in the four general characteristics of CZ elaborated by Lin [36]; in the CZEB, the material base refers to the five '*gua*'s (the "*five-xing*") and the non-material base refers to another three '*gua*'s of energy, force and information, which together form a dynamic '*Bagua*'. Each zone has the same five material and three non-material components with different rates, Figure 4.

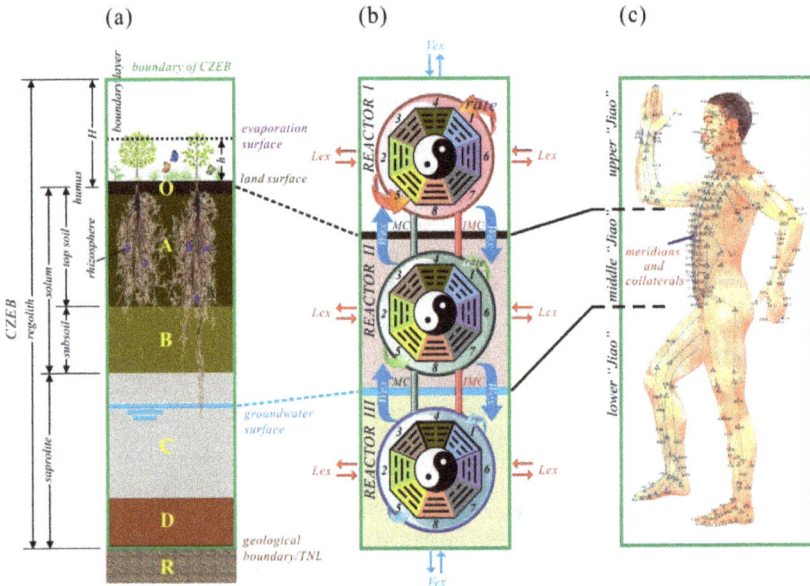

**Figure 4.** Schematic 'reactors' of the CZEB. The dynamic '*Bagua*' in each 'reactor' operates with its own rate; 1- 5 : the material base (1- air, 2- water, 3- soil, 4- rock, 5- organisms); 6-8: the immaterial base (6- energy, 7- force, 8- information).

Each zone will behave as a 'feed-through reactor' according to Anderson et al [47]. It follows then that there are three feed-through reactors coupled together in the CZEB (Figure 4), with probably different rates and residence time of materials.

*Reactor I.* The first zone is the above ground surface zone, which also contains above ground vegetation up to its interface with atmosphere (Figure 4a). Evaporation and plant transpiration may account for 50% or more of the total local precipitation and use up to 50% of the total solar energy.

*Reactor II.* The second zone is the unsaturated zone, extending from the soil surface to the upper surface of the groundwater table. It consists of the whole soil profile: The O horizon (humus), A horizon (topsoil), B horizon (subsoil) and C, and potentially D, horizons (Figure 4a). The unsaturated soil zone has been recognized as "the most complicated biomaterials on the planet" [36].

*Reactor III.* The third zone is the saturated zone, extending from the groundwater table down to bedrock, including the capillary fringe (Figure 4a). There are two general cases for CZEB, including phreatic groundwater and confined aquifers.

*Functioning of reactors*

1.  Reactor I includes hydrometeorological and ecohydrological processes, Reactor II includes hydropedological and hydroecological processes, while the Reactor III is more exclusively for hydrogeological processes.

2.  There is an overall trend of decreasing operation rates from Reactors I to III.

3.  Each Reactor has its own lateral flux exchanges $L_{ex}$ via the CZEB lateral boundaries (Figure 4b). The current hydrometric runoff data is only that from channel. Reactor I has vertical fluxes exchange $V_{ex}$ with atmosphere via the upper boundary of CZEB while Reactor III has $V_{ex}$ with deep aquifers and/or deep circulation via the bottom boundary of CZEB (Figure 4b). Fluxes include all material and immaterial components.

4.  Reactors are closely coupled within the boundaries of CZEB. There are flux exchanges $W_{ex}$ (Figure 5) between Reactor I and II and, and between II and III within the CZEB. There are flux channels *MC* (Figure 4) through I to III for materials, and flux channels *IMC* (Figure 4) through I to III for non-materials.

5.  The *MC* appears similar to blood vessel system and *IMC* to meridians and collaterals of the human body from the Chinese art of acupuncture. The three "reactors" are similar to the so-called human "*three Jiaos*" (Figure 4), a central concept of Chinese medicine philosophy.

# 4. A trial of the CZEB: The Chuzhou CZEB experimental system

### 4.1. The strategy: A two–way multi-scale approach

*Isolation of the natural system.* In order to investigate the natural watershed system, which is filled with complexity and heterogeneity, simplification is incorporated as much as possible (e.g., *"isolation"*) [9, 38]. Using detailed observation on ever-decreasing scales allows the investigator to focus and isolate processes from the naturally occurring heterogeneity (i.e., watershed, sub-watershed, unchanneled watershed, hillslope, monolith, a tree) (Figure 5). To some extent it is similar to dissection: splitting "problems into their smallest possible components"[48].

There are limitations in the field, however. Using a single tree as an example, the tree itself can be well instrumented, as well as the subsurface with pits up to meters deep, but still the fluxes from its bottom boundary would be unknown. In this sense, *isolation* refers to a system rather than the physical isolation of a natural unit, which can only be partially isolated in field. Strictly speaking, the field result from such natural systems can be explained only empirically, from a statistic or a stochastic view. Understanding causality mechanisms from observational studies alone is problematic. Thus, McDonnell et al (2007) noted that "most field experiments and observations in watershed science to date remain largely descriptive" [13]. They also discussed a shared 'philosophical path' which claims that "if we characterize enough hillslopes and watersheds around the world through detailed experimentations, some new understanding is bound to emerge eventually", the authors concluded that "what this approach to experimental design has succeeded in doing is to help characterize the idiosyncrasies of more and more watersheds, in different places and at different scales, but with little progress toward realizing the Dooge vision" of "hydrologic laws"[13]. This does not mean, however, that field studies have no place in hydrology. On the contrary, the problem becomes how to "put the pieces back together again" [48]. In our case it is how to put together hydrologic mechanisms for a monolith, a slope, a sub-watershed and, a watershed.

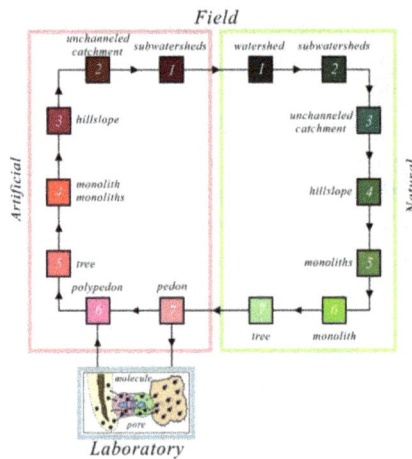

**Figure 5.** Concepts of the two-way multi-scale approach: natural and artificial

*Control and synthesis in the artificial system.* A series of artificial systems are suggested and defined for "control and synthesis" [9, 38]. They encompass the pedon, polypedon, monolith, hillslope, catchment, until subwatershed (Figure 5). Artificial systems are necessary in order to move beyond problems of boundary and parameter control found in natural systems. It is in fact inspired by the aforementioned empirical impasse of natural systems; artificial systems overcome heterogeneity by controlling all variables and leaving only test variables to fluctuate.

*The laboratory system.* The two-way multi-scale approach can only be applied at the macroscopic level of watersheds or, maybe the mesoscopic level (e.g., catena monolith). In fact many problems (e.g., the old water paradox) can be trace back to mechanisms occurring on a microscopic level (e.g., the molecular and pore related mechanisms as that of microbial processes of some contaminants transformation). The laboratory system connects with and supplements the artificial system (Figure 5).

*Could this be a new vision for watershed hydrology and a unified theory of hydrology?* Lin summarized three systems of different levels of complexity and a vast gap between the two extremes [36] as shown in Figure 6. The coupled natural and artificial systems are analogous to *'Yin'* and *'Yang'*, which is the successful philosophy of *'Taiji'* (Figure 7). Klmes suggested "a rational search for meaningful conceptualization in hydrology can proceed along two routes: upwards and downwards "[42], or, the "upward or bottom-up approach" and the "downward or top-down approach" [14]. The two-way multi-scale approach perhaps could also be upwards and downwards. Hopefully, it could be a way for a new vision for watershed hydrology [13] and, for the unified theory of hydrology [14].

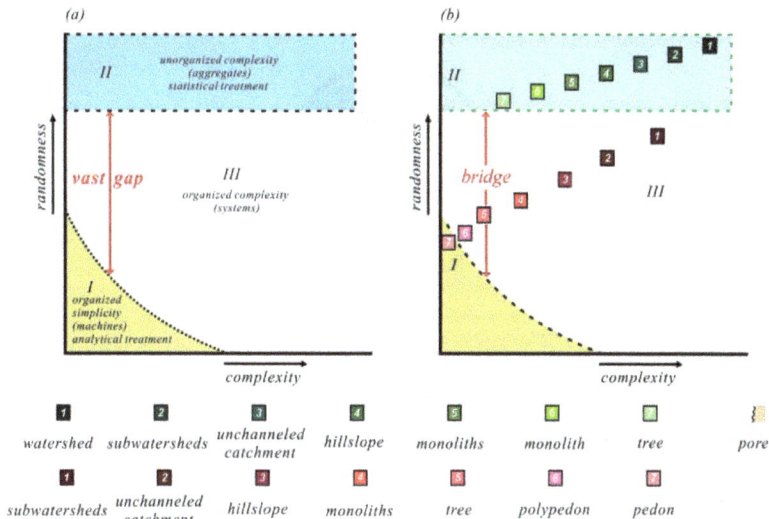

**Figure 6.** The gap between extremes of complex catchment systems. (a) Three types of complex catchment systems related to different treatment methods [14,19,36]; (b) The supposed field and artificial two-way multi-scale framework and the gap depending on the level of control. The gap becomes the bridge.

## 4.2. The trial practice of the CZEB: The Chuzhou CZEB experimental system

The Chuzhou CZEB can be attributed to the artistic conception of the *West Brook*, described in a poem of Tang Dynasty where intellectual exiles of Song Dynasty found their Arcadia ca 900 years ago. On the same brook in 1962, a hydrological experimental field station with three watersheds was founded. About fifteen years later, construction of the experimental watershed hydrology had barely survived from the cultural revolution of the 'Road to Perdition'. The Chuzhou CZEB was an adventurous plan for a two-way experimental watershed system including natural catchments and artificial catchments with controlled boundaries. The experiment was, unfortunately, interrupted again during the 'market-economy tides' of China and only partially completed. A turning point emerged a few years ago, with more efforts towards an innovative CZEB system. In the meantime, the components of the natural and artificial systems are: (1) the natural system with main infrastructure of hydrological monitoring and isotope biomonitoring encompasses watersheds with drainage area of 82.1 km$^2$ (hill 34%, forest 5%, ponds 1.2%), 17.5 km$^2$ (56%, 10%, 0.6% *idem*), 4.5 km$^2$ (100%, 2%, 0% *idem*), 3.3 km$^2$(100%, 100%, 0% *idem*), 2.0 km$^2$ (5%, 2%, 30% *idem*), 0.06 km$^2$ (80% paddy field), 4573 m$^2$ (Morning Glory Catchment, 100% weathered debris), and, a mini CZEB with directly observable surface and subsurface responses with a drainage area of 7897m$^2$; (2) the artificial system with measurable surface and subsurface responses encompasses a catchment with surficial and bottom area (horizontal projection) of 490m$^2$, a crop monolith of 150m$^2$, two grass monoliths (L1, L2) of 32 m$^2$ each, a saturated monolith (LS) of 1 m$^2$, and several weighing lysimeters. Here, the focus is on results from the two experimental watersheds - the mini CZEB watershed and the artificial catchment.

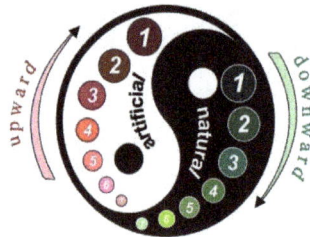

**Figure 7.** Reconciliation of two-way natural and artificial, *'Yin'* and *'Yang'* [14, 42].

*The natural CZEB, Nandadish*

Nandadish has a surficial drainage area of 7897 m$^2$ and rests on the consolidated bedrock of the concordant body of andesitic and tuffaceous facies with a thin weathered layer. The boundary condition of this mini CZEB belongs to the type shown in Figure 3a. The Quaternary regolith overlies the bedrock, its depth ranges from of 1 to 7 m with an average of 2.46 m; its bed rock topography was surveyed via 69 drillings (Figure 8). The regolith consists of brunisolic soil of heavy loam, medium and clay loams; saprolite with prismatic and block structures. Horizontal and vertical fissures and cracks developed in the upper regolith. This watershed has an altitude difference of 12.9 m with a surface slope of 6.7% to 17.1% at different directions. The brook gradient is 6.7%. The coverage during the watershed's construction in 1979 was natural grasses with small shrubs and a few Masson pines aged 5 to 6 years. Since then, coverage has shifted to a dense forest with canopy height ca 8 m.

The main infrastructure includes: (1) measurement for energy flux, water flux, geochemical flux, gas flux; (2) changes of watershed storages of energy, water, gas and that of geochemical ions; (3) CZ tree experiment; and (4) variations of precipitation isotope fingerprints in flux and in storage. The surface and subsurface runoff are monitored directly via a trench, which extends upslope to capture subsurface and surface flow. This CZEB block has deep soils near the divide but only 1-meter depths near the outlet, making the block easy to close via a concrete wall installed to the bedrock at the outlet point. In this way, all surface and subsurface flow drain into discharge measuring structures (Figure 9).

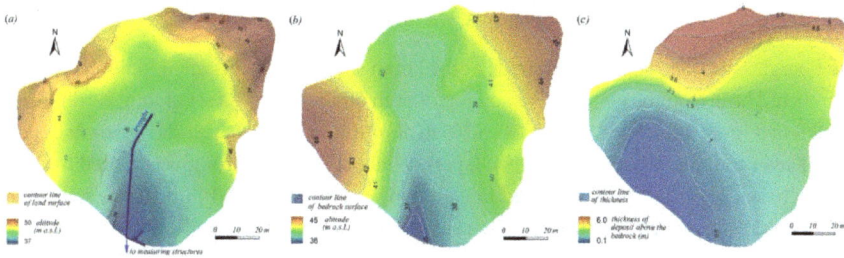

**Figure 8.** The mini CZEB Nandadish. (a) Its surficial watershed; (b) Its bedrock; (c) The depths of deposit above its bedrock.

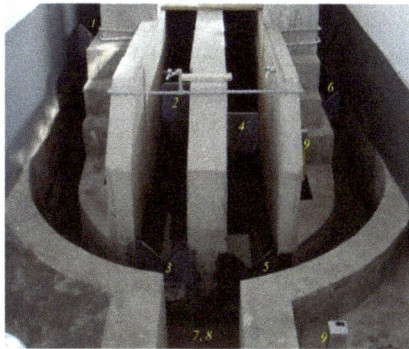

**Figure 9.** Discharge measurement structures for different runoff components from troughs with its location showing in Figure 8(a). 1- for rainfall; 2 and 3- for surface runoff; 4 and 5 – for interflow (30 cm below the soil surface); 6 – for interflow and groundwater flow (down to bedrock); 7 and 8 for total runoff (weirs are not shown). 1,3,5,6 and 8 are V-notch sharp-crested weir; 2, 4 and 7 are the full width rectangular sharp-crested weir; 9 is the tracking water head gauge. The picture design is the original concept, through renovations are now underway for gauge modernization.

*The artificial catchment, Hydrohill*

The Hydrohill catchment with a drainage area of 490 m² (horizontal projection by plane surveying) - 512 m² for its inclined surface [49, 50] - is sited on a small andesitic hill. The entire hillslope was first excavated to bedrock to create a bare catchment of ca 4700 m². A concrete aquiclude was created above it and consisted of two intersecting slopes dipping towards each other at 10°with an overall downslope gradients of 16.9°(Figure 10a). Impermeable concrete walls enclose the catchment on all sides to prevent any flow of water across the catchment boundaries. Silt-loam soil was removed, layer by layer, from an agricultural field near the Hydrohill site and installed on the artificial aquiclude, layer-by-layer again. Also the bulk density of the soil was adjusted during filling to approximate the natural soil profile of the original agricultural field. Hence, the final 1-m soil 'profile' was identical, at least in compo-sition, to the natural soil profile. Grass was then planted over the surface and soil allowed to settle for 3 years (Figure 10b), after which time a central drainage trench was then excavated at the intersection of the two slopes and the water-sampling instrumentation was installed.

Five fiberglass troughs, each 40 cm wide and 40 m in length, were installed longitudinally in the trench. These troughs were stacked on top of each other to create a set of long zero-tension lysimeters (Figure 11a). Each trough has a 20 cm aluminium lip that extends horizontally into the soil layer to prevent leakage between layers. Water collected in troughs is routed through V-notch based logarithm weirs located in a gauging room (Figure 11b) under the hill where discharge is continuously monitored. Water samples are collected manually above the ponding at the weirs.

As illustrated in Figure 11a, the uppermost trough collects rain; the next lower trough collects surface runoff. The next three troughs collect subsurface flow from soil layers spanning the depths of 0-30, 30-60, and 60-100 cm. These troughs will be referred to as the 30 cm, 60 cm, and 100 cm troughs. The source of the water in these troughs (i.e., whether the water is derived from interflow or saturated flow) varies locally and during storms. The lowermost trough collects either saturated flow or interflow, depending on the height of the water table. When the water table is high, saturated flow may be collected in both of the lower two troughs.

A network of 21 aluminium alloy access tubes for neutron moisture gauges [51], a tensiometer scanner, and 22 wells for water-table measurements and groundwater sampling were installed (Figure 12). The wells were drilled to the aquiclude and are slotted along the lowermost 20 cm. After installation, the spaces around the slotted lengths were packed with sands to allow movement of groundwater to the well, space above the slotted lengths were packed with clay to prevent vertical drainage along the pipes (Figure 11c). The neutron probe access tubes (Figure 11c) and soil water potential tensiometers were positioned adjacent to the wells for soil water monitoring. Catchment evaporation was monitored by methods of water balance, energy budget and, soil water variations above zero flux potential. The plan view of the surface topography, the locations of wells and access tubes, and, the central stacked lysimeter troughs and that of energy budget set are showing in Figure 12.

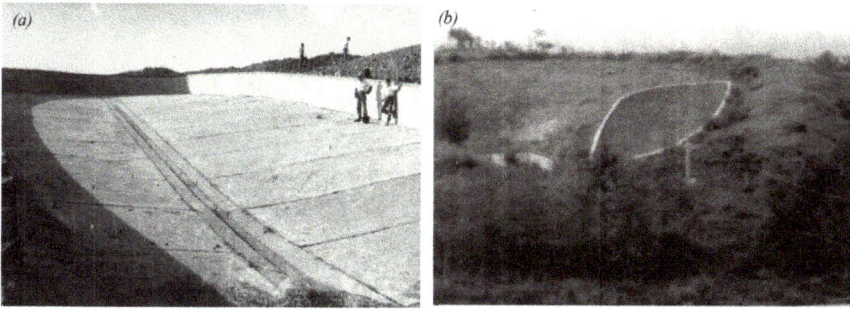

**Figure 10.** The artificial catchment Hydrohill: (a) Its concrete aquiclude; (b) Three years waiting for the settling of filled soil to try to match close to the natural soil profile.

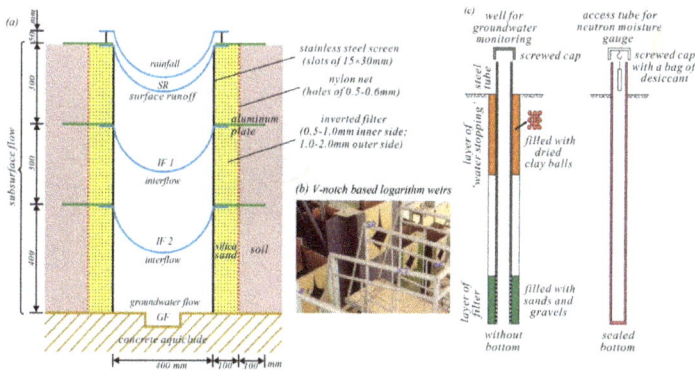

**Figure 11.** The artificial Hydrohill: (a) Schematic cross-section of rain, surface, and subsurface flow collectors at the catchment; (b) The V-notch based logarithm weirs for discharge measurement for these components; (c) The constructions of wells for groundwater monitoring and access tubes for neutron moisture gauge. (b) and (c) are the original design, now under renovation;

The renovation phase of the main infrastructures for this physically modeled CZEB includes: (1) measurement for energy flux, water flux, geochemical flux, gas flux; (2) changes of watershed storage of energy, water, gas and that of geochemical ions; (3) a small separate laboratory within the gauging room for high-frequency, real time measurement of isotope concentrations in all hydrologic components using of a laser spectrometer LWIA [52]; (4) a removable rainfall simulator on the tracks capable of covering an area of more than 500 m², capable of simulating a range of rainfall intensities; and (5) a system for carbon balance of this block.

# 5. Selected findings

## 5.1. Runoff composition

The catchment response to precipitation consists of surface and subsurface components. These components were directly measured at Hydrohill with both isotopic and hydrochemical evidences (Figure 13) [53, 54]. Surface runoff was not always the dominant component during a rainfall-runoff event. The hydrograph composition of surface and subsurface flows was discriminated into four broad categories (Figure 14): 1) S type which is dominated by surface runoff; 2) SS type which is dominated by subsurface flows; 3) Intermediate type M, with largely equal contributions of surface and subsurface flows; and 4) Variation type V, with switching between components during an event.

**Figure 12.** Hydrometric monitoring of Hydrohill. (a) Plan view of the surface topograph showing the locations of groundwater wells, access tubes for neutron moisture probe with potential scanner, the central stacked lysimeter troughs, the energy budget set. (b) Whole view of its surface watershed in operation during early years (1980's), under renovation now.

## 5.2. Spatial and temporal variability in hydrological parameters

The physical based distributed-parameter modeling appears to be a promising direction for watershed hydrology. However the scale of model elements, and their parameters, is a lingering problem. This has led to investigations into the establishment of a 'representative elementary area' (REP), 1 km² [55]. In fact, because of the spatiotemporal heterogeneity of hydrological parameters, little is known regarding hydrologic parameterization in scales less than 1 km². Previously unimagined heterogeneities were observed at Hydrohill:

*Intrastorm isotopic heterogeneity of shallow groundwater.* During a storm in July 1989, the newly developed groundwater showed considerable spatial and temporal variability in the $\delta^{18}O$ (Figure 15a), with values ranging from -12‰ to -6‰ while the groundwater derived from pre-event soil water ranged from 0 to 100% at different times and locations in this catchment [56].

The groundwater samples for δ¹⁸O were collected at three times during the storm from the wells distributed in whole area as shown in Figure 15a. At the end of the storm, the ground-waterδ¹⁸O values showed a 4‰ range in composition. This range of compositions during the storm appears to be caused by the combination effects of intrastorm variation in rainδ¹⁸O (Figure 13) and spatial and temporal variability in subsurface flowpaths [56] – including the downward displacement of pre-storm water versus delivery of new water to the aquiclude via macropores [54].

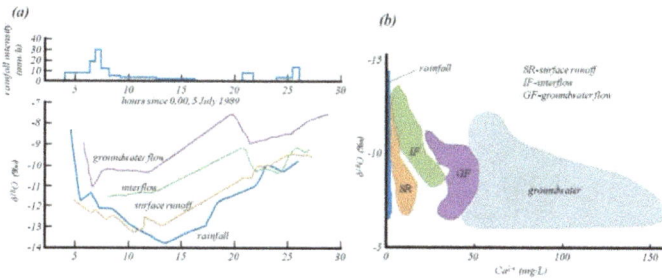

**Figure 13.** Isotopic and hydrochemical evidences of various runoff components. Samples for isotopes were analysed in Menlo Park laboratory of USGS by C. Kendall (same hereinafter)

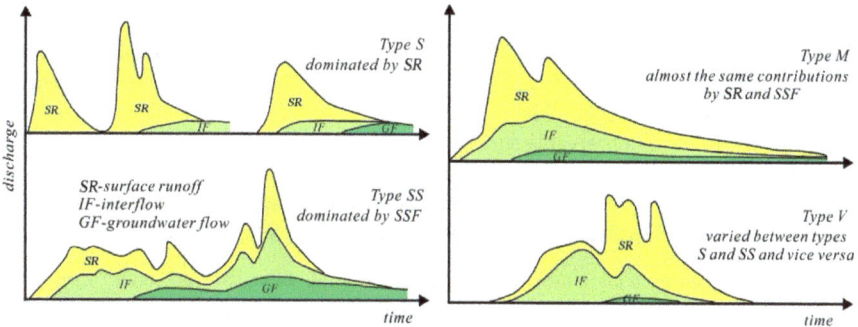

**Figure 14.** Types of runoff composition summarized from observed data of Hydrohill during 1981- 1992.

*Annual heterogeneity of land evaporation rate* Land evaporation was measured via the variations of unsaturated soil water at different positions in the soil profile during its matrix potential distribution in that profile showing a zero flux plane. Intensive measurements of volumetric soil moisture by neutron probe at 4 points within each of the 21 access tubes (84 in total) were made during the wet and dry seasons. Representative examples of heterogeneity in the daily evaporation rate are shown in Figure 15b [51].

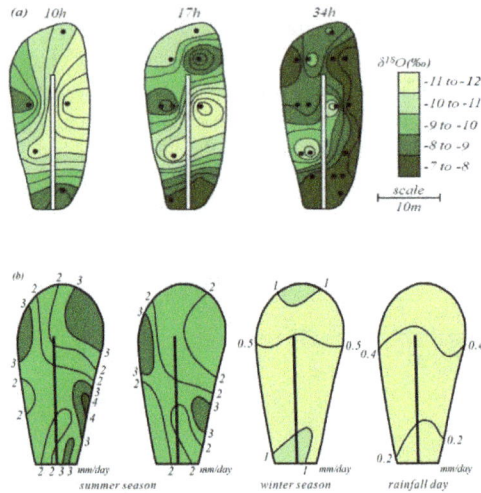

**Figure 15.** (a) Contour diagrams of δ¹⁸O in groundwater at three sampling times and (b) Typical spatial distribution of the daily land evaporation rate.

## 5.3. The double paradox

About ten years ago, Kirchner described a double paradox in catchment hydrology and geochemistry [31]. His objective was to raise questions rather than provide answers and a worthy challenge was then put forward (i.e., "need to take up the search for a unified theory", so that the paradoxes "would no longer seem paradoxes") [31]. For this challenge, perhaps catchments with directly observable surface and subsurface components could be useful in two ways: First, to identify how this double paradox will emerge in the subsurface and surface components individually, instead of in the integrated form at the stream outlet; and secondly, to trace surface and subsurface generation mechanisms using isotopic fingerprints.

*The delayed correspondency phenomena.* Three isolated rainfall events are presented from May 1987 at Hydrohill, and are ideal for establishing rainfall-runoff relationships as well as the unit hydrograph (Figure 16a). In short, the unit hydrography conceptualization is that rainfall event A produces the hydrograph peak A, rainfall events B and C produce the peaks B and C etc. It can be named as the concept of *'one-to-one correspondency'*. However, using isotope tracers, rainfall-runoff events at Hydrohill showed that rainfall event A corresponded to the hydrography peaks A1 and A2, though A2 only emerged later during event B. This pattern continued throughout subsequent storms with event B corresponding to hydrograph peaks B1 and B2, though B2 only emerged later during event C. This was termed as the phenomena of *delayed correspondency* [57]. In the case of Hydrohill, this delay appears very distinct. It is presumed that such a *delayed correspondency* will tend to diminish as drainage area increases. The diminishment of the *delayed correspondency* phenomena has been verified in a natural watershed with drainage area of 82.1 km² (Figure 16b). Rainfall event A produces a runoff hydrograph peak A, some of the constituents of rainfall event A were delayed to emerge during the rainfall event B. It is worth noting that the hydrograph produced by rainfall event A does not have the bell shape referred to in the unit hydrograph concept. In fact this *delayed correspondency* shows the formation of pre-event water ("old" water) during the runoff process. It follows that the current *'one-to-one correspondency'*, used to conceptualize rainfall-runoff relationships in applied hydrology, will be associated with large uncertainty.

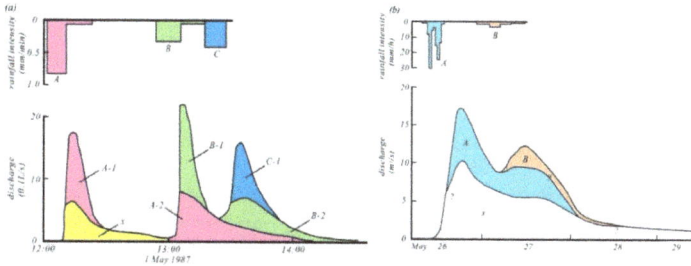

**Figure 16.** The delayed correspondency phenomena.

*"Old" water paradox identified from both surface and subsurface flows.* Data from the 'mini CZEB' watershed, Nandadish, suggest that pre-event water ("old water") appeared in all runoff components including surface runoff, interflow and groundwater or saturated flow (Figure 17). Figure 17a and Figure 17b refer to the surface runoff dominated type (type S, from above), and subsurface flow dominated type (type SS) respectively. The surface runoff and subsurface runoff processes are shown separately with their corresponding proportions of pre-event water. For the type S, the pre-event water accounts for 9% and 24% of the total amount of surface runoff and of subsurface flow respectively while for the type SS, it becomes 11%and 89% respectively [58]. This reveals that even in a catchment with an average soil depth of 2.46 m, large volumes of pre-event water ("old" water) are stored and released promptly by event input.

*Hydrochemistry distribution paradox* This was termed by Kirchner as the "'variable chemistry of old water' paradox: although baseflow and stormflow are both composed mostly of 'old' water, they often have very different chemical signatures" [31]. However, the artificial catchment at Hydrohill shows paradoxically the distribution of hydrochemical compositions in different runoff components, including event rainfall. This *'hydrochemistry distribution paradox'* results largely from: (1) inorganic ions in event rainfall input emerge in all runoff components; (2) a strong similarity between rainfall and surface runoff but less similarity in subsurface components; and (3) the fact that the total amount of ions of event rainfall is sometimes much smaller than the sum of all runoff components [59]. Figure 18 shows the processes of $Mg^{2+}$ and $Cl^-$ in event rainfall, surface runoff, interflow and groundwater flow (saturated flow).

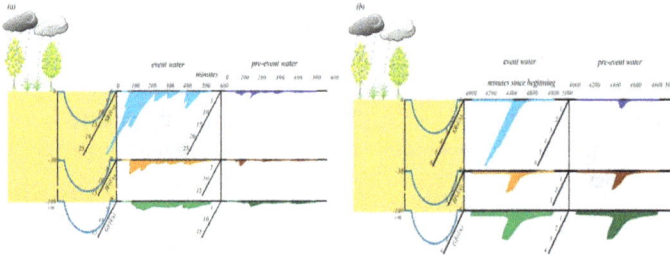

**Figure 17.** Processes of various runoff components, and that of their pre-event water of Nandadish. (a) for S type; (b) for SS type

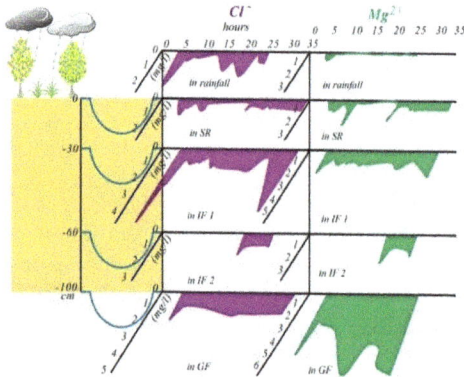

**Figure 18.** The $Mg^{2+}$ and $Cl^-$ processes in event rainfall and runoff components of Hydrohill. Samples were analysed in Atlanta laboratory of USGS by N.E. Peters (same hereinafter)

## 5.4. Does diel signal of hydrochemical constituents emerge linkage among multi–processes?

For the better understanding of the links between contaminants and multi-processes, exploration of diel signals in natural waters may yield "insight into the intricate linkages among hydrological, biological, and geochemical processes [23]". Diel variations of various hydrological constituents in surface and subsurface runoff responses during rainfall events were monitored in artificial catchments and monoliths with examples as following (Figure 19).

*Variations of pH.* The diel variations of pH in SR, IF, and GF of Hydrohill show their own individuations (solid lines in Figure 19a). Diel variation curve of SR to a large extent appears similar to that of rainfall. However, pH variation curve of IF is contrary to that of rainfall after 10 a.m., while the GF curve appears as the flattened IF curve. The pH variation curve of the interflow of monolith L1 and that of the flow of monolith LS are reasonably similar to that of

IF and GF of Hydrohill respectively (dot dash lines in Figure 19a). This reveals that the variation of pH in interflow and saturated flow is not driven only by rainfall input but also by "the biological processes of photosynthesis and respiration [22, 60]." This is ascribed to the coupling of the three "reactors" mentioned above via their MC and IMC for transformation and exchange (Figure 4).

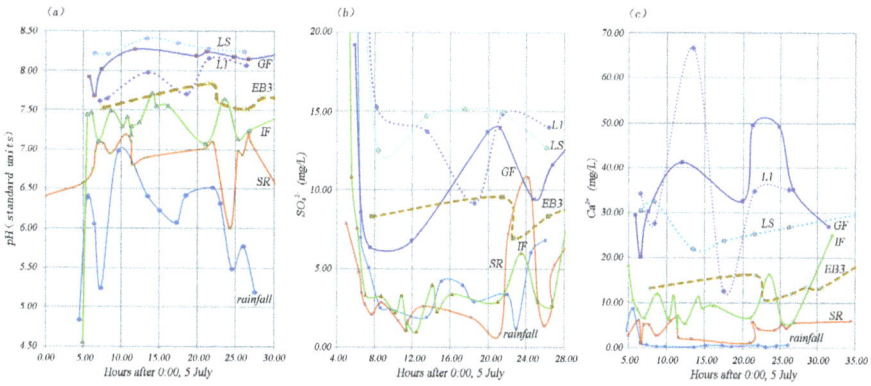

**Figure 19.** Diel variations of pH, and hydrochemical constituents $SO_4^{2-}$, $Ca^{2+}$ in event rainfall and runoff responses of Hydrohill and monoliths. SR, IF, GF- runoff responses from Hydrohill; L1- monolith of 32 m² serves as an element slope of Hydrohill; LS - saturated monolith of 1m² serves as solum in reduction environment; EB3- Morning Glory Catchment of 4573 m² serves as a catchment without saturated zone formed by debris only.

*Variation of ionic species.* (1) For anion $SO_4^{2-}$ : the diel variation curves of runoff responses of both Hydrohill and monoliths, with few exceptions, are contrary to that of rainfall. Mostly their peak concentrations happened during evening and midnight (Figure 19b). Different from the case of pH, $SO_4^{2-}$ of SR has a strong inversion with the event rainfall curve at night time. (2) For cation $Ca^{2+}$: the diel curve of rainfall (Fig.19c) shows a small variation after sunrise. However, it triggers variations in runoff responses with their peak concentrations at both a.m. and before midnight. Highest peak happens to LI at afternoon. These diel variations in runoff perhaps are metabolism related. (3) Nimick et al [60] discussed diel cycles of dissolved trace metal concentration in a Rocky Mountain stream. They found that the anionic species "have their highest dissolved concentrations in the late afternoon" while the cationic species "have their highest dissolved concentrations shortly before sunrise"[23,60].

*Role of soil.* Variations of pH, $SO_4^{2-}$, $Ca^{2+}$ in total runoff of Morning Glory Catchment, a special designed catchment without soil but debris (EB3 in Figure 19 with broken lines), are very similar to that of rainfall curves after 20:00 (EB3 data are not enough before this time). The role of soil in the formation of ionic species in runoff is apparent. Even the variations of pH, $SO_4^{2-}$, $Ca^{2+}$ in total runoff of EB3 are triggered by the event rainfall, but the resultant variations appear much simpler than various curves of Hydrohill and monoliths. This implies that only a simple process (i.e., mainly the hydrological process is involved in EB3). So, the complex diel variations of the runoff responses from Hydrohill and monoliths (Figure 19) can be reasoned as the results of multi-coupling among hydrological, biological, and geochemical processes. It shows that the reactor II (Figure 4) provides a key operator in contaminant hydrology.

## 5.5. Runoff generation

The measurements of various runoff components, within artificial systems containing controlled boundaries, provide the possibility to look inside the formation mechanisms of individual components. Isotopic and geochemical tracing can help to investigate these mechanisms but only if significant differences in isotopic compositions of components occur. The general mechanisms for these runoff components (i.e., the surface runoff, interflow and groundwater flow (saturated flow) ) are discussed in the following paragraphs [61,62].

*Surface runoff (SR).* Precipitation input is, of course, the essential condition for the generation of surface runoff. However, in order to actually generate runoff, there must be enough precipitation to form a thin, saturated soil layer ($L_{sat}$) at the ground surface. Saturation is key because unsaturated water movement is thought to be too slow to generate runoff. The thickness of $L_{sat}$ at catchments, monoliths and plots was found to vary between 5 to 50 mm throughout events, increasing downward during rainfall events and receding once rainfall stopped, although these findings were complicated by the irregularities of the soil surface. SR was not generated until a $L_{sat}$ was established at the surface. Once a $L_{sat}$ was developed, regardless of its thickness, SR was generated immediately. *Overland flow* was only observed on impermeable surface (*DO*) at artificial plot and, on saturated surface (*SO*) from a special lysimeter designed for $L_{sat}$ simulation. Intrastorm variation of isotopic composition of *DO* and *SO* can match that of event rainfall. In most cases however, SR is generated within the $L_{sat}$, with turbulent mixing of event and pre-event water stored in $L_{sat}$, i.e., the *saturated mixing surface flow (MS)*; Alternatively, small amounts of event water act on the surface of $L_{sat}$, to force out pre-event water in the $L_{sat}$, termed here as the *saturated expelled surface flow (ES)*. The isotopic composition of *MS* shows a mixing of event rainfall and pre-event water in $L_{sat}$. The isotopic composition of *ES* is similar to that of $L_{sat}$.

*Interflow (IF) in the unsaturated zone.* Three generation mechanisms are observed. (1) In cases there are soil layers with distinct bulk density and/or hydraulic conductivities, IF can be generated at the interface of soil layers. This was only observed in the 'mini CZEB', Nandadish, where IF occurred at the interface of the layers A (topsoil) and B (subsoil). This is termed as *layered interflow (LI)*. (2) During percolation, soil water moves downward and laterally towards the drainage interface and accumulated until saturation. The saturation will expand if the soil matrix potential $\psi < 0$. Once $\psi$ reaches zero, the accumulated saturated soil water will be released immediately at drainage interface. A small amount of infiltration water can trigger this process, to expel the accumulated soil water out as interflow. This process was observed for most IF events in Hydrohill and the 32 m² monolith. This is termed as *expelled interflow (EI)*. (3) *Macropore interflow (MI)* was also observed in a special designed catchment without saturated zone with area of 4573 m².

*Groundwater flow (GF) from saturated zone.* In most cases, groundwater flow was due to a rising groundwater table resulted from event water recharge via the capillary fringe zone, and was termed *recharged groundwater flow* (*RG*). RG was observed in most cases of both artificial and natural catchments. In addition, the *macropore-induced groundwater flow* (*MG*) is the only component not directly observed in this work but inferred from the isotope data in catchment *GF* during events.

## 5.6. Unreasonableness of current two-component isotope hydrograph separations

Hydrograph separation using conservative, two-component mixing models has been done in various natural basins. However, data from these two catchments, including both the natural and artificial systems indicate that this technique is unacceptable for natural basins and artificial catchments, yielding misleading results due to the unreasonableness of most of the assumptions involved and of the unrealistic operation procedures [63].

*Violation of assumptions.* In studying natural processes it is inevitable to make simplifying assumptions. However, as Kennedy et al [64] indicated, when simplifying assumptions are made, one runs the risk of drawing misleading conclusions. That may have been the case for the five assumptions [65] summarized for the use of isotopic techniques in precipitation-runoff studies.

The first assumption, that "Groundwater and baseflow are characterized by a single isotopic content" [65], is problematic. Baseflow in natural basins can be recharged by the active zone, passive zone, each with differing flowpaths for recharge resulting in differences in the isotopic composition of baseflow itself, which changes with time and discharge. It is also unreasonable to expect a single isotopic content for groundwater which is subject to both temporal and spatial variations (Figure 16).

The second assumption, that "Rain or snowmelt can be characterized by a single constant isotopic composition or, the variations are documented" [65], is problematic, because, as shown in Figure 14a the intra-storm variability of isotopic composition of rainfall is very significant. High isotopic variability has also been observed in experimental watersheds at Maimai, Panola, etc [56]. Additionally, the spatial variability of isotopic composition of rainfall in a watershed exists and shouldn't be ignored. Isotopic data from a 82.1 km$^2$ watershed showed that the largest difference in rainfall$\delta^{18}$O reached 8.2‰.

The third assumption, that "The isotopic composition of rain water is significantly different from that of groundwater/baseflow" [65], can be true. However it must be demonstrated by appropriate sampling [64].

The fourth assumption, that "Contributions from soil water are negligible, or the isotopic composition is identical to that of groundwater" [65] is problematic, because it is a misconception. In a natural watershed, this can only occur if there is no unsaturated zone.

The fifth assumption, that "Contributions from surface water-bodies (such as ponds) are negligible" [65], is reasonable for most upland watersheds [64]. However, in catchments of hilly area with land use including series of ponds and paddy fields, the contributions are large enough to significantly influence results [62].

Based on such problematic assumptions, one of the basic conclusions that "most stormflow is old water"[66], which resulted from two-component hydrograph separations in multiple catchments with different drainage areas, appears suspect.

*The physically ambiguous 'old' and 'new' water.* This two-component separation model in fact is based on the classic Horton infiltration theory with a very simple runoff generation concept that the soil surface acts as a sieve capable of separating rainfall into two basic components [67]. This 'old' and 'new' water separation model has led to many usages with a variety of defined components in addition to 'old' and 'new' water component of flow, e.g., 'pre-event and event component', 'pre-storm and storm component', 'pre-storm water and rain in storm run-off', 'pre-event and rainfall water', 'groundwater component and event water', 'groundwater and rainwater', 'groundwater discharge and surface runoff'[68], 'surface and subsurface runoff'[69]. The classic two-component method has also been extrapolated to incorporate three components (e.g. that of channel precipitation, soil water and groundwater [70]). Thus, it seems that success can be assured because there are no calibration constraints. As Kirchner [15] noted some models are "often good mathematical marionettes, they can dance to the tune of the calibration data".

The 'old'(O) and 'new'(N) water actually correspond to different generation patterns of runoff components. In fact, multiple runoff mechanisms can result in the same proportions of old versus new water, a problem known as equi-finality. As seen in Hydrohill, surface and subsurface runoff patterns emerge from multiple runoff mechanisms (e.g., macropore flow can result in a large proportion of event water or a large proportion of pre-event water depending on antecedent wetness in the catchment). The runoff components corresponding to old water and new water can be very ambiguous. Applied to surface and subsurface runoff [68, 69], surface runoff is labeled new water [62]. This is a misconception as shown in Figure 20.

*The mass balance equations for this model are untenable.* The two-component tracer mixing model is stated by two mass balance equations for the composition of stormflow at any time which is used for hydrograph separation: $Q_s = Q_n + Q_o$ and $Q_s \delta_s = Q_n \delta_n + Q_o \delta_o$ where Q is streamflow, $\delta$(Delta) is the D or $^{18}O$ content, the subscripts s, n and o represent the stream, new and old respectively[56]. At any time t, the first equation is correct (i.e., discharge $Q_{s(t)}$ at the outlet of a watershed is equal to the sum of the unknown $Q_{n(t)}$ and $Q_{o(t)}$ right at the same outlet at the time t). During time t, the $Q_{s(t)}$ at the outlet and its isotopic concentration $\delta_s$ are measurable and are the value of known without problem, however both the $Q_{n(t)}$ and $Q_{o(t)}$ at the outlet of time t, are the results of confluence from somewhere within the watershed before time t, i.e., the results of a convolution integration. The $Q_{n(t)}$ and $Q_{o(t)}$ at the outlet of time t are concentrated from the separated isochrone area somewhere within the watershed with different time of concentration. So, the second equation with algebraic sum is physically unrealistic, hinging on the assumptions outlined above. For this linear algebraic sum to be physically realistic, the isotopic concentrations in the reservoirs of new and old water must be constant, without mixing or fractionating during the time interval of event hydrograph. To be physically realistic and representative of the mixing and non-conservative nature of water in natural catchments, it would likely take on the form of an integral or a finite-difference isochrone [63]. No affluxion happens to both the new and old water in the watershed. This can only happen in a small pond, operationally for a watershed, it's "the Procrustean bed [17]".

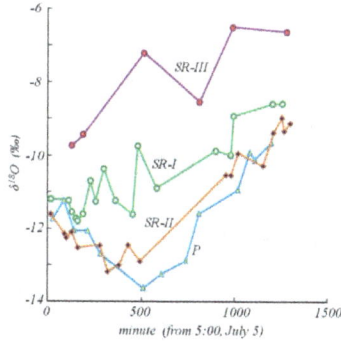

**Figure 20.** Intrastorm isotopic variations of an event rainfall, and isotopic variations of observed surface runoff in different watersheds. P–rainfall, SRI–surface runoff of the 'mini CZEB', Nandadish, with drainage area of 7897 m², SRII–surface runoff of Hydrohill of 490 m², SRIII- runoff of Morning Glory Catchment of 4573 m².

## 6. Conclusion

The river basin, watershed or catchment is a central concept in hydrology. Basin studies to assess watershed hydrology are approaching a period of transition and innovation. In fact, experimental watershed studies, the core of watershed hydrology, have tremendous and complicated challenges ahead.

The current challenges in basin experimental hydrology are mainly twofold. Advancing hydrologic science creates a fundamental challenge. Because the watershed system is a dynamic ecological system composed of a variety of biotic and abiotic processes driven by water and climatic processes, experimental watersheds should multi-couple these processes, organizing innovative measurements and approaches while continuing to support and test hydrological models. Anthropogenic hydrologic replumbing and natural climate oscillations are equally challenging. Field studies are the key to understanding and modeling the effect of hydrologic replumbing and climate change. There is a need for long term monitoring, systematic experimental facilities as well as data mining to reclaim historic data useful for determining baseline watershed metrics.

The main problems of the historical experimental basin approach are threefold: watershed surface laterality, hydrologic process laterality and downward components laterality. Lessons from fifty years of Chinese experimental basin studies are: (1) research facilities require natural conditions and also artificial controlled boundaries; (2) address not only the surface watershed, but also downward to the bedrock; (3) surface and subsurface runoff components should be directly monitored hydrometrically; (4) isotopic and hydrochemical tracers are key to understanding runoff generation mechanisms; and (5) account for interactions between the hydrosphere and multiple watershed processes.

Another kind of experimental basin is suggested going forward, namely the Critical Zone Experimental *Block* (CZEB), geologically a monolith-block within the Critical Zone. CZEB is a dynamic ecological and evolving system, coupled with various systems and united by hydrological process. The CZEB is a natural open system; both the energy and mass exchanges exist across its boundaries. It is a dissipative complex system with some degree of self-organization. The function of CZEB is threefold: mass, its material aspect; energy and force, its driven aspect and, organization and entropy, its thermodynamic/philosophical aspect.

To advance watershed hydrology and support development of a unified theory of hydrology, a two-way multi-scale experimental watershed system is suggested, including the natural system and the artificial system. Both have multi-scale subsystems from monolith, slope, subwatershed to watershed and follow the research idea of upwards and downwards routes. A trial for such a strategy is partly completed at the Chuzhou CZEB Experimental System.

To advance contaminant hydrology, the suggested two-way multi-scale experimental watershed system may provide a key to unravel the complex mechanisms coupling hydrological, biological, and geochemical processes. Contaminant transformation and fate and their effects on regional degradation of groundwater basin involve highly complex mechanisms. This two-way multi-scale system calls for new models using both natural and artificial experimental basin results and using upwards and downwards approaches to multidisciplinary techniques, including isotope tracing.

All hydrological knowledge ultimately comes from observations, experiments, and measurements [15]. Progress in hydrology results mainly from challenges to prevailing approaches and concepts [71]. Hydrological experimentation, including CZEB experimental watershed studies, is the building block for development of a unified theory of hydrology including contaminant hydrology. However, it is important to remember Werner Heisenberg's warning that, "what we observe is not nature herself, but nature exposed to our method of questioning".

## Acknowledgements

The renovation of the Chuzhou CZEB Experimental System is supported by the Hydrology Bureau of the Chinese Ministry of Water Resources and, the Nanjing Hydraulic Research Institutes. The adventurous plan for a two-way multi-scale hydrological experimental watershed system and its realization are led by Acadimician Jian-Yun Zhang.

Gu is deeply grateful to Jeffrey McDonnell and the group from USGS, Vance Kennedy, Carol Kendall, Norman (Jake) Peters for their kind help and support - the ever green cedars they planted by Hydrohill are now rooted and full of luxuriant foliage – as well as the wonderful water tracing methodology they taught, extending associations to new generation of colleagues and students. The authors are also deeply grateful to those who have visited Hydrohill for their teaching, help and support, they are: Sklash M. from Canada; Geyh M.A., Plate E. and Seiler K-P from Germany; Gat J. from Israel; Shiklomanov I.A. from Russia; Verhagen B. from South Africa, Littlewood I. from UK, Kinzelbach W. from Switzerland. Thanks to post-doctoral researchers Ma Tao, Xu J-T for their figures. Many thanks to the Editor of this book, Paul Bradley, for his kind help and, encouragements.

## Author details

Wei-Zu Gu[1*], Jiu-Fu Liu[1], Jia-Ju Lu[1] and Jay Frentress[2]

*Address all correspondence to: gweizu@163.com or gweizu@gmail.com

1 Institute for Hydrology and Water Resources, Nanjing Hydraulic Research Institutes, Nanjing, China

2 Oregon State University, Corvallis, USA

## References

[1] Rodda, J. C. Basin Studies. In: Rodda JC (ed.) Facets of Hydrology. London: John Wiley & Sons; (1976). , 257-297.

[2] Biswas, A. K. History of Hydrology. North-Holland Publishing Company; (1970).

[3] Unesco/WMO/IAHSThree Centuries of Scientific Hydrology. Paris: Unesco; (1974).

[4] Bates, C. G, & Henry, A. J. Forest and Stream-flow Experiment at Wagon Wheel Gap. Monthly Weather Review. Supplement.(1928). , 30, 1-79.

[5] Урываев ВА. Зкспериментальные Гидрологические Исследования на Валдае. Ленинград: Гидрометеорологическое Иэдательство; 1953

[6] Douglass, J. E, & Hoover, M. D. Forest Hydrology and Ecology at Coweeta. New York: Springer-Verlag. (1988).

[7] Delfs, J, Friedrich, W, Kiesekamp, H, & Wagenhoff, A. Der Einfluss des Waldes und des Kahlschlages auf den Abflussvorgane den Wasserhaushalt und den Bodenabtrag. Aus dem Walde. (1958). , 3, 223-325.

[8] Jacquet, J. Les Etudes d'Hydrologie Analytique sur Bassins Versant Experimentaux. Bull. Du Centre de Recherche et d'Essais de Chatou. (1962). , 2, 3-25.

[9] Gu, W-Z, Liu, C-M, Song, X-F, Yu, J-J, & Xia, J. Hydrological experimental system and environmental isotope tracing: a review on the occasion of the 50th Anniversary of Chinese basin studies and the 20th Anniversary of Chuzhou Hydrology Laborato-

ry. In: Xi R-Z, Gu W-Z, Seiler K-P (eds.) Research basins and hydrological planning: proceedings of the International Conference on Research Basins and Hydrological Planning, March 2004, Hefei, China. London: A.A. Balkema Publishers; (2004). , 22-31.

[10] Toebes, C, & Ouryvaev, V. editors. Representative and Experimental Basins. Paris: Unesco; (1970).

[11] Kelly, L. L, & Glymph, L. M. Experimental watersheds and hydrological research. In: Representative and Experimental Areas: proceedings of the International Symposium of Budapest, 28 September- 1 October. Paris: IASH; (1965).

[12] Kirby, C, Newson, M. D, & Gilman, K. editors. Plynlimon Research: The First Two Decades. Wallingford: Institute of Hydrology; (1991).

[13] Mcdonnell, J. J, Sivapalan, M, Vache, K, Dunn, S, Grant, G, Haggerty, R, Hinz, C, Hooper, R, Kirchner, J, Roderick, M. L, Selker, J, & Weiler, M. Moving beyond Heterogeneity and Process Complexity: A new Vision for Watershed Hydrology. Water Resources Research (2007). W07301

[14] Sivapalan, M. Pattern, Process and Function: Elements of a Unified Theory of Hydrology at the Catchment Scale. In: Anderson MG (ed.) Encyclopedia of Hydrological Science. John Wiley & Sons; (2005). , 193-219.

[15] Kirchner, J. W. Getting the Right Answers for the Right Reasons: Linking Measurements, Analyses, and Models to Advance the Science of Hydrology. Water Resources Research (2006). W03S04

[16] NRCBasic Research Opportunities in Earth Science. Washington D.C.: The National Academies Press; (2001).

[17] Klemes, V. Dilettantism in Hydrology: Transition or Destiny? Water Resources Research (1986). S-188S

[18] Bras, R, & Eagleson, P. S. Hydrology, the Forgotten Earth Science. Eos (1987).

[19] Dooge, J. C. Looking for Hydrologic Laws. Water Resources Research (1986). S- 58S

[20] Eagleson, P. S. Climate, Soil, and Vegetation, 1. Introduction to Water Balance Dynamics. Water Resources Research (1978). , 14, 705-712.

[21] Nash, J. E. Foreward. Journal of Hydrology (1988). v-viii.

[22] Grayson, R. B, Moore, I. D, & Mcmahon, T. A. Physically Based Hydrologic Modeling. 2. Is the Concept Realistic? Water Resources Research (1992). , 26(10), 2659-2666.

[23] National Research CouncilChallenges and Opportunities in the Hydrologic Sciences. Prepublication copy. Washington D.C.: The National Academies Press; (2012).

[24] Vince, G. An Epoch Debate. Science (2011). , 334, 32-37.

[25] Alley, R. B, Clark, P. U, Keigwin, L. D, & Webb, R. S. Making Sense of Millennial-Scale Climate Change. In: Clark PU, Webb RS, Keigwin LD (eds.) Geophys. Monogr. Ser., Mechanisms of Global Climate Change at Millennial Time Scale. Washington D.C.: AGU; (1999). , 112

[26] Alley, R. B, Mayewski, P. A, Sowers, T, Stuiver, M, Taylor, K. C, & Clark, P. U. Holocene Climatic Instability: A Prominent, Widespread Event 8200 yr ago. Geology (1997). , 25, 483-486.

[27] Adams, W. M. Green Development, Environment and Sustainability in the Third World. London: Routledge; (2001).

[28] Scudder, T. A Sociological Framework for the Analysis of New Land Settlements. In: Cernea M (ed.) Putting People First: Sociological Variables in Rural Development. Oxford: Oxford University Press; (1991). , 148-167.

[29] Gu, W-Z, & Lu, J-J. The Disposition of Water Resources of Arid Area with Special Reference to the Alxa Plateau, Inner Mongolia. In: Maldini D, Maher DM, Troppoli D, Studer M, Goebel J (eds.) Translating Scientific Results into Conservation Actions: New Roles, Challenges and Solutions for 21$^{st}$ Century Scientists. Boston: Earthwatch Institute; (2007).

[30] Hall, P. Great Planning Disasters. London: Weidenfeld and Nicolson; (1980).

[31] Kirchner, J. W. A Double Paradox in Catchment Hydrology and Geochemistry. Hydrological Processes (2003). , 17, 871-874.

[32] Gu, W-Z. Isotope Hydrology (in Chinese). Beijing: Science Press; (2011).

[33] Steffen, W, Crutzen, P. J, & Mcneill, J. R. The Anthropocene: Are Humans Now Overwhelming thr Great Forces of Nature? Ambio (2007). , 36, 614-642.

[34] Smakhtin, V. U. Low Flow Hydrology: A Review. Journal of Hydrology (2001).

[35] Li, F-X, & Yao, J-H. Synthetical Research on the Economic Development and the Environmental Rectification for the He-Xi Corridor area (in Chinese). Beijing: Environmental Science Press; (1998).

[36] Lin, H. Earth's Critical Zone and Hydropedology: Concepts, Characteristics, and Advances. Hydrology and Earth System Sciences (2010). , 14, 25-45.

[37] Falkenmark, M, & Rockstrom, J. Balancing Water for Humans and Nature: The New Approach in Ecohydrology. London: Earthscan; (2004).

[38] Gu, W-Z. Experimental and representative basin stuides in China. In: Hooghart JC, Posthumus CWS, Warmerdam PMM. (eds.) Hydrological research basins and the environment: proceedings of the international conference, September 1990, Wageningen, The Netherland. Hague: CHO; (1990). , 24-28.

[39] Brantley, S. L, White, T. S, White, A. F, Sparks, D, Richter, D, Pregitzer, K, Derry, L, Chorover, J, Chadwick, O, April, R, Anderson, S, & Amundson, R. Frontiers in explo-

ration of the Critical Zone: Report of a workshop sponsored by the National Science Foundation. (2005).

[40] Report prepared by the CZO CommunityFuture directions for Critical Zone observatory (CZO) science. (2010).

[41] National Critical Zone Observatory Program: http://criticalzoneorg/index.html

[42] Klemes, V. Conceptualization and Scale in Hydrology. Journal of Hydrology (1983).

[43] Seiler, K-P, & Lindner, W. Near Surface and Deep Groundwater. Journal of Hydrology (1995). , 165, 33-44.

[44] Arnold, R. W, Szabolcs, I, & Targulian, V. O. Global Soil Change. Laxenburg: International Institute for Applied Systems Analysis; (1990).

[45] Bateson, G. Mind and Nature: A Necessary Unity. New York: Dutton; (1979).

[46] Weinberg, S. Dreams of a Final Theory: The Scientist's Search for the Ultimate Laws of Nature. London: Vintage Books; (1994).

[47] Anderson, S. P, Von Blanckenburg, F, & White, A. F. Physical and Chemical Controls on the Critical Zone. Elements (2007). , 3, 315-319.

[48] Toffler, A. Forward: Science and Change. In: Prigogine I, Stengers I (eds.). Order out of Chaos. Toronto: Bantam Books; (1984).

[49] Gu, W-Z. Field research on surface water and subsurface water relationships in an artificial experimental catchment. In: Dahlblom P, Lindh G. (eds.) Interaction between groundwater and surface water: proceedings of the international symposium, 30 May- 3 June, Ystad, Sweden. Lund: Lund University; (1988).

[50] Gu, W-Z, & Freer, J. Patterns of surface and subsurface runoff generation. In: Leibundgut C (ed.) Tracer technologies for hydrological systems: proceedings of symposium H4 at the XXI General Assembly of the International Union of Geodesy and Geophysics, July 1995, Boulder, USA. IAHS Publ. (1995). , 1995(229), 265-273.

[51] Gu, W-Z. Measurements of spatial evapotranspiration characteristics of an experimental basin using a neutron probe. In: Isotope techniques in water resources development 1987: proceedings of the International Symposium on the Use of Isotope Techniques in Water Resources Development, 30 March- 3 April, Vienna. Vienna: IAEA; (1987). , 1987, 789-793.

[52] Mcdonnell, J. Personal communications.

[53] Kendall, C, Mcdonnell, J. J, & Gu, W-Z. A Look Inside 'Black Box' Hydrograph Separation Models: A Study at the Hydrohill Catchment. Hydrological Processes (2001). , 15, 1877-1902.

[54] Kendall, C, & Gu, W-Z. Development of isotopically heterogeneous infiltration waters in an artificial catchment in Chuzhou, China. In: Isotope techniques in water re-

sources development 1991: proceedings of the International Symposium on Isotope Techniques in Water Resources Development, 11-15 March, Vienna. Vienna: IAEA; (1992). , 1992, 61-73.

[55] Wood, E. F, Sivapalan, M, Beven, K. J, & Band, L. Effects of Spatial Variability and Scale with Implications to Hydrologic Modeling. Journal of Hydrology (1988). , 102, 29-47.

[56] Kendall, C, & Mcdonnell, J. J. Effect of intrastorm isotopic heterogeneities of rainfall, soil water, and groundwater on runoff modeling. In: Tracers in hydrology: proceedings of the International Symposium on Tracers in Hydrology, 11-23 July 1993, Yokohama, Japan. IAHS Publ. (1993). , 1993(215), 41-48.

[57] Gu, W-Z. Experimental Research on Catchment Runoff Responses Traced by Environmental Isotopes. Advances in Water Science (1992). in Chinese with English abstracts); , 3(4), 246-254.

[58] Gu, W-Z, Shang, M-T, Zhai, S-Y, Lu, J-J, Jason, F, Mcdonnell, J. J, & Kendall, C. Rainfall-runoff Paradox from A Natural Experimental Catchment. Advances in Water Science (2010). in Chinese with English abstracts); , 21(4), 471-478.

[59] Gu, W-Z, Lu, J-J, Zhao, X, & Peters, N. E. Responses of Hydrochemical Inorganic Ions in the Rainfall-runoff Processes of the Experimental Catchments and Its Significance for Tracing. Advances in Water Science (2007). in Chinese with English abstracts); , 18(1), 1-7.

[60] Nimick, D. A, Cleasby, T. E, & Mccleskey, R. B. Seasonality of Diel Cycles of Dissolved Trace Metal Concentrations in a Rocky Mountain Stream. Environmental Geology (2005). , 47(5), 603-614.

[61] Gu, W-Z. Challenge on some rainfall-runoff conceptions traced by environmental isotopes in experimental catchments. In: Hotzl H, Werner A. (eds.) Tracer hydrology: proceedings of the 6th International Symposium on Water Tracing, 21-26 September 1992, Karlsruhe, Germany. Rotterdam: A.A. Balkema; (1992). , 1992, 397-403.

[62] Gu, W-Z. Various Patterns of Basin Runoff Generation Identified by Hydrological Experimentation and Water Tracing using Environmental Isotopes. Journal of Hydraulic Engineering (1995). in Chinese with English abstracts); , 5, 9-17.

[63] Gu, W-Z. Unreasonableness of current two-component isotopic hydrograph separation for natural basins. In: Isotope techniques in water resources development 1995: proceedings of the International Symposium on Isotopes in Water Resources Development, 20-24 March, 1995, Vienna. Vienna: IAEA; (1996). , 1996, 261-264.

[64] Kennedy, V. C, Kendall, C, Zellweger, G. W, Wyerman, T. A, & Avanzino, R. J. Determination of the Components of Stormflow using Water Chemistry and Environmental Isotopes, Mattole River Basin, California. Journal of Hydrology (1986). , 84, 107-140.

[65]  Sklash, M. G, & Farvolden, R. N. The Use of Environmental Isotopes in the Study of High-runoff Episodes in Streams. In: Perry EC, Montgomery CW. (eds.) Isotope Studies of Hydrologic Processes. Dekalb: N. Illinois University Press; (1982). , 65-73.

[66]  Bishop, K. H. Episodic increases in stream acidity, catchment flow pathways and hydrograph separation. PhD thesis. University of Cambridge, Dep. Geol., Jesus College, Cambridge; (1991).

[67]  Chorley, R. J. The Hillslope Hydrological Cycle. In: Kirkby MJ. (ed.) Hillslope Hydrology. Chicheter: John Wiley & Sons; (1979). , 1-42.

[68]  Crouzet, E, & Hubert, P. Oliver Ph, Siwertz E. Le Tritium dans les Mesures d'Hydrologie de Surface. Determination Experimentale du Coefficient de Ruissellement. Journal of Hydrology (1970). , 11, 217-229.

[69]  Wels, C, Cornett, R. J, & Lazerte, B. D. Hydrograph Separation: A Comparison of Geochemical and Isotope Tracers. Journal of Hydrology (1991). , 122, 253-274.

[70]  Dewalle, D. R, Swistock, B. R, & Sharpe, W. E. Three Component Tracer Model for Stormflow on a Small Appalachian Forested Catchment. Journal of Hydrology (1988). , 104, 301-310.

[71]  Yevjevich, V. Misconceptions in Hydrology and Their Consequences. Water Resources Research (1968). , 4(2), 225-232.

# Permissions

The contributors of this book come from diverse backgrounds, making this book a truly international effort. This book will bring forth new frontiers with its revolutionizing research information and detailed analysis of the nascent developments around the world.

We would like to thank Paul M. Bradley, Ph.D., for lending his expertise to make the book truly unique. He has played a crucial role in the development of this book. Without his invaluable contribution this book wouldn't have been possible. He has made vital efforts to compile up to date information on the varied aspects of this subject to make this book a valuable addition to the collection of many professionals and students.

This book was conceptualized with the vision of imparting up-to-date information and advanced data in this field. To ensure the same, a matchless editorial board was set up. Every individual on the board went through rigorous rounds of assessment to prove their worth. After which they invested a large part of their time researching and compiling the most relevant data for our readers. Conferences and sessions were held from time to time between the editorial board and the contributing authors to present the data in the most comprehensible form. The editorial team has worked tirelessly to provide valuable and valid information to help people across the globe.

Every chapter published in this book has been scrutinized by our experts. Their significance has been extensively debated. The topics covered herein carry significant findings which will fuel the growth of the discipline. They may even be implemented as practical applications or may be referred to as a beginning point for another development. Chapters in this book were first published by InTech; hereby published with permission under the Creative Commons Attribution License or equivalent.

The editorial board has been involved in producing this book since its inception. They have spent rigorous hours researching and exploring the diverse topics which have resulted in the successful publishing of this book. They have passed on their knowledge of decades through this book. To expedite this challenging task, the publisher supported the team at every step. A small team of assistant editors was also appointed to further simplify the editing procedure and attain best results for the readers.

Our editorial team has been hand-picked from every corner of the world. Their multi-ethnicity adds dynamic inputs to the discussions which result in innovative outcomes. These outcomes are then further discussed with the researchers and contributors who give their valuable feedback and opinion regarding the same. The feedback is then collaborated with the researches and they are edited in a comprehensive manner to aid the understanding of the subject.

Apart from the editorial board, the designing team has also invested a significant amount of their time in understanding the subject and creating the most relevant covers. They scrutinized every image to scout for the most suitable representation of the subject and create an appropriate cover for the book.

The publishing team has been involved in this book since its early stages. They were actively engaged in every process, be it collecting the data, connecting with the contributors or procuring relevant information. The team has been an ardent support to the editorial, designing and production team. Their endless efforts to recruit the best for this project, has resulted in the accomplishment of this book. They are a veteran in the field of academics and their pool of knowledge is as vast as their experience in printing. Their expertise and guidance has proved useful at every step. Their uncompromising quality standards have made this book an exceptional effort. Their encouragement from time to time has been an inspiration for everyone.

The publisher and the editorial board hope that this book will prove to be a valuable piece of knowledge for researchers, students, practitioners and scholars across the globe.

# List of Contributors

Celeste A. Journey and Paul M. Bradley
U.S. Geological Survey, Columbia, SC, USA

Karen M. Beaulieu
U.S. Geological Survey, Hartford, CT, USA

Paul M. Bradley and Dana W. Kolpin
U.S. Geological Survey, USA

Prem B. Parajuli
1 Department of Agricultural and Biological Engineering at Mississippi State University, Mississippi State, USA

Ying Ouyang
USDA-Forest Service Center for Bottomland Hardwoods Research, Mississippi State, USA

Julia L. Barringer, Zoltan Szabo and Pamela A. Reilly
U.S. Geological Survey, USA

Francis H. Chapelle, Bruce G. Campbell and Mathew K. Landon
U.S. Geological Survey, Columbia, SC, USA

Mark A. Widdowson
Virginia Tech University, Blacksburg, VA, USA

Zulfiqar Ahmad and Gulraiz Akhter
Department of Earth Sciences, Quaid-i-Azam University, Islamabad, Pakistan

Arshad Ashraf
National Agricultural Research Center,Islamabad, Pakistan

Iftikhar Ahmad
College of Earth and Environmental Sciences, Punjab University, Lahore, Pakistan

Matjaž Glavan, Rozalija Cvejić, Matjaž Tratnik and Marina Pintar
University of Ljubljana, Biotechnical Faculty, Agronomy Department, Chair for Agrometeorology,
Agricultural Land Management, Economics and Rural Development, Ljubljana, Slovenia

**Luc Descroix, Moussa Malam Abdou and Kadidiatou Souley Yéro,**
IRD / UJF, Grenoble, France

**Ibrahim Bouzou Moussa, Oumarou Faran Maiga, Aghali Ingatan and Ibrahim Noma,**
UAM University, Niamey, Niger

**Pierre Genthon and Gil Mahé**
IRD-HSM, Montpellier, France

**Daniel Sighomnou**
Niger Basin Authority, Niamey, Niger

**Ibrahim Mamadou**
University of Zinder, Niger

**Jean-Pierre Vandervaere**
UJF-LTHE, Grenoble, France

**Emmanuèle Gautier**
Université Paris 8, France

**Jean-Louis Rajot**
IRD-BIOEMCO, Créteil, France

**Nadine Dessay**
IRD-ESPACE-DEV, Montpellier, France

**Harouna Karambiri**
2iE International high School, Ouagadougou, Burkina Faso

**Rasmus Fensholt**
University of Copenhague, Denmark

**Jean Albergel**
IRD-LISAH, Montpellier, France

**Jean-Claude Olivry**
IRD, France

**Arshad Ashraf**
Water Resources Research Institute, National Agricultural Research Center, Islamabad, Pakistan

**Peter C. Sakaris**
Department of Biology and Chemistry, Southern Polytechnic State University, Marietta, GA, USA

**Wei-Zu Gu, Jiu-Fu Liu and Jia-Ju Lu**
Institute for Hydrology and Water Resources, Nanjing Hydraulic Research Institutes, Nanjing, China

**Jay Frentress**
Oregon State University, Corvallis, USA

www.ingramcontent.com/pod-product-compliance
Lightning Source LLC
Chambersburg PA
CBHW070727190326
41458CB00004B/1071